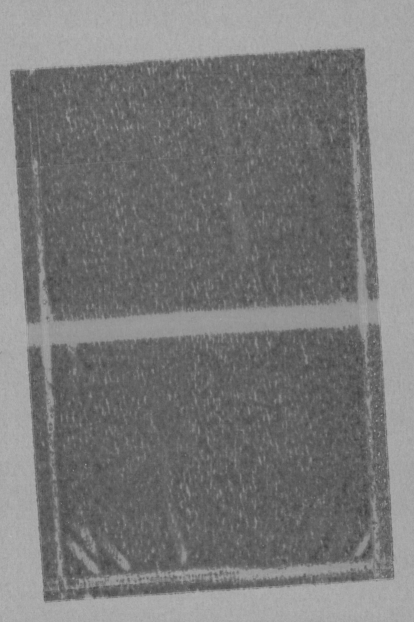

9.7.72

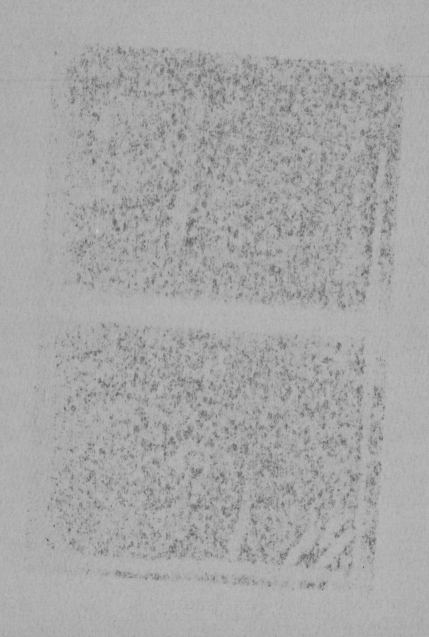

MAN
AND THE
ENVIRONMENT

MAN
AND THE
ENVIRONMENT

A Bibliography of Selected Publications

of the United Nations System 1946-1971

Compiled and edited by

Harry N. M. Winton

UNIPUB, INC. / R.R. BOWKER COMPANY

New York & London, 1972

Published by R. R. Bowker Co. (A Xerox Company)
1180 Avenue of the Americas, New York, N.Y. 10036

Copyright © 1972 by Xerox Corporation
All rights reserved.

International Standard Book Number: 0-8352-0536-3
Library of Congress Catalog Card Number: 72-739

Printed and bound in the United States of America.

This book has been printed on recycled paper, to
conserve our diminishing stock of natural resources.
The use of recycled paper contributes
to ecological balance while maintaining
high quality in paper production.

Contents

1704784

Preface

In recent years public concern has been dramatically aroused over two problems confronting mankind — the so-called "population explosion" and the deterioration of the environment. Neither problem is new and neither has been unforeseen as a possible consequence of economic and social development. Intensified industrialization has brought about the exploitation of natural resources on a vast scale accompanied by profound changes in the natural environment and in human society. We are beginning to understand, uncomfortably, that our natural resources are indeed not inexhaustible, that our ever-growing numbers require even further exploitation of these resources and severe readjustments of society, and that our exploitative activities have often been destructive to the natural balances in the biosphere. We are now coming to realize that if we cause too many irreversible changes in the ecological system of Earth, we imperil our own survival. The words "ecology," "environmental pollution," "population explosion," are heard everywhere. Scientists, technicians, development planners, and administrators are now exploring more fully the nature of the biosphere and the relation of man to the environment, and are considering the possible further ecological effects of programs of economic and social development.

Public opinion in the developed countries has been more concerned with local evidences of environmental deterioration than with their consequences for the nation as a whole, and has not fully grasped the truly international proportions of the problem. In the developing countries, planners of major projects of industrialization and agricultural development have often failed to consider adequately the possible adverse effects such projects may have on the environment. Some environmental and demographic problems are not merely local or national in scope, but international, interregional, or even global in their causes and effects; other problems, more local in character, arise in various parts of the world from similar causes and have similar effects. There is value, therefore, in the international approach to the study and treatment of these problems. The most useful scientific and technical approach to them must be interdisciplinary in character because the study and full understanding of complex environmental and demographic questions not only involves several of the sciences, but also requires the teamwork of specialists. Planners of economic and social development programs should encourage environmental research as part of rational planning. Study of the environment and planning for the rational use of natural resources calls for the united efforts of specialists from a variety of fields and from different countries, as well as the cooperation of many governments and international organizations. This bibliography marshals a selection of publications of the intergovernmental organizations in the United Nations system which treat of man, the resources available to him, the environment, and the population problem.

The publications of the international agencies of the United Nations system — the recommendations of expert committees, the results of seminars, the papers and proceedings of conferences and symposia, the collections of statistics or of texts of laws, the technical manuals, the directories of national and international institutions and documentation centers, the bibliographies, the reports of conditions in the developing countries — are of great value in the fields of the agencies' special competence for assessing the state of knowledge and consensus in the world at large or in particular regions and countries. Unfortunately, however, these publications are not so widely known in academic, governmental, and business communities as they deserve to be. Unlike the publications of commercial and academic presses, the publications of the intergovernmental organizations are not adequately covered by the standard current bibliographic sources, and they are often subject to inadequacies of distribution. The aim of this specialized bibliography, then, is to call attention to the generally valuable kinds of information and publications emanating from the United Nations system.

This bibliography of more than 1200 entries includes four indexes, which are provided for the convenience of the user. All index references are to entry numbers and not to page numbers. The Author Index includes both personal and corporate names. Because of the confusing variety of names for certain types of corporate entries, names of committees, conferences, groups, panels, seminars, symposia, and working parties are grouped under the terms "Committee," "Conference," "Group," etc. The Series and Serials Index lists the titles of series and serials alphabetically under the name of each international organization of issue. Titles of serials (yearbooks and other periodicals) are underlined to distinguish them from titles of monograph series. The Title Index lists alphabetically the titles of monographs. Finally, the alphabetical Subject Index covers the principal subjects treated in the selected publications, and emphasizes regional areas as well.

Since revised editions of this bibliography are planned, the compiler will be grateful to receive any suggestions of additional titles or corrections of errors in the present text. Future coverage of the bibliography may be expanded to include selected publications of international organizations outside the United Nations system.

HARRY N.M. WINTON
Former Chief, Documents Reference Section
Dag Hammarskjold Library, United Nations

The United Nations System

The United Nations family of intergovernmental organizations embraces the United Nations and seventeen related but autonomous agencies, each of which publishes more or less extensively. The United Nations itself is an international organization with broad political responsibilities and is active in the promotion of economic and social development, especially through its programs of technical cooperation with the developing countries. The related and specialized agencies promote international cooperation in the fields of their special competence. The following list, including the United Nations and all the related agencies by name and acronym, and giving the year of establishment and the headquarters of each, may be helpful to an understanding of the scope of the interests and activities of the United Nations system. Some subsidiary organs of the United Nations responsible for the preparation of various selections in this bibliography are included in the list, but these are not autonomous organizations.

United Nations (UN). 1945. New York
 Subsidiary organs (nonautonomous) responsible for some selected publications:
  International Trade Centre UNCTAD/GATT (ITC UNCTAD/GATT). 1964. Geneva. Under joint operation from 1968.
  United Nations Children's Fund (UNICEF). 1946. New York
  United Nations Conference on Trade and Development (UNCTAD). 1964. Geneva
  United Nations Development Programme (UNDP). 1965. New York
  United Nations Industrial Development Organization (UNIDO). 1967. Vienna
  World Food Programme (WFP). 1963. Rome (FAO). Under joint UN/FAO operation.

AUTONOMOUS SPECIALIZED AGENCIES AND BODIES HAVING SPECIAL STATUS

Food and Agriculture Organization of the United Nations (FAO). 1945. Rome
General Agreement on Tariffs and Trade (GATT). 1948. Geneva
Inter-Governmental Maritime Consultative Organization (IMCO). 1958. London
International Atomic Energy Agency (IAEA). 1956. Vienna
International Bank for Reconstruction and Development (IBRD or World Bank). 1945. Washington

Affiliates (forming, with the World Bank, the World Bank Group):
  International Finance Corporation (IFC). 1956. Washington
  International Development Association (IDA). 1960. Washington
  International Center for Settlement of Investment Disputes (ICSID). 1966.
    Washington.
International Civil Aviation Organization (ICAO). 1945. Montreal
International Labour Organisation (ILO). 1919. Geneva
International Monetary Fund (IMF or Fund). 1945. Washington
International Telecommunication Union (ITU). 1865. Geneva
United Nations Educational, Scientific and Cultural Organization (UNESCO).
    (UNESCO). 1946. Paris
Universal Postal Union (UPU). 1874. Berne
World Health Organization (WHO). 1946. Geneva
  WHO's Regional Office for the Western Hemisphere from 1949, but an auton-
  omous organization:
    Pan American Health Organization (PAHO). 1902. Washington
World Meteorological Organization (WMO). 1951. Geneva

The international point of view is the distinguishing characteristic of the publications of the United Nations family. The membership of each organization comprises most of the countries of the world and the professional staff of each secretariat also forms a wide spectrum of nationalities. Studies and surveys carried out
by secretariat staff express a consensus of work of several experts of different
nationalities, or when prepared by an individual staff expert may reflect his experience in several countries.

Two principal objectives of these organizations are the improvement of the
quality of technical standards and methods (including definitions and statistics of
international comparability) and assistance to the developing countries in the reali-
zation of their programs of economic and social development.

The organizations bring together experts from many countries to serve on
technical commissions, committees, panels, and working groups in their specific
fields in order to determine standards and methods of international validity and
to make recommendations for national action. The recommendations of these
groups are sometimes discussed in seminars and workshops throughout the world.
By such means the results of the international collaboration of experts may be
disseminated in the developing countries, tested as to their applicability to specific
national and regional situations, and modified by comment and criticism from
the field. There is also collaboration between the international agencies (as between experts of different disciplines) when a subject so requires — e.g., nutrition,
a field in which FAO, UNICEF and WHO work closely together.

The United Nations agencies also sponsor many scientific and technical conferences and symposia at which scientists and technical experts from all parts of
the world are able to exchange views in discussion and in personal contact and
to share their highly specialized knowledge in papers which are later made available to the rest of the world in published reports and proceedings.

In the area of technical aid to the developing countries, the United Nations
and the related organizations are doing a great deal to help governments to organ-
ize or improve national planning programs and statistical, scientific, and technical
services; to train administrative, scientific and technical staff; to plan and carry out
censuses and surveys of manpower and of natural resources; to develop educational
and public health programs; and to carry out agricultural, industrial and trade
development schemes. Many publications reflect this emphasis on the problems
of the "Third World" of the developing countries.

Introduction

TYPES OF MATERIAL SELECTED

Most publications in this bibliography are publications offered for sale, and include monographs, dictionaries and glossaries, bibliographies, directories, yearbooks and other periodicals, and filmstrips. In the case of yearbooks and periodic reports, the current issues have been selected for inclusion, and the annotations indicate in each case the number of the current issue and when the first issue was published.

A number of newsletters and public information pamphlets available free on request to the organization of issue, and a few unpriced monographs, have been included in the bibliography. Since the availability of unpriced documents is difficult to determine, and because most of them are transitory in character, they have been excluded.

Obsolete material has been excluded, but out-of-print titles of value have been listed, not only because they are frequently cited in current publications without indication of the fact that they have gone out of print, but also because this material may be available in depository and other libraries.

LANGUAGE EDITIONS

Most selections have been published in English — in monolingual English editions, in bilingual (English/French), in trilingual (English/French/Spanish) or in quadrilingual (English/French/Russian/Spanish) editions. Titles of these selections, however, are given only in English. For bilingual or multilingual editions, the languages involved are identified as "E" (English), "F" (French), "R" (Russian), or "S" (Spanish). The term "composite" has been used to describe publications in more than one language in which part of the text is in one language without being translated into the other language(s) used therein — e.g., directories in which each country section is published only in the language officially used in correspondence with the international organization of issue, such as French for sections on France and on Italy, and proceedings of conferences and symposia in which papers are published in full only in the language of submission with abstracts only in one or more languages. A few monolingual publications have been included which were issued solely in a language other than English.

PRICES

Quoted prices (including annual subscriptions for periodicals) are those listed in sales catalogs available when this bibliography was compiled, but they are subject to change without notice. Present fluctuations in the international monetary situation must also be borne in mind. Prices are for paperbound editions unless otherwise noted. Clothbound editions are identified by "cl." When both clothbound and paperbound editions are noted for the same title, the paperbound edition is identified by "pa."

AVAILABILITY OF PUBLICATIONS — NATIONAL DISTRIBUTORS

In a number of countries a publishing/distribution agency has been appointed by some of the international organizations as the exclusive distributor of their publications, and orders should therefore be addressed to the exclusive national distributor. In the case of other international organizations, their publications are obtainable directly from the organizations themselves, or through booksellers. The exclusive national distributors can provide information on the availability of earlier issues of yearbooks and periodic reports and back numbers of periodicals. These distributors may still have in stock some titles which have gone out of print at the headquarters of the issuing organizations. Although exclusive national distributors do not usually handle unpriced material other than sales catalogs, they may be able to supply information on the availability of such material.

The addresses of sales offices of the international organizations issuing publications included in this bibliography are given in the following list, along with national distributors of these publications in the United States and Canada.

SALES OFFICES AND EXCLUSIVE NATIONAL DISTRIBUTORS

FAO Food and Agriculture Organization of the United Nations
Via della Terme di Caracalla
01000 Rome, Italy   Distribution and Sales Section

Exclusive national distributors:

UNIPUB, Inc.
P.O. Box 433
New York, New York 10016

Information Canada
171 Slater Street
Ottawa, Ontario   KIA 0S9

IAEA International Atomic Energy Agency
P.O. Box 590
A-1011 Vienna, Austria   Publishing Section

Exclusive national distributors:

UNIPUB, Inc.
P.O. Box 433
New York, New York 10016

Information Canada
171 Slater Street
Ottawa, Ontario   KIA 0S9

ILO   International Labour Organisation
International Labour Office
CH-1211 Geneva 22, Switzerland   ILO Publications

ILO branch offices:

ILO Branch Office   *or*   UNIPUB, Inc.
917 Fifteenth Street, N.W.   P.O. Box 433
Washington, D.C. 20005   New York, New York
                 10016
ILO Branch Office
178 Queen Street (Room 307)
Ottawa 4, Ontario

IMCO   Inter-Governmental Maritime Consultative Organization
104 Piccadilly
London, W1V OAE, United Kingdom   Publications Section

PAHO   Pan-American Health Organization
525 Twenty-third Street, N.W.
Washington, D.C. 20037   Pan American Sanitary Bureau

UN   United Nations
New York, New York 10017   Sales Section

National distributors:

UNIPUB, Inc.
P.O. Box 433
New York, New York 10016

Information Canada
171 Slater Street
Ottawa, Ontario   KIA 0S9

UNESCO   United Nations Educational, Scientific and Cultural Organization
Place de Fontenoy
Paris 7e, France   Librairie de l'UNESCO

Exclusive national distributors:

UNIPUB, Inc.
P.O. Box 433
New York, New York 10016

Information Canada
171 Slater Street
Ottawa, Ontario   KIA 0S9

WHO   World Health Organization
CH-1211 Geneva 27, Switzerland   Distribution and Sales Services

National distributors:

American Public Health Association, Inc.
1015 Eighteenth Street, N.W.
Washington, D.C. 20036

Information Canada
171 Slater Street
Ottawa, Ontario   KIA 0S9

WMO   World Meteorological Organization
P.O. Box 1
CH-1211 Geneva 20, Switzerland   Secretary-General
Exclusive national distributors:
     UNIPUB, Inc.
     P.O. Box 433
     New York, New York 10016

     Information Canada
     171 Slater Street
     Ottawa, Ontario   KIA 0S9

SALES NUMBERS

Most priced publications of the United Nations bear sales numbers which indicate the year of publication, the subject category, and the number of the publications in that category during the year of issue. In recent issues the language edition is indicated by a letter prefix (e.g., Sales No. E.71.II.A.18, in entry 26). The bibliographic entry notes the sales number in the line giving the price of the publication. Acquisition orders for United Nations should always include the sales number in order to ensure prompt service. Publication numbers for World Meteorological Organization material, also noted in the line giving the price for each publication, should be included in any order for WMO material (WMO-No. 254.TP.142, in entry 5, is an example). Titles are not satisfactory for arranging publications in stock because of the frequency of very similar titles in one language and of the multiplicity of titles when three or more languages are used. The sales and publication numbers are constants.

DOCUMENT SYMBOLS

Document series symbols are used extensively for most United Nations material to identify the organ or subsidiary body for or by which the material has been prepared. Most publications of the International Atomic Energy Agency are issued with document symbols. The document symbol is always noted in a bibliographic entry after the series title — or after the pagination when there is no series title. The symbol is not required for ordering publications with sales numbers, but because of the large number of very similar titles issued by the International Atomic Energy Agency it is useful to include the symbol in an acquisition order to obviate misunderstanding. Many depository libraries for United Nations material (both mimeographed documents and sales publications) have used the symbol series to file or shelve the items deposited.

DEPOSITORY LIBRARIES

In addition to the many libraries which purchase the publications of international organizations as individual items or by standing orders for all or selected categories of publications, there are also a number of depository libraries. The United Nations deposits its publications in some 42 libraries in the United States and 13 in Canada. These depositories are named in the *List of Depository Libraries Receiving United Nations Material* (document ST/LIB/12/Rev.1, 1 January 1971), which is available from the Public Inquiries Unit, Office of Public Information, United Nations. The other organizations in the United Nations system have fewer depositories or make no deposits. The depository libraries for United Nations material in many cases locate this material in government document collections, arrange it in symbol series order, and rely for guidance to these documents and publications on the monthly *United Nations Documents Index* (entry 1144) rather than on library cataloging, especially if the material is checked in and shelved by symbol.

The acquisition librarian in a library which is a depository for United Nations publications would do well to check with the librarian in charge of the deposit collection before placing orders for United Nations material. Faculty members, students and librarians of departmental libraries are often unaware of the deposit collection, in which the item desired may have already been received. Such checking preliminary to an order is especially useful in the case of older material which may have gone out of print.

List of
Acronyms and Abbreviations

ACAST	Advisory Committee on the Application of Science and Technology to Development (UN)
ACMRR	Advisory Committee on Marine Resources Research (FAO)
Add.	Addendum, in United Nations document symbols, e.g., A/7622/Add.1 (See entry 147a)
ASGA	Association of African Geological Surveys
BCIS	Bureau Central International de Séismologie
BRGM	Bureau de Recherches Géologiques et Minières
CAB	Commonwealth Agricultural Bureaux
CAC	Codex Alimentarius Commission (FAO/WHO)
CBW	Chemical and biological weapons
CELADE	Centro Latinoamericano de Demografía
CES	Conference of European Statisticians (UN)
CIB	International Council for Building Research, Studies and Documentation
CICAR	Co-operative Investigations of the Caribbean and Adjacent Regions
CIG	Comité International de Géophysique
cl.	clothbound or hard cover
Corr.	Corrigendum, in United Nations document symbols, e.g., ST/SOA/SER.A/10/Corr.1 (See entry 784)
CSK	Co-operation Study of the Kuroshio and Its Adjacent Region
CNS	Centre International de la Recherche Scientifique (France)
COSPAR	Committee on Space Research (ICSU)
CTCA	Commission for Technical Co-operation South of the Sahara
E	English language
ECA	Economic Commission for Africa (UN)
ECAFE	Economic Commission for Asia and the Far East (UN)
ECE	Economic Commission for Europe (UN)
ECLA	Economic Commission for Latin America (UN)
ENEA	European Nuclear Energy Agency
F	French language
FAO	Food and Agriculture Organization of the United Nations
FFHC	Freedom from Hunger Campaign (FAO)
GARP	Global Atmosphere Research Programme (ICSU/WMO)

GATT	Agreement on Tariffs and Trade
GTS	Guinea Trawling Survey
IAEA	International Atomic Energy Agency
IAGC	International Association of Geochemistry and Cosmochemistry
IAH	International Association of Hydrogeologists
IAMAP	International Association of Meteorology and Atmospheric Physics
IAMS	International Association of Microbiological Societies
IAPO	International Association of Physical Oceanography
IASH	International Association of Scientific Hydrology
IAU	International Astronomical Union
IBE	International Bureau of Education (UNESCO)
IBRO	International Brain Research Organization
ICITA	International Co-operative Investigations of the Tropical Atlantic, 1963–1964
ICPM	International Commission on Polar Meteorology
ICSI	International Commission of Snow and Ice
ICSU	International Council of Scientific Unions
IEA	International Economic Association
IGOSS	Integrated Global Ocean Station System (IOC)
IGU	International Geographical Union
IHB	International Hydrographic Bureau
IHD	International Hydrological Decade, 1965–1975
IIEP	International Institute of Educational Planning
IIOE	International Indian Ocean Expedition, 1959–1965
IIR	International Institute of Refrigeration
illus.	illustrations
ILO	International Labour Organisation
IMCO	Inter-Governmental Maritime Consultative Organization
INCAP	Instituto de Nutrición de Centro América y Panamá
INQUA	International Union for Quaternary Research
IOC	Intergovernmental Oceanographic Commission (UNESCO)
ISA	International Sociological Association
ISSS	International Soil Science Society
ITC	International Trade Centre UNCTAD/GATT, Geneva
IUGG	International Union of Geodesy and Geophysics
IUGS	International Union of Geological Sciences
IUCN	International Union for Conservation of Nature and Natural Resources
IUSSP	International Union for the Scientific Study of Population
IWP	Indicative World Plan for Agricultural Development (FAO)
ODAS	Ocean Data Acquisition Systems
OECD	Organisation for Economic Co-operation and Development
pa.	paperbound
PAG	Protein Advisory Group of the United Nations System
PAHO	Pan American Health Organization
R	Russian language
Rev.	Revision, in United Nations document symbols, e.g., E/4608/Rev.1 (See entry 25)
S	Spanish language
SCAR	Scientific Committee on Arctic Research (ICSU)
SCOR	Scientific Committee on Oceanic Research (ICSU)
Supp.	Supplement
UN	United Nations

UNCTAD	United Nations Conference on Trade and Development
UNDP (SF)	United Nations Development Programme (Special Fund)
UNESCO	United Nations Educational, Scientific and Cultural Organization
UNICEF	United Nations Children's Fund, formerly (1946–1953) called the United Nations International Children's Emergency Fund
UNIDO	United Nations Industrial Development Organization
VAP	Voluntary Assistance Programme (WMO)
WHO	World Health Organization
WMO	World Meteorological Organization
WWW	World Weather Watch (WMO)

THE ENVIRONMENT

THE BIOSPHERE AND THE ECOSYSTEMS
_____

1   A Survey of HUMAN BIOMETEOROLOGY. Ed. by Frederick Sargent II
        and Solco W. Tromp. 113 pp., illus. (Technical Notes, 65)
        1964, WMO   WMO-No.160.TP.78 $5.00
            Chapters on the definition and importance of biometeorology; physio-
            logical regulations in man; influence of meteorological factors on
            physiological processes; influence of weather and climate on man,
            farm animals and insects, and on the treatment of diseases; urban
            biometeorology. References. The survey is the result of collabora-
            tion between WMO and the International Society for Biometeorology
            (ISB).

2   "Human-engineering the Planet." Impact of Science on Society, vol.
        19, no. 2, April/June 1969, pp. 105-219.
        1969, UNESCO   Out of print
            Articles on the ecosystem view of human society; space research
            and a better Earth; world's water resources, present and future;
            controlling the planet's climate; the city; the land: its future-endan-
            gering pollutants, and ecological farming.

3   USE AND CONSERVATION OF THE BIOSPHERE. Proceedings of the
        Intergovernmental Conference of Experts on the Scientific Basis for
        Rational Use and Conservation of the Resources of the Biosphere,
        Paris, 4-13 September 1968. 272 pp. (Natural Resources Research, 10)
        1970, UNESCO $6.00
            Final report and ten background papers, as revised in the light of
            discussion. Bears mainly on the terrestrial parts of the biosphere,
            including inland waters and coastal areas, but excluding oceanic re-
            sources, which are covered by other international conferences. Pa-
            pers cover: contemporary scientific concepts relating to the bio-
            sphere; impacts of man on the biosphere; soil resources; water re-
            sources; nonoceanic living aquatic resources; vegetation resources;
            animal resources; natural areas and ecosystems; problems of the

deterioration of the environment; man and his ecosystems. An excellent introduction to the study of the interrelationship of man and the environment.

4   URBAN CLIMATES. Proceedings of the WMO Symposium on Urban Climates and Building Climatology, Brussels, October 1968 (Volume I). xxv, 390 pp. (Technical Notes, 108)
1970, WMO   WMO-No.254.TP.141 $14.00
Composite: E/F
Papers (in full or abbreviated text or in abstracts) on urban climatology in general (3 papers) and special aspects; winds (3), radiation and temperature (10), atmospheric pollution (17), rainfall (5), and climatology and town planning (3).

5   BUILDING CLIMATOLOGY. Proceedings of the WMO Symposium on Urban Climates and Building Climatology, Brussels, October 1968 (Volume II). xxiv, 260 pp., illus. (Technical Notes, 109)
1970, WMO   WMO-No.255.TP.142 $7.00
Composite: E/F
Papers (30, in full or abbreviated text or in abstracts) on climatological aspects of building, effects of wind, ventilation, radiation, thermal behavior of buildings; effects of precipitation and snow; moisture absorption and transfer; indoor climate; air conditioning.

6   A Brief Survey of the Activities of the WORLD METEOROLOGICAL ORGANIZATION relating to HUMAN ENVIRONMENT. 22 pp. and fold. map. ([Special Environmental Reports, 1])
1970, WMO $1.00
Describes briefly WMO activities in monitoring the atmosphere, atmosphere pollution, urban climatology, weather and climate modification, and pollution of the oceans; with a short list of WMO publications.

7   "The Quantity-Quality Relationship in Environmental Management." By Gerardo Budowski. Impact of Science on Society, vol. 20, no. 3, July/September 1970, pp. 235-246.
1970, UNESCO $1.00
Considers the detrimental impacts upon the environment of the quantionship between quality and increasing quantity, citing the Aswan High Dam in Egypt and other examples of the quantity approach without regard for possible long-term consequences.

8   "Three Thousand Million People—Only One Biosphere." By K. Langlo. WMO bulletin, vol. 20, no. 4, October 1971, pp. 240-244.
1971, WMO $1.50
Brief review of the complex problems arising from the changes in the delicate balance of biological or physical processes in the biosphere occasioned by man's activities: population, food, and energy; chemical cycles in the biosphere; and effects on climate.

8 bis   ECE Symposium on PROBLEMS RELATING TO THE ENVIRONMENT. Proceedings and Documentation of the Symposium organized by the United Nations Economic Commission for Europe and held in Prague (Czechoslovakia) with a study tour to the region of Ostrava (Czechoslovakia) and Katowice (Poland). vii, 386 pp. (ST/ECE/ENV/1)
1971, UN   Sales No. E.71.II.E.6 $6.50
In two parts: (1) proceedings, including the report, of the Seminar; and (2) documents (59 papers). Subjects dealt with include environ-

mental conditions and problems in the metal, chemical, and petro-
chemical industries, buildings and related industries, agriculture,
soil erosion, forests and forestry, and water resources management;
problem areas such as recreational and tourist areas, zones of his-
toric value and interest, the metropolitan area of Marseilles-Aix-
Fos, the Potomac River Basin, the St. Louis interstate area, the
Yugoslav Adriatic coast, the industrial areas of Upper Silesia and
the Ruhr; environmental policy at various levels of government and
its socioeconomic and legal aspects; education and training; and
the work of the ECE, the International Council for Building Research, Studies and Documentation (CIB), the Council for Mutual
Economic Assistance (CMEA), the WHO Regional Office for Europe,
and the Organisation for Economic Co-operation and Development
(OECD). Included is a feasibility study and draft plan of action (first
stage) on international arrangements for the exchange and coordination of environmental information in Europe.

Alpine Region

9 RURAL PROBLEMS IN THE ALPINE REGION; an International Study.
By Michel Cépède and E.S. Abensour with the collaboration of Paul and
Germaine Veyret. 201 pp., illus.
1961, FAO $1.50
Comprehensive survey of 583 communities in the Alpine regions of
Austria, France, Germany, Italy, and Switzerland, depicting problems related to all areas where rural life is dominated by natural
forces beyond man's control. Reviews land use, occupations, institutions, social conditions, and population, among other subjects.

Arctic and Antarctic

10 CONFERENCE ON MEDICINE AND PUBLIC HEALTH IN THE ARCTIC
AND ANTARCTIC. Report. 29 pp. (WHO Technical Report Series,
253)
1963, WHO $0.30
Reviews health problems specific to polar areas and dwells upon
sanitation difficulties, the high rate of accidents, nutrition, mental
illness, and infectious diseases with particular reference to zoono-
ses. (See also entry 11.)

11 MEDICINE AND PUBLIC IN THE ARCTIC AND ANTARCTIC: SE-
LECTED PAPERS FROM A CONFERENCE. By K.L. Andersen and
others. 169 pp. (Public Health Papers, 18)
1963, WHO $2.00
Selection of papers presented at the conference of 1962 (entry 10).
Deals with demographic, social, and economic factors; health services; disease control; nutritional requirements; sanitary engineering; human adaptation to station life; physical working capacity of
arctic people; psychological aspects of transient populations.

Arid Zone

12 HUMAN AND ANIMAL ECOLOGY. Reviews of Research.... 244 pp.,
illus. (Arid Zone Research, 8)
1957, UNESCO Bilingual: E/F Out of print
Two reports on human ecology in arid zones; influence of human en-

vironment on human communities and their adaptation to it (with
special reference to housing, clothing, food and other social conditions, and comparing stationary with nomadic and long-established
with recently immigrant communities); and anatomy, physiology,
biochemistry, and pathology of human beings—four reports on animal ecology—animals, birds, locusts, other insects. Extensive
bibliographies.

13   NOMADES ET NOMADISME AU SAHARA. Éd. par Claude Bataillon.
195 pp., illus. (Recherche sur la Zone aride, 19)
1963, UNESCO   In French only $3.00
Contains seven chapters on traditional nomadism; seven chapters on
nomadism and the modern world; a list of tribes; glossary of local
terms; annotated, select bibliography.

14   ENVIRONMENTAL PHYSIOLOGY AND PSYCHOLOGY IN ARID CONDI-
TIONS. Reviews of Research. 345 pp., illus. (Arid Zone Research, 22)
1963, UNESCO $13.50 cl.
Review papers (10) on physiology and the arid zone; physiological
anthropology and climatic variations; nutrition and nutritional diseases in the arid zone; principles of renal function with reference to
the role of the kidney in salt and water metabolism; circulatory adjustments; endocrine functions in hot environments; modification of
the action of drugs by heat; work, sleep, comfort; tropical neuras-
thenia; field test methods. Extensive bibliographies. Prepared for
the Lucknow Symposium, 1962 (entry 15).

15   ENVIRONMENTAL PHYSIOLOGY AND PSYCHOLOGY IN ARID CONDI-
TIONS. Proceedings of the Lucknow Symposium.... 400 pp., illus.
(Arid Zone Research, 24)
1964, UNESCO   Composite: E/F $9.00 cl.
Papers (49) on medical climatology; water and electrolytes; nutrition
and heat; physiological anthropology; performance and comfort standards; comparative animal physiology; significance of solar radiation in the heat balance; neurophysiology of heat exposure; psychological aspects of life in hot climates. (See also entry 14.)

16   Studies on Selected DEVELOPMENT PROBLEMS IN VARIOUS COUN-
TRIES IN THE MIDDLE EAST, 1970. 158 pp. and corrigendum slip; tables, maps. (ST/UNESCOB/7)
1970, UN   Sales No. E.70.II.C.1 $2.50
Among other studies are: "Demographic Characteristics of Youth in
the Arab Countries of the Middle East: Present Situations and
Growth Prospects, 1970-1990" (pp. 71-103); and "Nomadic Populations in Selected Countries in the Middle East and Related Issues of
Sedentarization and Settlement," by Salah M. Yacoub (pp. 105-117).
The latter study describes the general features of nomadism as a
social phenomenon, nomadism in the context of social development,
the process of sedentarization and settlement, and notes some policy
considerations concerning nomadism. Study covers Iraq, Jordan,
Kuwait, Saudi Arabia, and Syria.

Subarctic Regions

17   ECOLOGY OF THE SUBARCTIC REGIONS. Proceedings of the Helsinki
Symposium.... 364 pp., illus. (Ecology and Conservation, 1)
1970, UNESCO   Composite:E/F $19.00 cl.
Papers (36) on subarctic definition, plant and animal ecology, meteo-

rology, snow cover, weathering and geomorphological processes, permafrost, soil-forming processes, conservation of nature, and rational use of renewable natural resources.

Study and Teaching

18   THE STUDY OF ENVIRONMENT IN SCHOOL; Research in Comparative Education. xli, 187 pp. (UNESCO/IBE Publications, 314)
1968, UNESCO/IBE $6.50
Concerns the aims of and place assigned to the study of environment in pre-school, primary, and secondary education; reference to such study in curricula and syllabuses; methods, teaching techniques and aids; training of teaching staff and information available to them. A comparative study based on individual studies of 79 countries which replied to a questionnaire.

19   INTERNATIONAL CONFERENCE ON PUBLIC EDUCATION, XXXIst Session, Geneva, 1968. SUMMARY REPORT. 172 pp. (UNESCO-IBE Publications, 320)
1968, UNESCO/IBE $4.00
Summary report of the plenary meetings; text of general reports and of the Recommendations adopted by the Conference. Its agenda included: (1) education for international understanding as an integral part of school curricula; (2) environmental studies; (3) brief reports from the Ministries of Public Education on educational development in 1967-1968.

20   OUT-OF-SCHOOL SCIENCE ACTIVITIES FOR YOUNG PEOPLE. By R.A. Stevens. 129 pp.
1969, UNESCO $2.00
Contains a chapter on the encouragement of nature conservation as well as chapters on the organization of science clubs, science fairs, camps, meetings, museums, activities of nonspecialized agencies, and the role of administration in encouraging these activities.

NATURAL RESOURCES
AND THE EARTH SCIENCES

GENERAL

21   Proceedings of the United Nations Scientific Conference on the CONSERVATION AND UTILIZATION OF RESOURCES, 17 August-6 September 1949, Lake Success, New York. 8 vol. (E/CONF.7/7, vol. 1-8)
1950-51, UN
Vol. 1: Plenary Meetings. 431 pp., illus. (E/CONF.7/7, vol. 1)
Sales No. 50.II.B.2   Out of print
Papers (57) include a review of world resources, critical shortages, and interdependence, and deal with soils and forests; fuels and energy; metals and minerals; development of new resources by applied technology; resource appraisal; conservation and development programs; education for conservation; labor and public health techniques; integrated river basin development—Tennessee Valley Authority (3 papers and a symposium).

Vol. 2: Mineral Resources. See entry 67.
Vol. 3: Fuel and Energy Resources. See entry 88.
Vol. 4: Water Resources. See entry 279.
Vol. 5: Forest Resources. See entry 459.
Vol. 6: Land Resources. See entry 407.
Vol. 7: Wildlife and Fish Resources. See entry 497.
Vol. 8: Index. 134 pp. (E/CONF.7/7, vol. 8)
Sales No. 50.II.B.9   Out of print

22   CURRENT TRENDS IN SCIENTIFIC RESEARCH. Survey of the Main
     Trends of Inquiry in the Field of the Natural Sciences, the Dissemina-
     tion of Scientific Knowledge and the Application of Such Knowledge for
     Peaceful Ends. By Pierre Auger. 245 pp.
     1963, 3rd printing, UNESCO $6.75 cl.
        Surveys (1) trends of research in the fundamental sciences (mathe-
        matics, automatics, physical sciences, chemical sciences, biological
        sciences), the earth and space sciences, the medical sciences, the
        food and agricultural sciences, fuel and power, and industry; (2)
        trends affecting the organization of scientific research and the dis-
        semination of results; and (3) makes recommendations. First issued
        in 1961.

23   SCIENCE AND TECHNOLOGY FOR DEVELOPMENT. Report of the
     United Nations Conference on the Application of Science and Technology
     for the Benefit of the Less Developed Areas. 8 vol. (E/CONF.39/1,
     vol. 1-8)
     1963, UN
        A narrative account of the Conference in nontechnical style.

     a. Vol. 1: World of Opportunity. 267 pp. (Sales No. 63.I.21) $6.00 cl.
        In three parts: The changing world; the Conference at work (sum-
        maries of the sessions); what has to be done.
     b. Vol. 2: Natural Resources. 243 pp. (Sales No. 63.I.22) $7.50 cl.
        In five parts: Landmarks and outlooks; the waters; minerals; en-
        ergy; scope for cooperation.
     c. Vol. 3: Agriculture. 309 pp. (Sales No. 63.I.23)   Out of print
        In five parts: Background; developing plant resources (soil im-
        provement, plant breeding, plant protection, forestry); developing
        animal and fish resources; developing the institutional framework;
        industrialization of agriculture (including handling, storing, pro-
        cessing of agricultural products, nutrition policy, agricultural
        mechanization, agricultural planning).
     d. Vol. 4: Industry. 265 pp. (Sales No. 63.I.24) $7.50 cl.
     e. Vol. 5: People and Living. 217 pp. (Sales No. 63.I.25) $7.00 cl.
        In three parts: Population trends; public health and nutrition;
        rural development and urbanization.
     f. Vol. 6: Education and Training. 197 pp. (Sales No. 63.I.26) $7.50 cl.
     g. Vol. 7: Science and Planning. 167 pp. (Sales No. 63.I.27)
                                                                 Out of print
     h. Vol. 8: Plenary Proceedings, List of Papers and Index. 295 pp.
        Sales No. 63.I.28) $9.00 cl.

24   AERIAL SURVEYS AND INTEGRATED STUDIES. Proceedings of the
     Toulouse Conference.... 575 pp., illus. (Natural Resources Research,
     6)
     1968, UNESCO   Composite: E/F $26.00 cl.
        Contains papers (57), organized panel discussions on survey princi-
        ples (6), and a general report. Appraises the last two decades of ex-
        perience in aerial photography for the study of natural resources.

Considers methods of ensuring the best use of aerial photography in the study of geology, vegetation, hydrology, ecology, geomorphology, and soils, and the application of these methods to integrated surveys throughout the world, especially in Australia, Africa, and Latin America. The conference was jointly organized by UNESCO, the French Centre National de la Recherche Scientifique (CNRS), and the University of Toulouse, and held at Toulouse 21-28 September 1964.

25   NATURAL RESOURCES OF DEVELOPING COUNTRIES: INVESTIGA-TION, DEVELOPMENT AND RATIONAL UTILIZATION. Report of the Advisory Committee on the Application of Science and Technology to Development. x, 174 pp. (E/4608/Rev.1-ST/ECA/122)
1970, UN   Sales No. E.70.II.B.2 $2.00
Policy guidelines for the inventory and assessment of the natural resources of developing countries and for undertaking the rational utilization of these resources; the role of the United Nations system in assistance to developing countries; areas for specific action. Selected references.

26   WORLD PLAN OF ACTION FOR THE APPLICATION OF SCIENCE AND TECHNOLOGY TO DEVELOPMENT. Prepared by the Advisory Committee on the Application of Science and Development for the Second United Nations Development Decade. 286 pp. (E/4962/Rev.1-ST/ECA/146)
1971, UN   Sales No. E.71.II.A.18 $4.00
In two parts. Part One (chap. I-V) lists priority areas as being primarily important and in which science and technology can make a resounding impact through national and international action during the Second United Nations Development Decade (1971-1980) — in research: high-yielding varieties of staple foods, edible protein, fish, pest and vector control, tropical hardwoods and fibers, groundwater, desalination, arid lands, natural disasters warning systems, indigenous building and construction material, industrial research and design, schistosomiasis, and human fertility — in the application of existing knowledge: storage and preservation of agricultural products, control of livestock diseases, human disease control, housing construction methods, glass and ceramics, improvement and strengthening of science-teaching in secondary schools, industrial extension, and natural resources — building up of an indigenous scientific and technological capacity in the developing countries. It also outlines the Committee's proposals for the implementation and financing of the Plan. Part Two (chap. VI-XV) contains more detailed proposals concerning science and technology policies and institutions, science and technology education, natural resources, food and agriculture, industry, transport and communications, housing, building and urban development, health, population, and application of new technologies.
The Committee had previously presented, in summary form, its selection of its findings and recommendations that should be explicitly embodied in the international development strategy for the Decade, which was reissued in Science and Technology for Development: Proposals for the Second United Nations Development Decade (1970; 47 pages; ST/ECA/133; sales no. E.70.I.23; $0.75). The text of General Assembly resolution 2626 (XXV), of 24 October 1970, proclaiming the Decade and adopting an International Development Strategy: Action Programme of the General Assembly for the Second United Nations Development Decade (1970 [i.e. 1971]; 20 pages; ST/ECA/139; $0.50).

Africa

27   A Review of the NATURAL RESOURCES OF THE AFRICAN CONTI-
      NENT. 437 pp., maps, fold. map in pocket. (Natural Resources Re-
      search, 1)
      1963, UNESCO $15.00 cl.
        Contains critical studies covering topographic mapping; geology,
        geophysics, and mineral resources; seismicity, climate and meteo-
        rology; hydrology; soils; flora and fauna, including freshwater and
        marine biology, conservation of wildlife, national parks and equiva-
        lent reserves, and economic aspects of entomology. Bibliographies;
        list of authorities consulted by UNESCO.

28   International Conference on the Organization of Research and Training
      in Africa in Relation to the Study, Conservation and Utilization of Natu-
      ral Resources, Lagos, Nigeria, 28 July to 6 August 1964:
      a. Final Report of the Lagos Conference. 102 pp.
         1964, UNESCO $0.75
      b. Outline of a Plan for Scientific Research and Training in Africa.
         25 pp.
         1964, UNESCO $0.30
      c. Lagos Conference Selected Documents. 214 pp., illus.
         1965, UNESCO $2.00
      d. Scientific Research in Africa: National Policies, Research Institu-
         tions. 214 pp.
         1966, UNESCO $4.00
         The conference examined the present state of scientific research
         and training of scientific and technical personnel in Africa, the
         organization of science policy and preparation of programs at the
         national level, the needs of research institutions, and the costs of
         training research programs. The fourth publication contains
         statements made at the conference on African national scientific
         policies and an inventory of African scientific and technical
         institutions.

Arid Zones

29   Guidebook to RESEARCH DATA FOR ARID ZONE DEVELOPMENT.
      Ed. by B.T. Dickson. 191 pp., illus. (Arid Zone Research, 9)
      1957, UNESCO   Out of print
        Sketches the scope of resource surveys and the type of data required
        to be collected before decisions are taken on development projects in
        arid and semiarid lands. Contains nine chapters on physical and
        biological factors—mapping, climatology, geology and geophysics,
        geomorphology, hydrology (including irrigation), soil resources,
        vegetation and forests; five chapters on human factors—population,
        land tenure, animal husbandry; nutrition, health and sanitation;
        sources of energy.

30   ARID LANDS: A GEOGRAPHICAL APPRAISAL. Ed. by H.S. Ellis.
      xviii, 461 pp., illus., map
      1966, UNESCO   Out of print
        Results of collaboration of 17 scientists from eight countries on
        problems including water supply and its use and management through
        industrialization and social life. Maps out future research in devel-
        opment of arid lands.

31   GEOGRAPHY OF COASTAL DESERTS. By Peveril Meigs. 140 pp.,
        maps. (Arid Zone Research, 28)
        1966, UNESCO   Out of print
        In two parts: (1) General aspects of coastal desert problems, and (2)
        regional surveys of 37 coastal deserts.

                            Asia and the Far East

32   ATLAS OF PHYSICAL, ECONOMIC AND SOCIAL RESOURCES OF THE
        LOWER MEKONG BASIN. Prepared under the direction of the United
        States Agency for International Development, Bureau for East Asia, by
        the Engineer Agency for Resources Inventory and the Tennessee Valley
        Authority for the Committee for Coordination of Investigations of the
        Lower Mekong Basin (Cambodia, Laos, Republic of Viet-Nam and Thai-
        land), UN Economic Commission for Asia and the Far East .... x,
        257 pp., including maps.
        1968, UN   Bilingual: E/F
        A collection of base maps (common scale, 1:2,000,000), with sepa-
        rate texts and statistical data. Topics include physical resources
        (physiography, climatology, geology, land resources, water re-
        sources development); human resources; soil and economic infra-
        structure; transportation and communications; development activi-
        ties; mapping and photography.

LEGAL STATUS OF NATURAL RESOURCES

33   I. The Status of PERMANENT SOVEREIGNTY OVER NATURAL
        WEALTH AND RESOURCES. II. Report of the Commission on Perma-
        nent Sovereignty over Natural Resources. 245 pp. and corrigendum
        sheet. (A/AC.97/5/Rev.2-E/3511, and Corr.1)
        1962, UN   Sales No. 62.V.6 $3.00
        Problem of control by states over natural wealth and resources in
        relation to foreign individuals or corporate enterprises. The
        greater part of the study centers on measures of control as ex-
        pressed in laws, decrees, treaties, and agreements.

RECREATIONAL AREAS

34   United Nations List of National Parks and Equivalent Reserves. 2nd ed.
        Prepared and published by the IUCN International Commission on Na-
        tional Parks pursuant to United Nations Economic and Social Council
        Resolution 810 (XXXI) [under the direction of Prof. Jean-Paul Harroy].
        601 pp. and corrigendum slip, illus., maps, tables.
        1971, Hayez (Brussels)
        List of national parks and equivalent reserves in 140 countries and
        territories completed and edited for publication under the responsi-
        bility of the International Union for Conservation of Nature and Na-
        tural Resources (IUCN). The first edition had been compiled and re-
        vised and extended as documents of the United Nations Economic and
        Social Council in 1961 and the World Conference on National Parks,
        Seattle, in 1962. The French version of the second edition was pub-
        lished by Hayez in 1967. This English version, prepared under the
        direction of Sir Hugh Elliott, Secretary of the IUCN Commission on
        Ecology, in consultation with Professor Harroy, incorporates some
        amendments and additional information received up to 30 June 1970.

35   PLANNING AND DEVELOPMENT OF RECREATIONAL AREAS IN-
CLUDING THE DEVELOPMENT OF NATURAL ENVIRONMENT. Proceedings of the Seminar . . . held in France and Luxembourg, 27 April-
10 May 1969. 2 vol. (182; [246] pp., including annexes; maps)
(ST/ECE/HOU/40, vol. 1-2)
1969, UN   Sales No. E.69.II.E/Mim.34, vol. 1-2 $4.40 the set
Vol. 1 identifies the main issues arising from the discussions in the
seminar and contains papers (4) dealing with factors governing recreational policies and the planning of recreational areas. Vol. 2
contains case studies (6) on the establishment of recreational areas
(U.S.A., U.S.S.R., Bulgaria, Czechoslovakia, Sweden, Yugoslavia);
papers (3) on recreational areas in France, related to the study tour
in France; and background statistics.

GEOGRAPHY

36   Unesco Source Book for GEOGRAPHY TEACHING. xvi, 254 pp.
1965, UNESCO/Longmans (London) $4.50 cl.
Contains practical advice from international experts on how to improve the teaching of geography at primary and secondary levels.
Emphasis is on methods and teaching materials rather than on specific recommendations for a syllabus.

37   UNESCO GEOGRAPHY SERIES
1967- , UNESCO
Set 1: Europe (Denmark, France, Hungary, Netherlands, Spain,
Switzerland, U.S.S.R., the United Kingdom). Set of 16 wall charts
size 30 x 40 in. (2 for each country), each consisting of 4 full-color
photographs size $12\frac{1}{4}$ x $18\frac{1}{2}$ in.; and a booklet, Teachers' Notes (90
pp.).
1967, UNESCO/Educational Productions Ltd. $19.00 the set
A practical aid for geography teaching in primary and secondary
schools, designed to give children a better understanding of the
life and culture of other peoples.

GEOLOGY

38   RADIOACTIVE DATING. Proceedings of the Symposium on Radioactive
Dating held by IAEA in co-operation with the Joint Commission on Applied Radioactivity (ICSU) in Athens, 19-23 November 1962. 440 pp.,
illus. (Proceedings Series; STI/PUB/68)
1963, IAEA   Composite: E/F/R $8.50 cl.
Papers (31) on new methods and possibilities of the radioactive dating method and its applications in geochemistry and geophysics, geology and meteorites.

39   RADIOACTIVE DATING AND METHODS OF LOW-LEVEL COUNTING.
Proceedings of a Symposium organized by IAEA in co-operation with
the Joint Commission on Applied Radioactivity (ICSU) and held in Monaco, 2-10 March 1967. 744 pp., illus. (Proceedings Series; STI/PUB/
152)

1967, IAEA   Composite: E/F $15.00
Papers (67) on: geochemistry and cosmochemistry of radiocarbon
and radiocarbon dating; radiocarbon dating and dating by cosmic-
ray-produced isotopes; study of meteorites; dating by primeval isotopes; low-level gas counting and methods of tritium enrichment;
low-level counting with scintillation and solid state detectors.

40   METEORITE RESEARCH. Proceedings of a Symposium on Meteorite
Research held in Vienna, Austria, 7-13 August 1968. Ed. by Peter M.
Millman. 941 pp., illus. (Astrophysics and Space Science Library, 12)
1969, IAEA/Reidel (Dordrecht)   Composite: E/F/R $48.00
Papers (73) on early history of meteorites, composition and struc-
ture, isotope studies and chronology, and orbits. Symposium orga-
nized by IAEA in co-operation with UNESCO, ICSU, IAU, IUGS,
IAGC, and the Meteoritical Society.

41   GONDWANA STRATIGRAPHY. IUGS Symposium, Buenos Aires, 1-15
October 1967.... xvi, 1,173 pp., illus., maps. (Earth Sciences, 2)
1969, UNESCO   Composite: E/S $16.00
Papers (80) of specialists in Gondwana stratigraphy arranged in sections on: biochronology, correlation and boundary of units, palaeo-
geography and palaeoclimates, sedimentary environments, regional
geology of the Southern Hemisphere—South America, Southern
Africa, India, Australia, and Antarctica.

42   Field Manual for MUSEUMS. 171 pp., illus. (Museums and Monuments,
12)
1970, UNESCO $7.00
Contains one chapter on scientific research on items in museum collections and nine chapters on the collecting of objects which will ex-
pand and enrich museum collections. Among other subjects are
chapters on prospecting methods (including geophysical methods) in
archaeology, and field-work techniques in geology and mineralogy,
in botany, and in zoology.

43   "Information géologique." Par P. Laffitte et A. Marelle. Natural Resources Forum, vol. 1, no. 1 (ST/ECA/142), 1971, pp. 59-66.
1971, UN   Sales No. E/F/S.71.II.A.13 $2.00
The authors discuss the problems posed by geological data processing which may be resolved by combinations of simpler programs,
such as (a) bank programs (data file storage, automatic documentation); (b) computing programs; (c) cartography programs, in which
the computer produces a map on the basis of the data; and (d) correlation programs. In the latter part of the article, they consider correlation programs involving semantic data, such as the GEOSEMAN-
TICA 70 project initiated by the École des mines (Paris) and the
Royal School of Mines (London).

Africa

44   GEOLOGICAL MAP OF AFRICA 1/5 000 000.... 8 sheets in 20 basic
colors, 1 sheet with full legends in English and French, size 44 x 30 in.;
and explanatory brochure (40 pp.) (Natural Resources Research, 3)
1964, UNESCO/ASGA   Bilingual: E/F $55.00 the complete set (9
sheets and brochure), $6.50 each sheet, $0.65 the brochure
Prepared by the Association of African Geological Surveys (ASGA),
Paris. One sheet contains the general legend. The Explanatory
Note: Geological Map of Africa (1/5 000 000), by R. Furon and J.

Lombard (40 pp., bilingual: E/F), gives the general background of and general observations on the map, describes the geological formations, lists the basic documents for each sheet, and contains a bibliography. Described in detail (with plan) in UNESCO map catalogue (entry 1143).

45   [Map of] AFRICA: GEOLOGICAL BACKGROUND TO SERVE AS A BASIS FOR VARIOUS SYNTHESIS OR THEMATIC MAPS. 2 maps in 3 colors. Map a: scale 1:10,000,000, size 37 x 37 in.; map b: scale 1:20,000,000, size 49 x 49 in.
1967, UNESCO/ASGA   Bilingual: E/F $0.75 each map. Special prices for schools: 5 copies of map a and 1 copy of map b $4.00; 10 copies of map a and 2 copies of map b $7.00
Prepared by the Association of African Geological Surveys (ASGA), Paris.

46   INTERNATIONAL TECTONIC MAP OF AFRICA 1/5 000 000.... 9 sheets using 20 to 30 colors; size $36\frac{1}{2}$ x $26\frac{3}{4}$ in.; and explanatory brochure (54 pp.) (Earth Sciences, 4)
1968, UNESCO   Bilingual: E/F $50.00 the complete set (9 sheets and brochure), $8.50 each sheet
Compiled under the aegis of the International Geological Congress and the Commission for the Geological Map of the World (Sub-Commission for the Tectonic Map of the World). Published jointly by UNESCO and the Association of African Geological Surveys (ASGA), Paris. The Explanatory Note: International Tectonic Map of Africa (1/5 000 000), by R. Furon and J. Lombard (54 pp., bilingual: E/F) is in four chapters: The basement of Africa; the young mountain chains; the post-Precambrian platform covers; miscellaneous phenomena. Bibliography. Appendix: Principal corrigenda and addenda to the map. Described in detail (with plan) in UNESCO map catalogue (entry 1143).

47   Proceedings of the Symposium on the GRANITES OF WEST AFRICA: Ivory Coast, Nigeria, Cameron, March 1965. 162 pp., illus., maps. (Natural Resources Research, 8)
1968, UNESCO   Composite: E/F $10.00 cl.
Reports and papers (8) by geologists taking part in an itinerant symposium in the field and the laboratory, and a special report by F.F.M. Almeida on Precambrian geology of Northeastern Brazil and Western Africa and the theory of continental drift.

48   MINERAL MAP OF AFRICA 1/10 000 000.... Ed. by Jean Lombard. One sheet (in ten colors; size 47 x 36 in.) and Explanatory Note (18 pp.). (Earth Sciences, 3)
1969, UNESCO/ASGA   Bilingual: E/F $10.00 the set
Prepared by the Association of African Geological Surveys (ASGA), Paris, with source material supplied by the Geological Surveys of 39 African countries, and jointly published by UNESCO and ASGA. Map shows geological background and mineral deposits (morphology; substances; hydrocarbons). The Explanatory Note contains two annexes: mineral production in Africa for the year 1966, and a bibliography.

Asia and the Far East

49   OIL AND NATURAL GAS MAP OF ASIA AND THE FAR EAST. Scale 1:5,000,000. 4 sheets (in color), size (two) $28\frac{1}{2}$ x 43 in., (two) $32\frac{5}{8}$ in., and explanatory brochure.
1962, UN   Sales No. 62.I.16 $12.50

50   MINERAL DISTRIBUTION MAP OF ASIA AND THE FAR EAST. Scale
1:5,000,000. 4 sheets (in color), size 30 x 42 in., and explanatory brochure.
1963, UN   Sales No. 63.I.18 $12.50
    Three maps compiled under the auspices of the United Nations Eco-
    nomic Commission for Asia and the Far East from data supplied by
    countries in the region.

51   STRATIGRAPHIC CORRELATION BETWEEN SEDIMENTARY BASINS
FOR THE ECAFE REGION. (Report of Special Working Group and Documents Submitted at the Third Symposium on the Development of Petroleum Resources of Asia and the Far East, Tokyo, November 1965.)
131 pp., illus., maps. (ECAFE. Mineral Resources Development Series, 30; E/CN.11/803)
1969, UN   Sales No. E.68.II.F.11 $2.50
    The working group of stratigraphers and palaeontologists deals with
    problems encountered in the preparation of an atlas of the stratigra-
    phy of each sedimentary basin of the ECAFE region in standard
    form, tied into a common framework of geochronologic units to be
    used in all basins, and the procedure for establishing detailed corre-
    lation. Seventeen papers cover specific areas. Supplements the
    Proceedings of the Third Symposium (entry 101).

52   ENERGY ATLAS OF ASIA AND THE FAR EAST. v, 25 pp., including 24
maps in color, $14\frac{3}{4}$ x 18 in. (E/CN.11/900)
1970, UN   Sales No. E.70.I.11 $12.50
    Prepared by the secretariat of the Economic Commission for Asia
    and the Far East, the atlas incorporates data as at the end of 1968
    concerning oil, gas and coal fields, oil refineries, electric power
    generating stations, hydroelectric potential sites, high voltage elec-
    tric power lines, and oil or gas pipelines. Maps are included for 20
    ECAFE countries but not for Brunei, Burma, Mongolia, and Western
    Samoa. A table provides statistical data for 1967 on annual con-
    sumption of solid fuels, gas, and hydroelectric energy as well as the
    generating capacity and generating electricity supply for individual
    countries in the region.

53   GEOLOGICAL MAP OF ASIA AND THE FAR EAST. Scale 1: 5,000,000.
2nd ed. (rev.) 4 sheets, 39 x 27 in. in 18 basic colors, with full legend
in English; explanatory brochure (in E/F)
1971, UNESCO/UN $12.50
    Prepared under the sponsorship of the United Nations Economic
    Commission for Asia and the Far East with the cooperation of mem-
    ber countries. First edition was issued in 1960; the revised edition
    reflects progress in geological mapping in the area. Described in
    detail (with plan) in UNESCO map catalogue (entry 1143).

Europe

54   GEOLOGICAL SURVEY AND MINING DEVELOPMENT IN EUROPE
AND THE USSR. Report of a Study Group of Geologists and Mining Engineers from Asia and the Far East 4 August-5 November 1955. 215
pp., illus. (ECAFE. [Mineral Resources Development Series, 6]; ST/
TAA/SER.C/27
1958, UN   No sales number   Out of print
    On organization and planning of geological surveys, training of geol-
    ogists and mining engineers, survey and prospecting methods and
    equipment, exploration, and development of fuel and mineral
    resources.

.. 55   INTERNATIONAL GEOLOGICAL MAP OF EUROPE. Scale 1:1,500,000.
49 sheets (in color), size 25 x 25 in., legend in French; a series of bro-
chures with translation of legends in English; a supplementary sheet
containing the general legend. New ed.
1965- , UNESCO/Bundesanstalt für Bodenforschung
Sheets already published:

A3, C1	$1.50 per sheet
C2, D1	$6.00 per sheet
B3, B4, C3, D2, D3, D4, D5	$8.00 per sheet

Prepared by the Committee for the Geological Map of Europe of the
Commission for the Geological Map of the World, to be completed by
1980. Described in detail (with plan) in UNESCO map catalogue (entry 1143).

56   INTERNATIONAL QUATERNARY MAP OF EUROPE. Scale 1:2,500,000.
15 sheets (in color), size 34 x 27 in., and one additional sheet for the
general legend; explanatory brochure with each sheet (with text of legends in German, E, F, R); general explanatory brochure.
1968- , UNESCO/Bundesanstalt für Bodenforschung
Published: Sheets 1, 2, 5, 6 $8.00 including brochure
Prepared by the Bundesanstalt für Bodenforschung, Hannover, in
cooperation with a special commission of the International Union for
Quaternary Research (INQUA), to be completed in 1973. Described
in detail (with plan) in UNESCO map catalogue (entry 1143).

57   CARTE MÉTALLOGÉNIQUE DE L'EUROPE AU 1/2 500 000. 1ère éd.
1968-1970. 9 sheets (30 to 36 colors, size 27 x 38 in.; 9 corresponding
Listes des Gites Minéraux; 1 explanatory brochure; 1 companion volume, The Metallogeny of Europe. In French except for explanatory
brochure (bilingual E/F) and monolingual editions of the companion
volume.
1968- , UNESCO/BRGM   Subscription for whole work: $158.00 with
companion volume, $143.00 without; $19.00 per sheet with corresponding list of deposits and explanatory brochure.
Published: Sheets 1, 4, 6
Compiled by the Drafting Committee on the Metallogenic Map of
Europe under the auspices of the Sub-Commission for the Metallogenic Map of the World of the Commission for the Geological Map
of the World. Published jointly by UNESCO and the Bureau de
Recherches Géologiqeus et Minières (BRGM), Paris. Sheets 2, 3,
5-9 and the corresponding lists of deposits, the explanatory brochure, and the companion volume are in preparation. Described in
detail (with plan) in UNESCO map catalogue (entry 1143).

GEOPHYSICS

58   RADIOISOTOPES IN THE PHYSICAL SCIENCES AND INDUSTRY. Proceedings of the Conference on the Use of Radioisotopes in the Physical
Sciences and Industry held by IAEA, with the co-operation of UNESCO,
at Copenhagen, 6-17 September 1960. 3 vol. (542; 554; 633 pp.); illus.
(Proceedings Series; STI/PUB/20, v.1-3)
1962, IAEA   Composite: E/F/R $8.00 per vol. cl.
Volume 1 includes twelve papers on radioisotopes in geophysics.

59   RADIOISOTOPE INSTRUMENTS IN INDUSTRY AND GEOPHYSICS.
Proceedings of the Symposium on Radioisotope Instruments in Industry
and Geophysics held by IAEA in Warsaw, 18-22 October 1965. 2 vol.
(577; 477 pp.); illus. (Proceedings Series; STI/PUB/112, vol. 1-2)
1966, IAEA   Composite: E/F/R   Vol. 1 $12.00; Vol. 2 $10.00
Volume 2 contains papers (26) on geophysical applications; neutron
methods in geophysics; gamma-gamma methods; density/moisture
gauges.

60   RADIOISOTOPE TRACERS IN INDUSTRY AND GEOPHYSICS. Proceedings of the Symposium on Radioisotope Tracers in Industry and Geophysics held by IAEA in Prague, 22-25 November 1966. 710 pp., illus.
(Proceedings Series; STI/PUB/142)
1967, IAEA   Composite: E/F/R $14.50
Papers on geophysical applications (6), dispersion of effluents (pollutants) in the sea, rivers, lakes, and air (4), among others.

61   Notes on GEOMAGNETIC OBSERVATORY AND SURVEY PRACTICE.
By K.A. Wienert. 217 pp., illus. (Earth Sciences, 5)
1970, UNESCO $11.00 cl.
An up-to-date practical handbook intended for the use of the scientific and technical personnel engaged both in observatory and in field
work.

SEISMOLOGY

62   ASEISMIC DESIGN AND TESTING OF NUCLEAR FACILITIES. Recommendations of a Panel on Aseismic Design and Testing of Nuclear Facilities, held in Tokyo, Japan, 12-16 June 1967. 49 pp. (Technical Reports Series, 88; STI/DOC/10/88)
1968, IAEA $1.00
Contains recommendations on site selection or evaluation (taking
into account seismicity and ground motions, fault displacements, soil
and rock stability, and—for coastal sites—the effects of tsunamis),
aseismic design, and future action including investigations and research problems.

63   Proceedings of the International Seminar on EARTHQUAKE ENGINEER-
ING held under the auspices of the Federal Government of Yugoslavia
and of UNESCO, Skopje, 29 September to 2 October 1964 .... 137 pp.,
illus.
1968, UNESCO   Bilingual: E/F $3.00
Papers (24) on earthquake engineering studies and geological and
seismological investigations in the Skopje area, and on recent developments in earthquake engineering research in eight other countries.

64   The SKOPJE EARTHQUAKE OF 26 JULY 1963. Report of the UNESCO
Technical Assistance Mission .... 152 pp.; illus., including 2 fold.
maps. (UNESCO Earthquake Study Missions)
1968, UNESCO   Composite: E/F $8.50
Studies by four experts of the engineering seismology aspects of the
earthquake, and of the seismicity of the Skopje area and other parts
of Macedonia.

65   SKOPJE RESURGENT. The Story of a United Nations Special Fund
       Town Planning Project. 383 pp., illus., maps. (DP/SF/UN/17.Yugo-
       slavia)
       1970, UN   No sales number $25.00 cl.
       General report prepared for the Government of Yugoslavia acting as
       executing agency for the UNDP. Includes a chapter on the earthquake
       of 1963. The other chapters and the annexes concern the planning
       and execution of the Skopje Urban Plan Project.

66   The SEISMICITY OF THE EARTH, 1953-1965.... By J.P. Rothé.
       336 pp., maps. (Earth Sciences, 1)
       1969, UNESCO   Bilingual: E/F $16.00 cl.
       Contains a summary of seismic activity (1953-1965), statistical data
       and a general bibliography, and a catalogue of 5,329 earthquakes ar-
       ranged in 51 regional tables, with notes and bibliographies. The
       text, tables, and maps were compiled by the Bureau Central Interna-
       tional de Séismologie (BCIS), of which Professor Rothé is Director.
       The volume is intended to represent a continuation of the classic
       work Seismicity of the Earth and Associated Phenomena by B.
       Gutenberg and C.F. Richter.

MINERAL RESOURCES

GENERAL
---

67   Proceedings of the United Nations Scientific Conference on the CON-
       SERVATION AND UTILIZATION OF RESOURCES, 17 August-6 Septem-
       ber 1949, Lake Success, New York. 8 vol. (E/CONF.7/7, vol. 1-8)
       1950-51, UN
       Vol. 2: Mineral Resources. 303 pp., illus. (E/CONF.7/7, vol. 2)
       Sales No. 50.II.B.3   Out of print
       Contains papers on: mineral supplies and their measurement (7);
       discovery of mineral resources (15); conservation in mining and
       milling (13); in manufacture (8), by corrosion control (7), and by
       substitution (5); inorganic fertilizers in conservation (7).

68   NON-FERROUS METALS IN UNDER-DEVELOPED COUNTRIES.
       129 pp., maps. (E/2798-ST/ECA/36)
       1956, UN   Sales No. 55.II.B.3   Out of print
       Technical, fiscal, and other problems that face developing countries
       in making the best use of their nonferrous mineral resources to fur-
       ther economic development.

69   The Establishment of the BRICK AND TILE INDUSTRY IN DEVELOP-
       ING COUNTRIES. By H.W.H. West. 122 pp., illus. (ID/15)
       1969, UN   Sales No. E.69.II.B.19 $1.50
       In seven chapters: raw materials; winning and haulage of clays; clay
       preparation and product manufacture; drying; firing; productivity and
       efficiency; establishment of the heavy clay industry in developing
       countries. Prepared for the Interregional Seminar on the Develop-
       ment of Clay Building Materials Industries in Developing Countries,
       Copenhagen, August 1968, and also presented as a background docu-
       ment at the Workshop on Organizational and Technical Measures for
       the Development of Building Materials, Moscow, 25 September-19

October 1968. The report of the seminar (UN, 1969, 50 pages, (ID/ 28-ID/WG.16/8), sales number E.69.II.B.18, $1.00) is also available.

70   NUCLEAR TECHNIQUES AND MINERAL RESOURCES. Proceedings of the Symposium on the Use of Nuclear Techniques in the Prospecting and Development of Mineral Resources, held by IAEA in Buenos Aires, 5-9 November 1968. 546 pp., illus. (Proceedings Series; STI/PUB/198)
1969, IAEA   Composite: E/F $13.00
   Papers on ore geophysics and processing; uranium exploration; radioisotope x-ray fluorescence; oil field geophysics; activation analysis.

71   MINERAL RESOURCES DEVELOPMENT WITH PARTICULAR REFER- ENCE TO THE DEVELOPING COUNTRIES. 74 pp. and corrigendum sheet. (ST/ECA/123 and Corr.1)
1970, UN   Sales No. E.70.II.B.3 $1.00
   Defines objectives of a mineral development policy, states what is required to attain them, and suggests principles and options for the guidance of those concerned with the framing of such policy. Relates to exploitable mineral resources, metallic and nonmetallic, other than petroleum, natural gas, ground water, and mineral water. An appendix lists United Nations Development Program (Special Fund) projects in the field of geology, hydrogeology, and mining executed by the United Nations, 1959-1969.

72   NUCLEAR TECHNIQUES FOR MINERAL EXPLORATION AND EX- PLOITATION. Proceedings of a Panel on Nuclear Techniques and Min- eral Resources in Developing Countries, organized by IAEA and held in Cracow, 8-12 December 1969. 187 pp., illus., fig., tables. (Panel Pro- ceedings Series; STI/PUB/279)
1971, IAEA   Composite: E/F $5.00
   Papers (15) and summary of the findings of the panel, which re- viewed the present status of and identified those nuclear techniques most suitable for adoption by developing countries.

73   "Geochemical Exploration in United Nations Mineral Surveys." By Claude Lepeltier. Natural Resources Forum, vol. 1, no. 1 (ST/ECA/ 142), 1971, pp. 11-14.
1971, UN   Sales No. E/F/S.71.II.A.13 $2.00
   Briefly reviews the main features of ground exploration projects during 1961-1969 under United Nations auspices.

74   "An Approach to Mineral Policy Formation." By W. Keith Buck and R.B. Elver. Natural Resources Forum, vol. 1, no. 1 (ST/ECA/142), 1971, pp. 16-20.
1971, UN   Sales No. E/F/S.71.II.A.13 $2.00
   Discusses the possible elements of a national mineral policy and the four main avenues by which such a policy may be implemented. Can- ada is used as an example.

Asia and the Far East

75   DEVELOPMENT OF MINERAL RESOURCES IN ASIA AND THE FAR EAST. (Report and Documents of the ECAFE Regional Conference on Mineral Resources Development held in Tokyo, Japan, from 20 to 30 April 1953) 366 pp. (ECAFE: [Mineral Resources Development Series, 2]; E/CN.11/374)

1953, UN   Sales No. 53.11.F.5   Out of print
Report and papers on geological surveys and mineral resources development, including solid mineral fuels, iron ore, ferroalloymetal-
lic deposits, and nonferrous ore resources.

76A   MINING DEVELOPMENTS IN ASIA AND THE FAR EAST 1961. 68 pp.
       (ECAFE. Mineral Resources Development Series, 19; E/CN.11/632)
       1963, UN   Sales No. 64.II.F.2 $1.00

76B   ———1962. 74 pp. (ECAFE. Mineral Resources Development Series,
       22; E/CN.11/666)
       1964, UN   Sales No. 64.II.F.13 $1.00

76C   ———1963. 67 pp. (ECAFE. Mineral Resources Development Se-
       ries, 24; E/CN.11/694)
       1965, UN   Sales No. 65.II.F.14 $1.00

76D   MINING DEVELOPMENTS IN ASIA AND THE FAR EAST; a Twenty-
       year (1945-1965) Review to Commemorate the Twentieth Anniversary of
       ECAFE. 136 pp. (ECAFE. Mineral Resources Development Series,
       27; E/CN.11/751)
       1967, UN   Sales No. 67.II.F.12 $2.50

76E   MINING DEVELOPMENTS IN ASIA AND THE FAR EAST 1966. 92 pp.
       (ECAFE. Mineral Resources Development Series, 31; E/CN.11/832)
       1968, UN   Sales No. E.68.II.F.19 $2.00

76F   ———1967. 134 pp. (ECAFE. Mineral Resources Development Se-
       ries, 33; E/CN.11/893)
       1968 [i.e. 1970], UN   Sales No. E.70.II.F.10 $2.50

76G   ———1968. 93 pp. (ECAFE. Mineral Resources Development Se-
       ries, 35; E/CN.11/949)
       1970 [i.e. 1971], UN   Sales No. E.71.II.F.7 $2.50
       Surveys of mineral production in the ECAFE region: fuels, iron ore
       and mineral resources for ferroalloys, nonferrous and precious
       metals, light metals, radioactive minerals, nonmetallics other than
       fuels; mineral exploration; mineral policy. Country summaries.
       Surveys for the years 1953/54 through 1960 are out of print; they
       were issued during 1954-1962 (Mineral Resources Development Se-
       ries, 4, 5, 8, 11, 13, 15, 16; E/CN.11/393, 421, 459, 509, 537, 565,
       596).

77   Proceedings of the United Nations Seminar on AERIAL SURVEY METH-
       ODS AND EQUIPMENT (held at Bangkok, 4 January to 5 February 1960).
       [204] pp., illus. (ECAFE. Mineral Resources Development Series, 12;
       E/CN.11/536)
       1960, UN   Sales No. 60.II.F.5   Out of print
       Seminar report and papers. Includes country reports on aerial sur-
       veys in the ECAFE region, and papers on aerial photography, photo-
       grammetric mapping, photo-interpretation, and airborne geophysical
       surveys—some reporting work in countries outside the ECAFE
       region.

78   COPPER, LEAD AND ZINC ORE RESOURCES OF ASIA AND THE FAR
       EAST. 63 pp., charts, maps. (ECAFE. Mineral Resources Develop-
       ment Series, 14; E/CN.11/538)
       1960, UN   Sales No. 60.II.F.8   Out of print
       Position of the region in relation to total world output and reserves;
       survey of resources and production country by country in the ECAFE
       region.

79 BAUXITE ORE RESOURCES AND ALUMINIUM INDUSTRY OF ASIA
    AND THE FAR EAST. 51 pp., illus., maps. (ECAFE. Mineral Re-
    sources Development Series, 17; E/CN.11/598)
    1963, UN   Sales No. 63.II.F.2 $0.75
    Study of bauxite resources in the region, aluminum technology, re-
    gional consumption and production of aluminum, and future trends.

80 Proceedings of the Seminar on GEOCHEMICAL PROSPECTING METH-
    ODS AND TECHNIQUES (held at Bangkok, 5-14 August 1963). 202 pp.,
    illus. (ECAFE. Mineral Resources Development Series, 21; E/CN.11/
    634)
    1963, UN   Sales No. 64.II.F.7 $2.50
    Seminar report and papers on prospecting for minerals and the inte-
    gration of geochemical prospecting with other methods of mineral
    exploration, and geochemical prospecting for petroleum and natural
    gas.

81 TIN RESOURCES OF ASIA AND AUSTRALIA. 58 pp., illus. (ECAFE.
    Mineral Resources Development Series, 23; E/CN.11/671)
    1965, UN   Sales No. 65.II.F.4 $0.75
    Summaries by country of tin concentrate and smelter production, tin
    consumption, tin ore resources.

Latin America

82 "Mining in Latin America." Economic Bulletin for Latin America, vol.
    14, no. 2, 1969, pp. 78-109.
    1969, UN   Sales No. E.69.II.G.5 $1.50
    Presents some preliminary information on known mineral reserves
    and resources and the present situation and recent trends in mining
    in Latin America, exclusive of hydrocarbons.

LAWS AND REGULATIONS

83 Survey of MINING LEGISLATION WITH SPECIAL REFERENCE TO
    ASIA AND THE FAR EAST. 111 pp. (ECAFE. Mineral Resources De-
    velopment Series, 9; E/CN.11/462)
    1957, UN   Sales No. 57.II.F.5   Out of print
    In three parts: I. Mineral legislation in the ECAFE region; II.
    Comparative survey of recent petroleum legislation in China, India,
    Pakistan, Philippines, and in countries outside the ECAFE region
    (Guatemala, Israel, Italy, Libya, Mexico, Peru, Turkey, Venezuela);
    III. Comparative survey of provisions on proprietary rights, pros-
    pecting and mining licenses, leases and concessions, governmental
    supervision and control of mining operations, and government rev-
    enue, selected from mining laws of countries outside the ECAFE
    region (Victoria, Australia; Belgian Congo; Brazil; Manitoba, Can-
    ada; Egypt; Mexico; Peru).

84 Proceedings of the Seminar on MINING LEGISLATION ADMINISTRA-
    TION (held at Manila, Philippines, 19 to 23 October 1969). viii, 350 pp.
    (ECAFE. Mineral Resources Development Series, 34; E/CN.11/919)
    1970, UN   Sales No. E.70.II.F.12 $5.00
    In two parts: (1) Report of the Seminar, which reviewed basic poli-
    cies, laws, and regulations governing the exploration and exploita-
    tion of mineral resources, mining investment policies in the ECAFE

region, the classification of minerals adopted for determining the different types of mining licenses, leases and concessions, governmental revenues from mineral resources, government incentives to encourage exploration and development of mineral resources, and government authority in the administration of mineral laws and mineral lands; (2) documentation (60 papers) on these subjects, and a summary of mining laws in the ECAFE area covering Australia (5 States and the Northern Territory) and 17 other countries and territories.

IRON ORES

85   WORLD IRON ORE RESOURCES AND THEIR UTILIZATION, with special reference to the Use of Iron Ores in Under-developed Countries. (ST/ECA/6)
      1950, UN   Sales No. 50.II.D.3   Out of print
      Appraisal of the economic feasibility of developing iron and steel production in the developing countries based on a survey of their reserves of iron ore, coking coal, location of sources and markets, technology of production.

86   Survey of WORLD IRON ORE RESOURCES: Occurrence, Appraisal and Use. Report of a Committee of Experts appointed by the Secretary-General. 345 pp., illus., maps. (E/2655-ST/ECA/27)
      1955, UN   Sales No. 54.II.D.5   Out of print
      Analysis of problems arising in the appraisal of iron ore resources; survey of chief reserves and exploitations in various countries; production problems. Twelve detailed technical papers supporting the general analysis.

87   Survey of WORLD IRON ORE RESOURCES: Occurrence and Appraisal. Report of a Panel of Experts appointed by the Secretary-General. 479 pp., maps. (ST/ECA/113)
      1970, UN   Sales No. E.69.II.C.4 $6.50
      Contains the general report and recommendations of the panel, which met in 1966-1967, five technical reports, and eight regional appraisals of iron ore deposits throughout the world. Revises substantially the 1955 study, Survey of World Iron Ore Resources: Occurrence, Appraisal and Use (entry 86).

FUEL AND ENERGY RESOURCES

GENERAL

88   Proceedings of the United Nations Scientific Conference on the CONSERVATION AND UTILIZATION OF RESOURCES, 17 August-6 September 1949, Lake Success, New York. 8 vol. (E/CONF.7/7, vol. 1-8)
      1950-1951, UN
            Vol. 3: Fuel and Energy Resources. 333 pp., illus. (E/CONF.7/7, vol. 3)
            Sales No. 50.II.B.4   Out of print
                Contains papers on: techniques of oil and gas discovery and production (6); new techniques for increasing production of oil and

gas (8); petroleum chemistry (7); coal mining (4); coal preparation (4); underground gasification of coal (3); coal carbonization
(6); conservation in use of coal for space heating (7); the integrated power system (5); new developments in production and
utilization of fuel and energy, including thermal and wind power.

89 THERMAL POWER STATIONS; a Techno-economic Study. 69 pp. (E/
CN.11/891)
1970, UN   Sales No. E.70.II.F.2 $1.50
Reviews developments in the siting, layout and design, and cost studies of thermal power stations, operation and maintenance personnel,
cost control and operational efficiency, the problem of atmospheric
pollution, and the use of crude oil as boiler fuel.

COAL AND LIGNITE

90 COAL AND IRON ORE RESOURCES OF ASIA AND THE FAR EAST.
155 pp., maps. (ECAFE. [Mineral Resources Development Series, 1];
ST/ECAFE/5)
1952, UN   Sales No. 52.II.F.1   Out of print
Covers briefly the present status of surveys, the extent of known ore
resources, the status of production, and the possibilities of increasing production; country studies cover fourteen countries.

91 The AUSTRALIAN LIGNITE (BROWN COAL) INDUSTRY in relation to
the Development of Low Grade Coal Deposits in Asia. Report of the
Study Group of Lignite Experts on Their Visit to Australia (October-
November 1953). 175 pp., illus., fold. map. (ECAFE. [Mineral Resources Development Series, 3]; ST/TAA/SER.C/15)
1954, UN   No sales number   Out of print
Analysis of Australian organizations, operational methods, and utilization processes and their applicability to existing conditions in
other countries of the ECAFE region.

92 LIGNITE RESOURCES OF ASIA AND THE FAR EAST, Their Exploration, Exploitation and Utilization. 134 pp., illus. (ECAFE. Mineral
Resources Development Series, 7; E/CN.11/438)
1957, UN   Sales No. 57.II.F.3   Out of print
Surveys the availability and production of lignite in the region, methods of exploration for and development of deposits, processing methods, and the utilization of lignite and other low-grade coals in thermal power plants, the iron and steel industry, in households, handi-
crafts, and agriculture, and for other industrial purposes.

NUCLEAR RAW MATERIALS

93 Proceedings of the International Conference on the PEACEFUL USES
OF ATOMIC ENERGY, held in Geneva, 8 August-20 August 1955. 17
vol. (A/CONF.8/1-17)
1956, UN
a. Vol. 6: Geology of Uranium and Thorium. 825 pp., illus., maps.
(A/CONF.8/6)
Sales No. 56.IX.1, vol. 6   Out of print
Papers (128) on the natural occurrence of uranium and thorium
and prospecting for their ores.

   b. Vol. 8: Production Technology of the Materials Used for Nuclear
      Energy. 627 pp., illus. (A/CONF.8/8)
      Sales No. 56.IX.1, vol. 8   Out of print
      Papers (95) on the treatment of uranium and thorium ores and
      ore concentrates; production of metallic uranium and thorium;
      analytical methods in raw material production; heavy water;
      graphite; zirconium; beryllium.

94   Proceedings of the Second United Nations International Conference on
     the PEACEFUL USES OF ATOMIC ENERGY, held in Geneva, 1 Septem-
     ber-13 September 1958. 33 vol. (A/CONF.15/1, vol. 1-33)
     1958, UN
     a. Vol. 2: Survey of Raw Material Resources. 843 pp., illus., maps.
        (A/CONF.15/1, vol. 2)
        Sales No. 58.IX.2, vol. 2   Out of print
        Papers (102) on raw material resources, geochemistry (including
        isotopic composition and age determination), mineralogy and gen-
        esis of deposits, geology of deposits, prospecting (geochemical
        and radiometric).
     b. Vol. 3: Processing of Raw Materials. 607 pp., illus. (A/CONF.15/1,
        vol. 3)
        Sales No. 58.IX.2, vol. 3   Out of print
        Papers on mining aspects and instrumentation (17) and ore treat-
        ment and analytical methods (61).

95   SURVEYING AND EVALUATING RADIOACTIVE DEPOSITS. By A.H.
     Lang. 50 pp. (Review Series, 1; STI /PUB/15/1)
     1959, IAEA   Out of print
     Review of the literature in English 1955-1959 on practical problems
     involved in the exploration and evaluation of radioactive deposits of
     uranium and thorium.

96   Metallurgy of THORIUM PRODUCTION. By B. Praksh, S.R. Kantan,
     and N.K. Rao. 56 pp. (Review Series, 22; STI/PUB/15/22)
     1962, IAEA $1.00
     After dealing briefly with the history and occurrence of thorium ores
     (chiefly monazite), deals with the metallurgy of the production of
     thorium and with its applications. References.

97   Chemical Processing of URANIUM ORES. By E.T. Pinkney, W. Lurie
     and P.C.N. van Zyl.—Tratamiento de MINERALES DE URANIO. Por
     L.G. Jodra y J.M. Josa. 133 pp., illus. (Review Series, 23/24; STI/
     PUB/15/23/24)
     1962, IAEA $1.00
     Both monographs deal with the processing of uranium ores, with the
     first monograph giving a short outline of their mineralogy. The
     second monograph is in Spanish only. Both contain extensive ref-
     erences.

98   Proceedings of the Third International Conference on the PEACEFUL
     USES OF ATOMIC ENERGY, held in Geneva, 31 August-9 September
     1964. Multilingual ed. 16 vol. (A/CONF.28/1, vol. 1-16)
     1965, UN   Composite: E/F/R/S
     a. Vol. 12: Nuclear Fuels—III. Raw Materials. 496 pp., illus. (A/
        CONF.28/1, vol. 12)   Sales No. 65.IX.12 $12.50 cl.
        Papers on uranium and thorium resources and requirements (12),
        prospecting techniques and recovery from ores (20), and isotope
        separation (19).

99   Processing of Low-grade URANIUM ORES. Proceedings of a Panel held
       in Vienna, 27 June-1 July 1966. 247 pp., illus. (Panel Proceedings Se-
       ries; STI/PUB/146)
       1967, IAEA   Composite: E/F/R/S $5.00
       Status reports from 13 countries, technical reports (13), summaries
       of discussion, and the panel recommendations for an IAEA program
       of action.

100   URANIUM EXPLORATION GEOLOGY. Proceedings of a Panel on Ura-
       nium Exploration Geology held in Vienna 13-17 April 1970. 384 pp. and
       corrigendum slip; illus., maps. (Panel Proceedings Series; STI/PUB/
       277)
       1970, IAEA   Composite: E/F/S $10.00
       Contains introductory review of uranium reserves, future demand,
       and extent of the exploration problem; descriptive summaries (12) of
       uranium deposits; uranium deposit genesis; general geology (2 pa-
       pers), strata-bound deposits (4), favorability criteria (6); reports
       from five working groups; and recommendations.

101   URANIUM: RESOURCES, PRODUCTION AND DEMAND, SEPTEMBER
       1970. A joint report by the European Nuclear Energy Agency and the
       International Atomic Energy Agency. 54 pp., tables.
       1970, OECD $3.00
       Estimates of uranium resources, production and demand (in 19 coun-
       tries, exclusive of the USSR, Eastern Europe, and the People's Re-
       public of China), and brief estimate of thorium resources, prepared
       jointly by ENEA/IAEA experts. Updates two previous reports: Ura-
       nium Resources, Revised Estimates (1967, OECD) by a joint ENEA/
       IAEA working party, and World Uranium and Thorium Resources
       (1965, OECD) by an ENEA study group.

102   THE RECOVERY OF URANIUM. Proceedings of a Symposium on the
       Recovery of Uranium from Its Ores and Other Sources, organized by
       IAEA and held in São Paulo, 17-21 August 1970. 478 pp., illus., fig.,
       tables. (Proceedings Series; STI/PUB/262)
       1971, IAEA   Composite: E/F/R/S $13.00
       Papers (35) and discussions on the processing of uranium ores, pre-
       concentration and recent development in separation methods and re-
       covery of uranium from clear solutions and ore pulps, and recovery
       of uranium as a by-product.

                                  Africa

103   PEACEFUL USES OF ATOMIC ENERGY IN AFRICA. Proceedings of a
       Symposium on the Peaceful Uses of Atomic Energy in Africa organized
       by the Organization of African Unity in Kinshasa, 28 July-1 August 1969
       in collaboration with IAEA and the Nuclear Science Commission of the
       Democratic Republic of the Congo. 574 pp., illus. (Proceedings Series;
       STI/PUB/233)
       1970, IAEA   Composite: E/F $16.00
       Contains report and recommendations of the symposium and papers
       (72) on: evolution of nuclear science in Africa, nuclear techniques in
       the agricultural sciences (soil science, irrigation practices, plant
       breeding, plant and animal protection), hydrology, geology and ex-
       ploration for nuclear raw materials (9 papers), medical and biologi-
       cal sciences, physical and chemical sciences, reactors, and training
       of personnel.

PETROLEUM AND NATURAL GAS

104   TECHNIQUES OF PETROLEUM DEVELOPMENT. Proceedings of the
        United Nations Inter-regional Seminar on Techniques of Petroleum De-
        velopment, New York, 23 January to 21 February 1962. 345 pp., illus.
        (ST/TAO/SER.C/60)
        1964, UN   Sales No. 64.II.B.2 $4.00
        Lectures (34) by experts on the technical, administrative, and oper-
        ational aspects of the petroleum industry, including petroleum geol-
        ogy (1) and petroleum exploration and production (8).

105   Utilization of OIL SHALE: Progress and Prospects. 112 pp. and corri-
        gendum sheet; illus. (ST/EGA/101 and Corr.1)
        1967, UN   Sales No. 67.II.B.20 $2.00
        In three parts: (1) Progress in oil shale development and research;
        (2) oil shale development in selected areas (United Kingdom, France,
        Brazil, Chile, Uruguay, Bolivia); (3) some economic aspects of the
        utilization of low-grade oil shale for thermal power generation.

106   "Latest Advances in Off-shore Petroleum Technology." By Frank F.H.
        Wang. Natural Resources Forum, vol. 1, no. 1 (ST/ECA/142), 1971,
        pp. 22-30.
        1971, UN   Sales No. E/F/S.71.II.A.13 $2.00
        Reviews the latest technological improvements and achievements in
        offshore petroleum exploration and exploitation.

                            Asia and the Far East

107   Proceedings of the Symposium on the Development of PETROLEUM
        RESOURCES OF ASIA AND THE FAR EAST. 251 pp. (ECAFE. Mineral
        Resources Development Series, 10; E/CN.11/507)
        1959, UN   Sales No. 59.II.F.3   Out of print
        Report of the Symposium, held at New Delhi, December 1958, and the
        principal papers. There are summaries of papers on some coun-
        tries outside the ECAFE region (France, Federal Republic of Ger-
        many, Netherlands, U.S.S.R., and U.S.A.) and on recent petroleum
        legislation inside and outside the region.
        a. ———. Annex. 73 pp., illus., maps. (ECAFE. Mineral Resources
        Development Series, 10 (Annex); E/CN.11/507 (Annex))   Out of print
            Contains the 72 figures for the papers contained in the main vol-
            ume, and maps.

108   Proceedings of the Second Symposium on the Development of PETRO-
        LEUM RESOURCES OF ASIA AND THE FAR EAST. 2 vol. (556, 511
        pp.), illus. (ECAFE. Mineral Resources Development Series, 18, vol.
        1-2; E/CN.11/601, vol. 1-2)
        1963, UN   Sales No. 63.II.F.3 and 63.II.F.9 $5.50 each
        The symposium was held in Tehran, 1-15 September 1962. Vol. I
        contains the report of the symposium, papers reviewing recent pro-
        gress of the oil industry, and papers on petroleum geology and ex-
        ploration and exploitation techniques. Vol. II contains papers on
        natural gas production, transportation, and use; economics of the
        petroleum industry; and technical training in the industry. Most
        papers relate to countries in the ECAFE region, but others on each
        topic deal with countries outside the region.

109 Case Histories of OIL AND GAS FIELDS IN ASIA AND THE FAR EAST.
161 pp., illus. (ECAFE. Mineral Resources Development Series, 20;
E/CN.11/633)
1964, UN   Sales No. 64.II.F.4 $2.00
Papers deal with exploration and exploitation, geological background,
and reservoir characteristics of selected oil and gas fields in Brunei, China (Taiwan), Iran, Japan and Pakistan, and offshore exploration in British Borneo. Papers had been submitted to the Second
Symposium, Tehran, 1962. For the second and third series, see entries 112 and 113.

110 Proceedings of the SEMINAR ON THE DEVELOPMENT AND UTILIZA-
TION OF NATURAL GAS RESOURCES with special reference to the
ECAFE Region. xiii, 436 pp., illus. (ECAFE. Mineral Resources Development Series, 25; E/CN.11/711)
1965, UN   Sales No. 66.II.F.6 $5.00
Report of the seminar and papers on: review of recent progress of
natural gas development (15 papers); development, operation, and
conservation of natural gas fields (7); legal and administrative aspects of development of the natural gas industry (3); financial aspects (2); transmission, storage, and distribution of natural gas (12);
marketing problems of natural gas and its allied products (4); utilization of natural gas and its allied products—as materials for fertilizer and other chemical industries (5); natural gas for the iron and
steel industry (2); use of natural gas as a fuel (4); and contribution of
natural gas to agriculture (4); regional arrangements in natural gas
development and in the establishment of industries based on natural
gas (3). Most papers deal with developments in Asia with Far East
but some relate to Europe, Trinidad and Tobago, the U.S.S.R., and
the U.S.A.

111 Proceedings of the Third Symposium on the Development of PETRO-
LEUM RESOURCES OF ASIA AND THE FAR EAST. 3 vol. (Mineral
Resources Development Series, 26, vol. 1-3; E/CN.11/750, vol. 1-3)
1967, UN
a. Vol. 1: Report of the Symposium; Documents: Petroleum Geology and
Exploration Methods. xxx, 603 pp., illus., maps.
Sales No. 67.II.F.10 $8.00

b. Vol. 2: Documents: Petroleum Exploitation Methods; Storage, Transportation and Utilization of Oil and Natural Gas. xx, 369 pp., illus.
Sales No. 67.II.F.19 $5.00
c. Vol. 3: [Documents]: Economics, Technical Training, U.N. Assistance. xviii, 181 pp., illus. Sales No. 67.II.F.20 $3.00
The symposium was held in Tokyo, 10-20 November 1965. (See also
entry 51.)

112 Case Histories of OIL AND GAS FIELDS IN ASIA AND THE FAR EAST
(Second Series). Submitted to the Third Symposium on the Development
of Petroleum Resources of Asia and the Far East, Tokyo, November
1965. 96 pp., illus., maps. (ECAFE. Mineral Resources Development
Series, 29; E/CN.11/800)
1968, UN   Sales No. E.68.II.F.10   Out of print
Studies of selected fields in Afghanistan, Australia, China (Taiwan),
Japan, New Zealand, and Thailand, including an offshore structure in
the Sea of Japan. For the first and third series, see entries 109 and
113.

113 Case Histories of OIL AND GAS FIELDS IN ASIA AND THE FAR EAST
(Third Series). (Submitted to the Fourth Symposium on the Develop-

ment of Petroleum Resources in Asia and the Far East, Canberra,
Australia, October-November 1969.) 90 pp., illus., maps. (ECAFE.
Mineral Resources Development Series, 37; E/CN.11/952)
1971, UN   Sales No. E.71.II.F.8 $2.50
Studies of selected oilfields in Australia, China (Taiwan), Japan, and
Thailand, with an account of the discovery and development of the
Sarir oilfield in Libya. For the first and second series, see entries
109 and 112.

WIND, SOLAR, TIDAL, AND GEOTHERMAL ENERGY

114   WIND AND SOLAR ENERGY. Proceedings of the New Delhi Sympo-
       sium. 238 pp., illus. (Arid Zone Research, 7)
       1956, UNESCO   Composite: E/F/S   Out of print
       Papers (23) of this 1954 symposium deal with the application of wind
       and solar energy in arid regions.

115   NEW SOURCES OF ENERGY AND ECONOMIC DEVELOPMENT: Solar
       Energy, Wind Energy, Tidal Energy, Geothermic Energy and Thermal
       Energy of the Seas. 150 pp. (E/2997-ST/ECA/47)
       1957, UN   Sales No. 57.II.B.1   Out of print
       In three parts: (1) Comparative study of the five new sources of en-
       ergy in question; (2) five studies by an expert on each of these
       sources; (3) selected annotated bibliography (480 items).

116   NEW SOURCES OF ENERGY AND ENERGY DEVELOPMENT. Report
       on the United Nations Conference on New Sources of Energy: Solar En-
       ergy—Wind Power—Geothermal Energy, Rome, 21 to 31 August 1961.
       65 pp. and corrigendum sheet. (E/3577/Rev.1-ST/ECA/72, and Corr.
       1)
       1962, UN   Sales No. 62.I.21 $0.75
       Reviews briefly the Conference background and arrangements, sum-
       marizes the proceedings, and reviews the implications of the Confer-
       ence for international action. Synthesizes the papers and related
       discussions.

117   Proceedings of the United Nations Conference on NEW SOURCES OF
       ENERGY: SOLAR ENERGY, WIND POWER AND GEOTHERMAL EN-
       ERGY, Rome, 21-31 August 1961. 7 vol. (E/CONF.35/2-8)
       1964, UN   Composite: E/F $33.50 the set
       a. Vol. 1: General Sessions. 218 pp. (E/CONF.35/2)
          Sales No. 63.I.2 $2.50
          Papers and reports on new sources of energy and energy develop-
          ment (9), and combined use of various energy sources and energy
          storage problems (10).
       b. Vol. 2: Geothermal Energy: I. 420 pp., illus. (E/CONF.35/3)
          Sales No. 63.I.36 $5.00
          Prospection of geothermal fields and evaluation of their capacity
          (40).
       c. Vol. 3: Geothermal Energy: II. 516 pp., illus. (E/CONF.35/4)
          Sales No. 63.I.37   Out of print
          Harnessing of geothermal energy and geothermal electricity pro-
          duction (29), utilization of geothermal energy for heating pur-
          poses and combined schemes and by-products (11).
       d. Vol. 4: Solar Energy: I. 665 pp., illus. (E/CONF.35/5)
          Sales No. 63.I.38 $7.50
          Use of solar energy for mechanical power and electricity produc-

tion (22), solar energy availability and instruments for measurement (21), new materials in solar energy utilization (11).

 e. Vol. 5: Solar Energy: II. 423 pp., illus. (E/CONF.35/6)
  Sales No. 63.I.39 $4.50
  Use of solar energy for heating purposes (38).

 f. Vol. 6: Solar Energy: III. 454 pp., illus. (E/CONF.35/7)
  Sales No. 63.I.40 $5.00
  Use of solar energy for cooling purposes (10), for the production of fresh water (14), for high-temperature processing (solar furnaces) (14).

 g. Vol. 7: Wind Power. 408 pp., illus. (E/CONF.35/8)
  Sales No. 63.I.41   Out of print
  Contains 43 papers.

118 SITES FOR WIND-POWER INSTALLATIONS. (Report of a Working Group of the Commission for Aerology) 38 pp., illus. (Technical Notes, 63)
   1964, WMO   WMO-No. 156.TP.76 $2.00
   Summarizes theoretical and experimental studies of air flow over hills with particular attention to the estimation of wind speeds over the summits of such hills. Prepared in response to a request of the United Nations Conference on New Sources of Energy, Rome, 1961.

119 "The Economics of Geothermal Power." By J.C.C. Bradbury. Natural Resources Forum, vol. 1, no. 1 (ST/ECA/142), 1971, pp. 46-53.
   1971, UN   Sales No. E/F/S.71.II.A.13 $2.00
   Discusses the basic factors involved in assessing the costs of the different phases associated with exploration and development of geothermal energy resources.

120 "Multipurpose Exploration and Development of Geothermal Resources." By Joseph Barnes. Natural Resources Forum, vol. 1, no. 1 (ST/ECA/142), 1971, pp. 55-58.
   1971, UN   Sales No. E/F/S.71.II.A.13 $2.00
   Advocates the multipurpose approach to the development of geothermal resources from the very earliest stage of reconnaissance to that of reservoir management.

OCEANOGRAPHY

GENERAL

121 Proceedings of the UNESCO Symposium on PHYSICAL OCEANOGRAPHY, 1955, Tokyo. 292 pp., illus. (Marine Sciences Programme)
   1957, UNESCO/Japanese Society for the   Out of print
   Promotion of Science (Tokyo)
   Papers (46) on instrumentation, circulation, and information on various oceanographic research activities.

122 Draft of a General Scientific Framework for WORLD OCEAN STUDY.
   76 pp. (IOC/89/K)
   1964, UNESCO/IOC   Out of print
   Preliminary draft prepared for IOC by ICSU Scientific Committee on

Oceanic Research (SCOR). Revised on the basis of comments received and reissued as <u>Perspectives in Oceanography</u>, 1969 (<u>entry 126</u>).

123   INTERGOVERNMENTAL OCEANOGRAPHIC COMMISSION (Five Years of Work). 39 pp., illus., maps. (IOC. Technical Series, 2)
1966, UNESCO $1.00
Short history (1960-1965) and description of IOC and its current operations, its international expeditions, with list of IOC working and coordination groups, their meetings, membership, and terms of reference. Includes IOC Statutes and a list of documents and publications about IOC.

124   MORNING REVIEW LECTURES OF THE SECOND INTERNATIONAL OCEANOGRAPHIC CONGRESS (including three in the original French text), Moscow, 30 May to 9 June 1966. 256 pp., illus., maps.
1969, UNESCO $18.00 cl.
Contains lectures (22) covering ocean/atmosphere reaction, ocean circulation, sea surface, ocean food chain, productivity and distribution of marine organisms, submarine geology and geochemistry, and research in the Indian Ocean.

125   SEA-SURFACE TEMPERATURE. Lectures presented during the scientific discussions at the fifth session of the Commission for Maritime Meteorology. xv, 151 pp., illus. (Technical Notes, 103)
1969, WMO   WMO- No.247.TP.135 $6.00
Seven lectures on sea-surface temperature: ways of measuring it, uses of observations to increase the harvest of the sea, the persistency of ocean temperature patterns, and the uses of sea temperature measurements in long-range weather prediction.

126   PERSPECTIVES IN OCEANOGRAPHY. 90 pp., fig. (IOC. Technical Series, 6)
1969, UNESCO $3.00
Revision of the 1964 <u>Draft of a General Scientific Framework for World Ocean Study</u> (<u>entry 122</u>). A comprehensive outline of the possible future development of oceanographic research through international cooperation, it cites expected economic benefits and touches upon aspects of modern marine technology and marine chemistry, and opportunities and problems in physical oceanography, marine geology, geophysics, and marine biology. It omits aspects of fisheries research, which were dealt with in the earlier <u>Draft</u>.

127   GLOBAL OCEAN RESEARCH. Report prepared by a Joint Working Party of the Scientific Committee on Oceanic Research (ICSU), the Advisory Committee on Marine Resources Research (FAO) and the Advisory Group on Oceanic Research (WMO), Ponza and Rome, 29 April to 7 May 1969. xxxvii, 47 pp. (Reports on Marine Science Affairs, 1)
1970, WMO $2.00
Preliminary report identifying the most important oceanic research problems, suggesting the principal programs of research which should be undertaken and the areas of the ocean where they would be most usefully carried out, and estimating what supporting facilities, services and manpower would be necessary to implement an expanded program of oceanic research. The Joint Working Party established small, interdisciplinary discussion groups for closer examination of research problems and programs in ocean circulation and ocean-atmospheric interaction, life in the ocean, marine pollution, and dynamics of the ocean floor.

128 EXPLORING THE OCEAN. By Daniel Behrman. 89 pp., illus.
(Unesco and Its Programme)
1971, UNESCO $1.00
Description of recent international cooperation in exploring the
ocean, studying its influence on the weather, considering its possibilities as a source of food and mineral resources, and examining
the problem of marine pollution.

Tropical Atlantic

129 Proceedings of the Symposium on the OCEANOGRAPHY AND FISH-
ERIES OF THE TROPICAL ATLANTIC, Abidjan, Ivory Coast, 20-28
October 1966. Results of the ICITA and of the GTS, organized through
the joint efforts of UNESCO, FAO and OAU. Review Papers and Contributions.... 430 pp.
1969, UNESCO   Composite: E/F $11.00
Largely devoted to the results of the International Cooperative Investigations of the Tropical Atlantic (ICITA) and the Guinean Trawl-
ing Survey (GTS). Papers (38) on oceanography of the tropical Atlantic and the Gulf of Guinea, and plankton and fish distribution.

Caribbean Sea

130 SYMPOSIUM ON INVESTIGATIONS AND RESOURCES OF THE CARIB-
BEAN SEA AND ADJACENT REGIONS Preparatory to the Co-operative
Investigations of the Caribbean and Adjacent Regions (CICAR), orga-
nized jointly by Unesco and FAO, Willemstad, Curaçao, Netherlands
Antilles, 18-26 November 1968. Papers on Physical and Chemical
Oceanography, Marine Biology. 545 pp., fig., tables.
1971, UNESCO   Composite: E/S $16.00 cl.
Papers on physical and chemical oceanography (25), marine geology
and geophysics (13), marine biology (27). Note: The report of the
Symposium and abstracts of all papers have been issued as an unpriced item by FAO in FAO Fisheries Reports, 71.1 (1969, 165 pp.,
FRm/R.71.1), and the full texts in their original languages of the
papers contributed to the Fishery Resources Section of the Symposium are said to be appearing eventually in the same FAO series.

Indian Ocean

131 INTERNATIONAL INDIAN OCEAN EXPEDITION. Collected Reprints.
Published upon the recommendation of the Scientific Committee on
Oceanic Research (SCOR) and the Intergovernmental Oceanographic
Commission (IOC). 4 vols.
1965-67, UNESCO Composite: E/F/German/R
a. Vol. 1: Reprints No. 1-77. 1965. xv, 915 pp., illus.
      Physics and chemistry of water; marine geology; geophysics.
b. Vol. 2: Reprints No. 78-120. 1966. ix, 670 pp., illus.
      Marine biology.
c. Vol. 3: Reprints No. 130-204. 1966. ix, 712 pp., illus.
      Marine biology; marine chemistry; physical oceanography;
      marine geology and geophysics.
d. Vol. 4: Reprints No. 205- . 1967. x, 867 pp., illus.
      Marine biology; marine chemistry; physical oceanography;
      marine geology and geophysics; papers presented by title or
      abstract only.

132   METEOROLOGY IN THE INDIAN OCEAN. By C. Ramage. 31 pp., illus.
1965, WMO   WMO-No.166.TP.81 $1.00
Describes the climate of the region, the meteorological program of the International Indian Ocean Expedition (IIOE), and the work of the International Meteorological Centre, established in 1963 in Bombay.

Pacific Ocean

133   Proceedings of Symposium on the KUROSHIO, Tokyo, October 29, 1963. 66 pp. and errata sheet; illus., maps.
1965, Oceanographical Society of Japan/UNESCO (Tokyo) $1.50
Papers (6) on the Kuroshio current system in the Pacific Ocean, its physical, chemical, and fishery aspects; plankton in its waters; and thoughts about planning the cooperative study (CSK) of the Kuroshio and its adjacent region, 1965-

METHODOLOGY

134   DETERMINATION OF PHOTOSYNTHETIC PIGMENTS IN SEA-WATER. 69 pp. (Monographs on Oceanographic Methodology, 1)
1969, 2nd printing, UNESCO $3.00
Contains report (1964) of SCOR-UNESCO Working Group 17; a survey by T.R. Parsons of existing methods; and two Australian papers on methods of determination of phytoplankton pigments. First issued in 1966.

135   ZOOPLANKTON SAMPLING. 174 pp., illus. (Monographs on Oceanographic Methodology, 2)
1968, UNESCO $6.50
Review papers (7) on zooplankton sampling methods and relevant parts of reports of four working parties on standardization of methods for sampling zooplankton at sea.

136   OCEANIC PART OF THE HYDROLOGICAL CYCLE. A Review on Evaporation and Precipitation over the Oceans and Transport of Water Vapour. By T. Laevastu, L. Clarke, and P.M. Wolff. [111 pp., including charts] (Reports on WMO/IHD Projects, 11)
1969, WMO $2.00
Presents formulas for computation of evaporation and gives examples of synoptic and climatological evaporation charts. Discusses precipitation at sea and outlines the possibilities of extended hydrometeorological and hydrological forecasts.

ACQUISITION AND EXCHANGE OF DATA

137   RADIO COMMUNICATIONS REQUIREMENTS FOR OCEANOGRAPHY. 19 pp., illus. (IOC. Technical Series, 3)
1967, UNESCO $0.75
Information of oceanographers' needs for radio frequencies to be used for oceanographic data transmission. Includes a description of scientific research and technical development in this field.

138   Manual on INTERNATIONAL OCEANOGRAPHIC DATA EXCHANGE.
2nd ed. (rev.) 49 pp., charts. (IOC. Technical Series, 4)
1967, UNESCO $1.25
     Assembles in convenient form, for the use of practicing oceanogra-
phers, the various documents concerned with the exchange of oceanographic data of all kinds. Supersedes the 1965 Manual (IOC. Technical Series, 1) and the Guide to International Exchange through the
World Data Centres (Section on Oceanography, pp. 48-50) issued by
the Comité International de Géophysique (CIG) in November 1963,
and Supplement No. 1 thereto, issued in December 1964.

139   LEGAL PROBLEMS ASSOCIATED WITH OCEAN DATA ACQUISITION
SYSTEMS (ODAS). A Study of Existing National and International Legislation, prepared jointly by the secretariats of UNESCO and IMCO,
1962-68. Rev. under the authority of and with the assistance of the
IOC Group of Experts on the Legal Status of Ocean Data Acquisition
Stations. 40 pp. (IOC. Technical Series, 5)
1969, UNESCO $1.50
     In four parts: (1) History of the problem and relevant background
documentation; (2) domestic laws and regulations and comments by
states members of IOC on existing international legislation; (3) bibliographic information on international conventions; (4) extract from
the report of the second meeting (1966) of the IOC Working Group
on Ocean Data Stations.

140   INTEGRATED GLOBAL OCEAN STATION SYSTEM: The General Plan
and Implementation Programme for Phase I. Plan adopted by IOC and
WMO. 27 pp., tables. (Reports on Marine Science Affairs, 2)
1970, WMO $1.00
     Plan developed by the IOC Working Committee for IGOSS and the
WMO Executive Committee Panel on Meteorological Aspects of
Ocean Affairs. The IGOSS will provide from the oceanographic side
oceanographic and some meteorological data for forecast services
and research in a manner similar to meteorological programs, such
as the World Weather Watch, and in a complementary fashion.

141   INTERNATIONAL LIST OF SELECTED, SUPPLEMENTARY AND
AUXILIARY SHIPS.... 1970 ed. Loose-leaf.
1970, WMO   Bilingual: E/F WMO-No.47.TP.18 $7.00
     Eleventh edition. Lists by country the ships participating in the
WMO Voluntary Observing Scheme covering meteorological observations from ocean areas.

141   THE BEAUFORT SCALE OF WIND FORCE (Technical and Operational
bis   Aspects). Report submitted by the President of the Commission for
Maritime Meteorology to the WMO Executive Committee at its twenty-
second session. 22 pp., fig. (Reports on Marine Science Affairs, 3)
1970, WMO $1.00
     Study prepared by J.M. Dury, Vice-President of CMM, of the techni-
cal and operational aspects of the Beaufort scale of wind force and
its use for the issue of storm warnings and in statistical data needed
for research. The new set of equivalent wind speeds corresponding
to the Beaufort scale, proposed in 1964 at the fourth session of
CMM, is included as annex III.

141   REQUIREMENTS FOR MARINE METEOROLOGICAL SERVICES. Re-
ter   port prepared by the Working Group on Requirements for Marine Mete-

orological Services of the Commission for Marine Meteorology. 22 pp. (Reports on Marine Science Affairs, 4)
1971, WMO   WMO-No.288 $1.00
Identifies the requirements of various marine user groups—shipping, fisheries, coastal and offshore activities, recreational boating, and marine pollution control—for marine meteorological and subsurface information, and analyzes the present state of the provision of information.

TIDES, WAVES AND SEICHES

142   Methods of FORECASTING THE STATE OF THE SEA ON THE BASIS OF METEOROLOGICAL DATA. Lectures presented at the Scientific Conference during the Third Session of the Commission for Maritime Meteorology. By J.J. Schule, K. Terada, H. Walden, G. Verploegh. xii, 35 pp., illus. (Technical Notes, 46)
      1970, reprint, WMO WMO-No.124.TP.55 $4.50
      Lectures (1960) on the structure and spectrum of ocean waves; the present state of wave measurement and analysis in Japan; forecasting wind-generated waves; application of wave forecasting to ship operations. Published together with Precipitation Measurements at Sea (Technical Notes, 47) (entry 251). First issued in 1962.

143   PROCEEDINGS OF THE SYMPOSIUM ON TIDAL INSTRUMENTATION AND PREDICTIONS OF TIDES.... Paris, 3-7 mai 1965. 242 pp., illus. (IAPO. Publications scientifiques, 27)
      1967, UNESCO/IAPO Composite: E/F $8.50
      Technical papers (15) on tide gauge recording instruments and methods of analyzing tidal observations and predicting tides.

144   PROCEEDINGS OF THE SYMPOSIUM ON TIDES organized by the International Hydrographic Bureau, Monaco, 28-29 April 1967. 204 pp., illus.
      1969, UNESCO   Composite: E/F $11.00
      Papers (18) on tidal analysis; tide predictions; computer applications to tidal analysis and data exchange; data tele-announcing system for tsunami warning.

145   SEICHES ET DENIVELLATIONS CAUSÉES PAR LE VENT DANS LES LACS, BAIES, MERS, ESTUAIRES. By L.J. Tison and G. Tison, Jr. 59 pp. (Technical Notes, 102)
      1969, WMO   WMO-No.246.TP.134 In French only $4.50
      In two parts. First and major part deals with seiches and the horizontal theory of seiches, their cause and forecasting. Second part is a survey of variations in levels of lakes, bays, seas, and estuaries and the effects of wind speed. Extensive bibliography.

MARINE MINERAL RESOURCES

146   Report of the Ad Hoc Committee to Study the PEACEFUL USES OF THE SEA-BED AND THE OCEAN FLOOR BEYOND THE LIMITS OF NATIONAL JURISDICTION. 69 pp. (General Assembly. Official Records, 23rd session, 1968, document A/7230)
      1968, UN

147   Report of the Committee on the PEACEFUL USES OF THE SEA-BED AND THE OCEAN FLOOR BEYOND THE LIMITS OF NATIONAL JURISDICTION. 161 pp. (General Assembly. Official Records, 24th session, 1969, Supplement No. 22 (A/7622))
1969, UN   No sales number $4.00
a. ——Addendum.... 3 pp. (General Assembly. Official Records, 24th session, 1969, Supplement No. 22A (A/7622/Add.1))
1970, UN   No sales number $0.50

148   Report of the Committee on the PEACEFUL USES OF THE SEA-BED AND THE OCEAN FLOOR BEYOND THE LIMITS OF NATIONAL JURISDICTION. iii, 194 pp. (General Assembly. Official Records, 25th session, 1970, Supplement No. 21 (A/8021))
1970, UN   No sales number $4.00

149   Report of the Committee on the PEACEFUL USES OF THE SEA-BED AND THE OCEAN FLOOR BEYOND THE LIMITS OF NATIONAL JURISDICTION. vi, 266 pp. (General Assembly. Official Records, 26th session, 1971, Supplement No. 21 (A/8421))

The reports of the ad hoc and the standing Committees include the reports of their respective subsidiary bodies—in the 1968-1970 reports, on economic, technical, and legal aspects; in the 1971 report, on the proposed international regime and machinery, on the law of sea, and on the preservation of the marine environment—on the possibilities of reserving the area exclusively for peaceful purposes, exploiting its mineral resources under an international regime for the benefit of all mankind, while assuring protection of the marine environment. Each report also contains selected documents, such as studies and draft articles.

150   LIMITS AND STATUS OF THE TERRITORIAL SEA, EXCLUSIVE FISHING ZONES, FISHERY CONSERVATION ZONES AND THE CONTINENTAL SHELF (with Particular Reference to Fisheries). 32 pp, and Addendum sheet. (FAO Legislative Series, 8)
1969, FAO $1.00
Provides, in the form of a synoptical table, a summary of the claims of 102 coastal States as reflected in their laws or in international agreements. Survey was undertaken by the Legislation Branch and the Fishery Liaison Office of the FAO Secretariat.

151   MODERNIZATION AND MECHANIZATION OF SALT INDUSTRIES BASED ON SEAWATER IN DEVELOPING COUNTRIES. Proceedings of Expert Group Meeting, Rome, 25-29 September 1968. 161 pp., illus. (ID/32)
1970, UN   Sales No. E.70.II.B.25 $2.50
Contains papers (9) prepared for the meeting on experience in Peru, U.S.A. (solar salt), Kuwait, France, Portugal, Italy, India, Venezuela, and energy requirements and related costs for selected desalination processes.

152   MINERAL RESOURCES OF THE SEA. 49 pp. (ST/ECA/125)
1970, UN   Sales No. E.70.II.B.4 $1.00
Report, prepared by Dr. Frank F.H. Wang, in cooperation with the staff of the Resources and Transport Division, Department of Economic and Social Affairs, United Nations Secretariat, on the known mineral resources of the sea as well as the most recent technical developments regarding exploration and exploitation techniques, with an indication of problems relating to mineral resources development in the marine environment which require special attention.

153   "Legal Problems of the Exploitation of the Ocean Floor." By D.P.
       O'Connell. Impact of Science on Society, vol. 21, no. 3, July/September
       1971, pp. 253-264.
       1971, UNESCO $1.25
       Surveys the difficulties in internationalizing the deep-sea bed, espe-
       cially in relation to defining its limits with respect to the extent of
       national jurisdiction over the continental shelf and to the form of in-
       ternationalization to be agreed upon.

METEOROLOGY AND CLIMATOLOGY

GENERAL, METHODOLOGY AND PRACTICE

154   TECHNICAL REGULATIONS. General introduction and 3 vols. Loose-
       leaf, with binder. (Basic Document No. 2; WMO-No.49)
       1971- , WMO
       Vol. 1: General. 1971 ed. ix, 114 pp., loose-leaf. $9.00
       In three sections: (A) The World Weather Watch—the Global Observ-
       ing System; the Global Data-processing System; the Global Telecom-
       munication System; (B) Research Activities; (C) Interaction of Man
       and His Environment—chap. C.1: Meteorological Services to Marine
       Activities; chap. C.2: Meteorological Services for Agriculture.
       Vol. 2: Meteorological Services for International Air Navigation.
       In preparation
       When published will form chap. C.3 of Section C, and will replace the
       present vol. 2: Meteorological Services for International Air Naviga-
       tion, 3rd ed. 1970, vi, 125 pp., loose-leaf and binder (WMO-No.49.
       BD.3), which formed chap. 12 of the former ed. of Technical Regula-
       tions.
       Vol. 3: Operational Hydrology. 1st ed. viii, 14 pp., loose-leaf. $2.00
       Comprises chap. C.4.1: Hydrological Observing Networks and Sta-
       tions; chap. C.4.2: Hydrological Observations; chap. C.4.3: Hydro-
       logical Warnings and Forecast. This is a new set of regulations.
       ———Binder. $3.00
       The present Technical Regulations were adopted by the sixth WMO Con-
       gress (1971) and incorporate major changes from previous editions
       which reflect the new obligations upon WMO members stemming from
       the World Weather Watch plan adopted by the fifth Congress (1967), in-
       clude a new set of regulations concerning operational hydrology, and
       have resulted in a new layout of the volume. Their contents are now
       rearranged according to the major WMO programs—namely, the World
       Weather Watch, Research, and Interaction of Man and His Environment.
       The three volumes may be obtained separately.

155   Guide to CLIMATOLOGICAL PRACTICES. Loose-leaf. illus.
       1960- , WMO   WMO-No.100.TP.44 $5.50
       Provides information on climatological practices and procedures,
       including administrative organization; observation and data; use of
       statistics and data processing; descriptive climatology; microclima-
       tology; marine climatology; climatology of the free atmosphere;
       CLIMAT reports; publication of data; and application of climatologi-
       cal analysis. Amended by Supplements, issued irregularly.

156 Guide to METEOROLOGICAL INSTRUMENT AND OBSERVING PRAC-
TICES. 4th ed. Loose-leaf.
1971- , WMO   WMO-No.8.TP.3 $21.50
Contains advice on methods required to keep observing stations up to
international standards, recommendations for international observing practices for the taking of observations before coding, information about uniform procedures for applying correction with a view to
eliminating errors, and, in general, the best methods of obtaining
correct observations. Includes chapters on measurement of atmospheric pressure, temperature, atmospheric humidity and soil moisture, surface wind, precipitation, evaporation, radiation and sunshine
visibility, cloud observations, measurement of upper wind, radiosonde techniques, meteorological balloon techniques, observations of
atmospherics, instruments and methods of observation at aeronautical meteorological stations, marine observations, meteorological
observations from aircraft. Bibliography. Supplements issued
irregularly.

157 WEATHER AND MAN: the Role of Meteorology in Economic Development. 80 pp., illus.
1964, WMO   WMO-No. 143.TP.67 $1.00
Shows how the work of the meteorologist may contribute to economic
development in many fields, such as the development of water resources, town planning, agriculture, transport and tourism, and the
establishment of new industries. Describes the role of WMO.

158 AUTOMATIC WEATHER STATIONS. Proceedings of the WHO Technical Conference on Automatic Weather Stations, Geneva, 1966. 364 pp.,
and corrigendum slip; illus. (Technical Notes, 82)
1967, WMO   WMO-No.200.TP.104 $11.50
Lectures and papers (47) and background papers (2) on automatic
weather stations on the ocean, on land, and aboard ships, including
design, operation, data systems, etc.; description of national activities in France, India, Japan, Norway, Romania, South Africa,
U.S.S.R., and the U.S.A.

159 ASSESSING THE ECONOMIC VALUE OF A NATIONAL METEOROLOG-
ICAL SERVICE. 14 pp. (World Weather Watch. Planning Reports, 17)
1969, 2nd printing, WMO $1.50
Provides guidance on possible methods of conducting cost/benefit
analyses of national weather services, with some examples of economic studies of weather services. First issued in 1967.

160 The ECONOMIC BENEFITS OF NATIONAL METEOROLOGICAL SER-
VICES. Papers presented to the 20th Session of the WMO Executive
Committee. xiv, 55 pp., illus. (World Weather Watch. Planning Reports, 27)
1968, WMO $3.50
Examines economic benefits already gained by agriculture and various specific industries from the national meteorological services in
Australia, France, the United Kingdom, U.S.S.R., and the Federal
Republic of Germany. Country reports conclude that the estimated
overall benefits to the community amounted to at least 20 times the
annual budget of the meteorological service. Also includes reports
on weather and the construction industry; the application of cost/
benefit analysis in assessing the value of meteorological services;
less the potential economic benefits from improvements in meteorological information; and the economic benefits of meteorological
services to developing countries.

161   LECTURES ON NUMERICAL SHORT-RANGE WEATHER PREDICTION.
        WMO Regional Training Seminar, Moscow, 17 November-14 December
        1965. 706 pp., fig., tables.
        1969, Hydrometeozdat (Leningrad) for WMO $25.00
        Texts of lectures and papers (21) by scientists from five countries in
        the field of numerical weather prediction.

162   Proceedings of the WMO/IUGG Symposium on NUMERICAL WEATHER
        PREDICTION in Tokyo, November 26-December 4, 1968. Sponsored by
        the World Meteorological Organization and the International Union of
        Geodesy and Geophysics. [631] pp., illus.
        1969, Meteorological Society of Japan (Tokyo) $25.00
        Opening and closing addresses, papers (64) and discussions on nu-
        merical weather prediction—the physical basis, mathematical mod-
        els for short-, medium-, and long-range forecasting and for fore-
        casting in the tropics (typhoon and hurricane formation), numerical
        experiments, and mathematical and numerical procedures.

163   THE PLANNING OF METEOROLOGICAL STATION NETWORKS. By
        L.S. Gandin x, 35 pp., fig., tables. (Technical Notes, 111)
        1970, WMO   WMO-No.265.TP.149 $3.50
        Report by the Rapporteur to the WMO Commission for Climatology
        on the best methods for evaluating existing networks and for design-
        ing improved ones, with special attention to climatological stations
        for macrometeorological investigations and to reference climato-
        logical stations.

164   WMO HELPS THE DEVELOPING COUNTRIES. 85 pp.
        1971, WMO   WMO-No.307 $1.00
        Notes the role of meteorology and operational hydrology in economic
        development; gives illustrative examples of WMO technical coopera-
        tion projects; and outlines the programming procedures of the United
        Nations Development Programme (UNDP) and WMO's Voluntary
        Assistance Programme (VAP).

164   Selected Papers on METEOROLOGY AS RELATED TO THE HUMAN
bis   ENVIRONMENT. xv, 151 pp., fig. (Special Environmental Reports, 2)
        1971, WMO   WMO-No.312 $11.50
        Contains papers (22) dealing with meteorological aspects of air and
        water pollution (inland and ocean waters, monitoring), urban climate,
        natural resources planning, natural disasters, and the relation of
        man to his environment.

164   METEOROLOGY AND THE HUMAN ENVIRONMENT. 40 pp., illus.
ter   1971, WMO   WMO-No.313 $1.00
        General description of the role of meteorology in human environ-
        mental affairs, and explanation of the steps taken by WMO to adjust
        and extend its programs in meteorology (including operational hy-
        drology) in this field. Four chapters deals with the efficient use of
        natural resources, air pollution, water pollution, and modification of
        weather and climate—intentional and unintentional.

                        Africa and the Tropics

165   PROBLEMS OF TROPICAL METEOROLOGY (A Survey). By M.A.
        Alaka. 121 pp., illus. (Technical Notes, 62)
        1964, WMO   WMO-No.155.TP.75 $2.00
        Formulation of unsolved problems in tropical meteorology, reviewed

by a WMO/IUGG Symposium on Tropical Meteorology, 1963. Covers general circulation in the tropics; Asian summer monsoon; convection; hurricanes and typhoons; rainfall; observations; synoptic analysis; forecasting; numerical methods; weather modification.

166   The Role of METEOROLOGICAL SERVICES IN THE ECONOMIC DE-VELOPMENT OF AFRICA. Proceedings of the ECA Seminar, Ibadan (Nigeria), 23-28 September 1968. 145 pp.
1969, WMO/UN   • $4.50
Report of seminar, and papers (14) on meteorology and economic development; cost/benefit studies; application of meteorology to agriculture, water resources, and transport; organization of meteorological services. Issued in cooperation with the United Nations Economic Commission for Africa, ECA.

Europe

167   CLIMATIC ATLAS OF EUROPE.... I: Maps of Mean Temperature and Precipitation. Technical Supervisor, Prof. F. Steinhauser. vi pp., 28 fold., detachable map sheets in 5 or 6 colors.
1970, WMO/UNESCO/Cartographic (Budapest) $50.00 cl.
Contains foreword, introduction and 27 maps: monthly and annual mean temperature distribution (13 maps on scale of 1:10,000,000 with inset of Greenland at 1:20,000,000), mean annual range of temperature (one map with inset of Greenland, same scales), distribution of average monthly precipitation totals (12 maps with inset of Greenland, same scales), and annual precipitation (one map on scale 1:5,000,000 in two parts: western Europe, and eastern Europe with an inset of Greenland on scale 1:10,000,000, respectively). The observational data used cover the period 1931-1960. Described in detail (with plan) in UNESCO map catalogue (entry 1143).

EDUCATION AND TRAINING

168   The Problem of the PROFESSIONAL TRAINING OF METEOROLOGICAL PERSONNEL OF ALL GRADES IN THE LESS-DEVELOPED COUNTRIES. By J. Van Mieghem. x, 75 pp. (Technical Notes, 50)
1967, 2nd printing, WMO   WMO-132.TP.59 $2.50
Comprehensive report by a highly qualified consultant. Describes a system of grading meteorological personnel in four classes—from Class I, comprising university graduates holding at least a degree in mathematics or physics who have successfully completed a period of postgraduate training in meteorology, down to Class IV, comprising personnel who must have general knowledge of at least primary education standard and intermediate vocational training of at least three years—and proposes a detailed syllabus in basic and professional knowledge for each class. First issued in 1963.

169   PROBLEM WORKBOOK FOR THE TRAINING OF CLASS III METEO-ROLOGICAL PERSONNEL. By Pemmaraju S. Pant. x, 247 pp.
1968, WMO   WMO-No.223.TP.118 $3.50
Workbook for Class III meteorological personnel containing problems to be solved which are intended to familiarize a candidate with units, dimensions, and orders of magnitude of physical quantities, and to enable him to understand and apply the theory taught in lectures and textbooks.

170   Report on METEOROLOGICAL TRAINING FACILITIES. 196 pp.
1969, WMO   WMO-No.240.TP.131 $9.00
Third report. Gives information on current status of meteorological
training facilities in 72 countries and territories. Previous reports
were issued in 1959 and 1964.

171   Guidelines for the EDUCATION AND TRAINING OF METEOROLOGI-
CAL PERSONNEL. Prepared by the Executive Committee Panel of Experts on Meteorological Education and Training. xxiv, 161 pp.
1969, WMO   WMO-No.258.TP.144 $5.50
Comprehensive syllabi for basic and specialized training in all fields
of meteorology and for all classes of personnel in national meteorological services.

172   HOW TO BECOME A METEOROLOGIST; Meteorological Education and
Training. 16 pp., illus.
1970, WMO   WMO-No.257.TP.143 $1.00
Comments on meteorological education and training at secondary
school level and below.

173   PROBLEMS IN DYNAMIC METEOROLOGY. By D.L. Laikhtman and
others. English translation by G. Tarakanov, ed. by Aksel C. Wiin-
Nielsen. 245 pp.
1970, WMO   WMO-No.261.TP.146 $3.50
Workbook for Class I and II meteorological personnel published by
the Hydrometeorological Service of the U.S.S.R. in 1967. Problems
therein are intended as basic material for practical lessons and test
papers in dynamic meteorology.

174   Proceedings of the WMO/IAMAP SYMPOSIUM ON HIGHER EDUCA-
TION AND TRAINING (Rome, April 1970). viii, 313 pp.
1970, WMO   WMO-No.278.TP.156 $7.00
Introductory lectures (3), papers (33) and reports of chairmen (7) on
education and training of Class I and Class II meteorological personnel, training of research personnel, training methods and facilities,
and general problems of meteorological education.

175   Compendium of Lecture Notes for TRAINING CLASS IV METEORO-
LOGICAL PERSONNEL. Prepared by B.J. Retallack. 2 vols.
1970, WMO   WMO-No.266-TP.150 $7.00 the set
Vol. 1: Earth Science. xii, 179 pp., fig., tables.
Vol. 2: Meteorology. xv, 435 pp., fig., tables.
Volume 1 deals with earth science (in 14 chapters) and should be studied prior to the training course in meteorology. Volume 2 comprises
four units: (1) general meteorology, (2) surface instruments and methods of observation, (3) surface weather reports, and (4) aeronautical
meteorology.

176   Compendium of Lecture Notes for TRAINING CLASS III METEORO-
LOGICAL PERSONNEL. Prepared by B.J. Ratallack. xviii, 381 pp.,
fig.
1971, WMO   WMO-No.291 $7.50
Compendium in seven units: (1) general meteorology, (2) upper air
instruments and methods of observation, (3) upper wind computations, (4) radiosonde computations, (5) upper air meteorological reports, (6) meteorological transmissions, and (7) aeronautical meteorology.

WORLD WEATHER WATCH, WWW

177   WORLD WEATHER WATCH. 32 pp., illus.
1966, WMO   WMO-No.183,TP.92 $0.50
Popular account of the need for a better world weather observation system and the plans for WWW, with its use of artificial satellites and computers.

178   The POTENTIAL ECONOMIC AND ASSOCIATED VALUES OF THE WORLD WEATHER WATCH. By J.C. Thompson. 35 pp. (World Weather Watch. Planning Reports, 4)
1966, WMO $1.50
Preliminary assessment of economic and related benefits to be expected from WWW; suggestions for further study.

179   The ESSENTIAL ELEMENTS OF THE WORLD WEATHER WATCH. 28 pp., maps.
1966, WMO $1.50
Concise report of WMO plan for a new worldwide system for meteorological observations, services, and research. Summarizes expected economic benefits, describes former system and how its deficiencies may be remedied, and implementation and financing of WWW.

180   WORLD WEATHER WATCH: THE PLAN AND IMPLEMENTATION PROGRAMME, as approved by the Fifth World Meteorological Congress, Geneva, April 1967 and the Nineteenth Session of the WMO Executive Committee, Geneva, May 1967. 56 pp., including charts.
1968, 2nd printing, WMO $2.50
Text of WWW plan for 1968-1971 and of resolutions related to the plan and its implementation including WMO Voluntary Assistance program (VAP). Originally issued in 1967.

181   Activities and Plans of the WORLD METEOROLOGICAL CENTRES. 104 pp., illus., maps.
1967, WMO $2.50
Describes present activities and future plans of the three world meteorological centers of WWW in Melbourne, Moscow, and Washington.

182   The Role of METEOROLOGICAL SATELLITES IN THE WORLD WEATHER WATCH. 38 pp., illus., maps. (World Weather Watch. Planning Reports, 18)
1969, 2nd printing, WMO $5.00
Reports on present use of meteorological satellites in the global observing system and on their possible use through 1971. Describes use of television cameras, infrared sensors, and radiation subsystems. First issued in 1967.

183   SCOPE OF THE 1972-1975 PLAN WITH PARTICULAR REFERENCE TO THE METEOROLOGICAL SATELLITE SUB-SYSTEM. [136] pp., including annexes, illus. (World Weather Watch. Planning Reports, 30)
1969, WMO $4.50
Report of the informal planning meeting of 29 September-3 October 1969 to consider the future WWW satellite subsystem and to review the broad objectives and scope of WWW Plan for 1972-1975.

184   Development of the WWW OCEANIC OBSERVING SUB-SYSTEM FOR
        1972-1975. [91] pp., including appendices; charts. (World Weather
        Watch. Planning Reports, 31)
        1970, WMO $3.50
        Report of an informal planning meeting of February 1970 which con-
        sidered ways of remedying the deficiency of surface and upper-air
        meteorological observations over the oceans, especially over those
        parts rarely visited by ships. Reviews parts which may be played by
        five types of vehicles for meteorological sensors: ships, aircraft,
        satellites (see item 183 for fuller coverage), superpressure balloons,
        and buoys.

185   WORLD WEATHER WATCH: FOURTH STATUS REPORT ON IMPLE-
        MENTATION. vi, [289] pp., including annexes, fold. maps.
        1971, WMO   WMO-No.308 $9.00
        Report on the situation as of June 1971 and improvements to be ex-
        pected in the years to come with respect to the global observing sys-
        tem, the global data-processing system and the global telecommuni-
        cation system. The previous reports were issued in 1968 (at $4.50),
        in 1969 (at $4.50) and in 1970 (at $7.00).

ATMOSPHERIC METEOROLOGY

186   TIDAL PHENOMENA IN THE UPPER ATMOSPHERE. by B. Haurwitz.
        27 pp., illus. (Technical Notes, 58)
        1964, WMO   WMO-No.146.TP.69 $1.50
        Summarizes present knowledge of tidal phenomena in the strato-
        sphere and mesosphere, with recommendations for future research.

187   METEOROLOGICAL SOUNDINGS IN THE UPPER ATMOSPHERE. By
        W.W. Kellogg. 46 pp., illus. (Technical Notes, 60)
        1964, WMO   WMO-No.153.TP.73 $3.00
        Guide to techniques of investigations of the upper atmosphere, on
        parameters available from such investigations, with methods for
        charting and synoptic analysis of these data.

188   The CIRCULATION IN THE STRATOSPHERE, MESOSPHERE AND
        LOWER THERMOSPHERE. By R.J. Murgatroyd and others. 206 pp.,
        illus. (Technical Notes, 70)
        1965, WMO   WMO-No.176.TP.87 $6.50
        Comprehensive review of present meteorological knowledge and a
        synthesis of the data available, with emphasis on main features of
        general circulation. Concludes with an assessment of present abil-
        ity to provide general climatology for these regions and description
        of various Standard Atmospheres now available.

189   LOWER TROPOSPHERE SOUNDINGS. (Report of a Working Group of
        the Commission for Instruments and Methods of Observation.) Pre-
        pared by D.H. Pack and others. 33 pp., illus. (Technical Notes, 77)
        1966, WMO   WMO-No.192.TP.98 $2.00
        Report on means of obtaining measurements of temperature, pres-
        sure, and humidity at intervals not greater than 20 meters in the
        lower troposphere for calculating the refractive index of air, pollu-
        tion, etc.

190 UPPER AIR OBSERVATIONS IN THE TROPICS. By Herbert Riehl.
xviii, 16 pp. (World Weather Watch. Planning Reports, 1)
1966, WMO $1.50
Provides a general background to observational requirements in
tropic regions, with reference to optimum density of radiosonde and
radio weather stations, frequency of observations, observations from
aircraft and ships.

191 The Nature and Theory of the GENERAL CIRCULATION OF THE
ATMOSPHERE. By Edward N. Lorenz. xxv, 161 pp., illus.
1967, WMO   WMO-No.218.TP.115 $14.00 cl.
Basic text which covers dynamic equations, observed circulation,
former theories of general circulation, the energy cycle, laboratory
models of the atmosphere, numerical simulation, theoretical investigation, and a qualitative explanation for main features of general
circulation.

192 Methods in Use for the REDUCTION OF ATMOSPHERIC PRESSURE.
21 pp. (Technical Notes, 91)
1968, WMO   WMO-No.226.TP.120 $3.50
Surveys methods of atmospheric pressure reduction to mean sea
level used in 107 countries, classified into three categories: method
indicated by Angot; hypsometric equation; others—with complete
mathematical descriptions. Supersedes a similar survey of 1952
forming Part I of out-of-print WMO Technical Note No. 7, Reduction
of Atmospheric Pressure.

193 RADIATION INCLUDING SATELLITE TECHNIQUES. Proceedings of
the WMO/IUGG Symposium held in Bergen, August 1968. xxxii, 555 pp.,
illus. (Technical Notes, 104)
1970, WMO   WMO-No.248-TP.136 $17.50
Extended summaries of papers presented on satellite instrumentation (8 papers), satellite observations and results (15), spectroscopy
(15), radiation climatology (22), radiation in clouds and aerosols (35),
surface and airborne instrumentation (28), radiation and atmospheric
dynamics (12), and radiative transfer and planetary atmospheres (9).
(See also entry 201.)

194 UPPER-AIR INSTRUMENTS AND OBSERVATIONS. Proceedings of the
WMO Technical Conference, Paris, 8-12 September 1969. xiv, 661 pp.,
illus., fig.
1970, WMO   Composite: E/F   WMO-No.284 $23.00
Papers (39) on new marine and land radiosonde/radiowind systems
(including automation), high-altitude balloon sounding, constant-level
balloon instrumentation, sensors and telemetry for meteorological
rockets, meteorological satellites as objective sensors of meteorological parameters, low-level sounding systems, sensors and ancil-
lary equipment, and recent developments in reference radiosonde
and CIMO comparison program. Includes also three papers from the
scientific discussions at the fifth session of the WMO Commission
for Instruments and Methods of Observation (CIMO), Versailles, 15-
29 September 1969.

GLOBAL ATMOSPHERIC RESEARCH PROGRAM, GARP

195 An INTRODUCTION TO GARP. 22 pp., illus. (GARP Publications
Series, 1)
1969, WMO/ICSU $2.00
Semipopular account of the objectives and the scientific basis of

GARP. Presents current techniques of weather prediction based on physico-mathematical models of the atmosphere and an analysis of the problems yet to be solved.

196 SYSTEMS POSSIBILITIES FOR AN EARLY GARP EXPERIMENT. 55 pp., illus. (GARP Publications Series, 2)
1969, WMO/ICSU $2.00
Report of Working Group VI of the ICSU Committee on Space Research (COSPAR) to the Joint Organizing Committee on GARP (JOC) on proposals for the space-based observing system for the first GARP global experiment.

197 The PLANNING OF THE FIRST GARP GLOBAL EXPERIMENT. 36 pp., illus. (GARP Publications Series, 3)
1969, WMO/ICSU $3.50
Describes scientific requirements for data output, organization and implementation of the first GARP global experiment, including preliminary consideration of costs.

198 The PLANNING OF GARP TROPICAL EXPERIMENTS. 78 pp. and corrigendum slip; illus. (GARP Publications Series, 4)
1970, WMO/ICSU $3.50
Suggested program of experimental studies and experiments in the investigation of tropical disturbances.

199 Report of the PLANNING CONFERENCE ON GARP, Brussels, March 1970. 42 pp. (GARP Special Reports, 1)
1970, WMO/ICSU $2.00
Report of the conference on its review of the proposed cooperative experiments planned by the ICSU/WMO Joint Organizing Committee on GARP (JOC), and recommendations.

200 Report of INTERIM PLANNING GROUP ON GARP TROPICAL EXPERIMENT IN THE ATLANTIC, London, July 1970. 25 pp. (GARP Special Reports, 2)
1970, WMO/ICSU $1.00
Reviews proposals by the ICSU/WMO Joint Organizing Committee on GARP (JOC) for the GARP Tropical Experiment in the Atlantic and a supplement thereto prepared by JOC. Annexes include regulations applicable to the Tropical Experiment Council (TEC), and to the Tropical Experiment Board (TEB).

201 Problems of ATMOSPHERIC RADIATION IN GARP. By F. Möller and C.D. Rodgers. xiii, 18 pp. (GARP Publications Series, 5)
1970, WMO/ICSU $2.00
Study which attempts to explain to what degree radiation processes ought to be considered in dynamic models of the atmosphere of different scales, and to demonstrate from the point of view of radiation research what special investigations have to be performed so that sufficiently accurate radiation models can be designed. Summarizes and elaborates in more detail the discussions at the GARP Study Conference in Stockholm, 1967, and at the Symposium on Radiation including Satellite Techniques, Bergen, 1968 (entry 193),

202 NUMERICAL EXPERIMENTATION RELATED TO GARP. By B.R. Döös. xiii, 68 pp., fig. (GARP Publications, 6)
1970, WMO/ICSU $3.50
Synthesis of the information received from active research groups in various countries. One chapter is devoted to the development of

four-dimensional analysis schemes, another to the problems of determining the limiting factors for numerical forecasting. A catalogue of twenty numerical weather prediction models is appended.

203   The GARP PROGRAMME ON NUMERICAL EXPERIMENTATION. By R.V. García. 31 pp. (GARP Publications, 7)
1971, WMO/ICSU $4.00
Report on problem areas of numerical experimentation where it is desirable to stimulate increased research, resulting from the meeting of the Study Group on Numerical Experimentation (Oslo, 1970), as revised by the JOC Working Group on the subject, and unified and organized by Professor García in order to provide a scientific basis for further planning of the GARP Tropical and Global Experiments.

204   The GLOBAL ATMOSPHERE RESEARCH PROGRAMME; a Co-operative Effort to Explore the Weather and Climate of Our Planet. By Professor Bert Bolin, Chairman, Joint GARP Organizing Committee. 28 pp., illus.
1971, WMO/ICSU $1.00
Explanation of the aims of GARP in terms which will be understandable to the nonspecialist.

CLOUDS

205   INTERNATIONAL CLOUD ATLAS. Abridged Atlas. 210 pp., including 72 plates (30 in color)
1969, 2nd printing, WMO $9.00 cl.
Reference work containing a comprehensive study of clouds and techniques for observing and reporting them and a set of plates; pictures are 4 x 6 in. in size. Abridged from the complete two-volume edition of 1956, now out of print. First issued in 1956.

206   INTERNATIONAL CLOUD ALBUM FOR OBSERVERS IN AIRCRAFT. 32 plates (13 in color) and table.
1956, WMO $3.50 cl.
Contains 32 plates (pictures 4 x 6 in.) of basic cloud types from ground and air selected from the International Cloud Atlas.

207   [MARINE CLOUD ALBUM] 40 bare plates (17 in color).
1956, WMO $1.50
Selection of 40 plates (pictures 4 x 6 in.) from the International Cloud Atlas with the printed legends and references deleted for the use of meteorological services wishing to print explanatory legends in languages other than English and French.

208   [CLOUD SHEET] $19\frac{1}{2}$ x 26 in. 40 plates (17 in color)
1956, WMO $1.00
Selection of 40 plates (pictures 2 x 4 in.) from the International Cloud Atlas.

209   INTERNATIONAL NOCTILUCENT CLOUD OBSERVATION MANUAL. Prepared under the auspices of WMO. 39 pp., illus.
1970, WMO   WMO-No.250.TP.138 $3.00
Summarizes characteristics of noctilucent clouds (NLC) and provides instructions for their observation and international data exchange. References.

AGROMETEOROLOGY AND AGROCLIMATOLOGY

210   WEATHER AND FOOD. By L.P. Smith. 80 pp., illus. (Freedom from
       Hunger Campaign. Basic Studies, 1)
       1962, WMO   WMO-No.113.TP.50 $1.00
       Explains relationships between weather and food production and how
       meteorology can be and is being applied in agriculture and fisheries.
       Bibliography.

211   GUIDE TO AGRICULTURAL METEOROLOGICAL PRACTICES. Loose-
       leaf. 76 pp.
       1963-   , WMO   WMO-No.134.TP.61 $4.50
       Guide to practices, procedures, and specifications in agricultural
       meteorology for setting up observing stations, for recording meteo-
       rological and biological phenomena, handling agrometeorological
       data, applying climatological data to agriculture, and meeting de-
       mand for agrometeorological forecasts. Amendments are issued in
       Supplements, appearing irregularly.

212   WINDBREAKS AND SHELTERBELTS. (Report of a Working Group of
       the Commission for Agricultural Meteorology). 188 pp., illus. (Tech-
       nical Notes, 59)
       1964, WMO   WMO-No.147.TP.70 $3.50
       Reports on effects of windbreaks and shelterbelts on microclimate,
       soil climate, soil erosion, and on plants and animals; influence on
       airflow, heat balance, water balance, chemical composition of soil
       and air; biological effects.

213   HARVEST FROM WEATHER. 48 pp., illus.
       1967, WMO   WMO-No.220.TP.117 $1.00
       Explains in nontechnical language relationships of meteorology and
       agriculture, and application of the science in development of agricul-
       ture and fisheries. Issued in connection with the celebration of
       World Meteorological Day on 23 March 1968.

214   AGROCLIMATOLOGICAL METHODS. Proceedings of the Reading Sym-
       posium. 392 pp., illus. (Natural Resources Research, 7)
       1968, UNESCO   Composite: E/F $14.00 cl.
       Conclusions of the symposium of July 1966 and papers (39) on avail-
       able data; primary and secondary effects of weather including the
       epidemiology of plants and diseases; and surveys, both on macro-
       and meso-scale with special attention to FAO/UNESCO/WMO sur-
       veys in the Near East and Western Africa (entries 216, 217, 220,
       221).

215   The WORLD WEATHER WATCH AND METEOROLOGICAL SERVICE
       TO AGRICULTURE. By L.P. Smith. 14 pp. (World Weather Watch.
       Planning Reports, 22)
       1969, 2nd printing, WMO $1.50
       Report on how WWW can help national meteorological services in
       meeting their responsibilities to the agricultural community. First
       issued in 1967.

Arid Zones

216   A Study of AGROCLIMATOLOGY IN SEMI-ARID AND ARID ZONES OF
      THE NEAR EAST. By G. Perrin de Brichambaut and C.C. Wallén. 64
      pp., illus., maps. (Technical Notes, 56)
      1968, 2nd printing, WMO   WMO-No.141.TP.66 $4.50
      Shows how climatological data can best be used in helping to solve
      problems in agricultural planning in semiarid and arid zones. Gen-
      eral report based on the technical report of a joint FAO/UNESCO/
      WMO project. First issued in 1963.

217   Technical Report on a Study of AGROCLIMATOLOGY IN SEMI-ARID
      AND ARID ZONES OF THE NEAR EAST. By G. Perrin de Brichambaut
      and C.C. Wallén. 290 pp., illus., maps. (23061/2)
      1962, FAO   Bilingual: E/F
      A main document (reference no. 61608-65-MR).

218   CHANGES OF CLIMATE. Proceedings of the Rome Symposium orga-
      nized by UNESCO and WMO. 488 pp., illus. (Arid Zone Research, 20)
      1963, UNESCO   Composite: E/F   Out of print
      Papers (45) deal with climatic changes during the period of meteo-
      rological records, during the late geological and early historical
      periods, theories of changes of climate, and significance of the
      changes. Symposium was held in 1961.

219   BIOCLIMATIC MAP OF THE MEDITERRANEAN ZONE.... 2 sheets
      in 17 basic colors, scale 1:500,000, size $29\frac{1}{2}$ x 39 in.; and explanatory
      notes (58 pp.) (Arid Zone Research, 21)
      1963, UNESCO/FAO   Maps bilingual: E/F $10.00 the set
      The Ecological Study of the Mediterranean Zone: Bioclimatic Map of
      the Mediterranean Zone; Explanatory Notes explains the map, des-
      cribes its preparation, notes the distribution of Mediterranean-type
      bioclimates throughout the world, including 4 maps in pocket, scale
      1:10,000,000, in color, of southern South America, South Africa,
      southwestern Australia, and western North America; bibliography.
      Prepared jointly by UNESCO and FAO. Described in detail (with
      plan) in UNESCO map catalogue (entry 1143). (See also entry 222.)

220   An AGROCLIMATOLOGY SURVEY OF A SEMIARID AREA IN AFRICA
      SOUTH OF THE SAHARA. By J. Cochemé and P. Franquin. 136 pp.,
      illus. (Technical Notes, 86)
      1967, WMO   WMO-No.210.TP.110 $5.50
      Survey of a semiarid area with summer rainfall lying south of the
      Sahara and running from the Atlantic Ocean eastward to the Chad-
      Sudan frontier, consisting of parts of Senegal, Mauritania, Mali, Up-
      per Volta, Ghana, Togo, Dahomey, Niger, Nigeria, Cameroon, and
      Chad. The main purpose was to take an inventory of the climatic
      resources of the area, and, to that extent, of its food-producing po-
      tentials. This is the general report of an FAO/UNESCO/WMO inter-
      agency project.

221   Technical Report on a Study of the AGROCLIMATOLOGY OF THE
      SEMI-ARID AREA SOUTH OF THE SAHARA IN WEST AFRICA.... By
      J. Cochemé and P. Franquin. 325 pp., illus., maps.
      1967, FAO   Bilingual: E/F
      A main document (reference no. 00572-67-MR).

222   VEGETATION MAP OF THE MEDITERRANEAN REGION.... 2 sheets
in color, scale 1:5,000,000, size $29\frac{1}{2}$ x $39\frac{3}{8}$ in.; and explanatory notes
(90 pp.) (Arid Zone Research, 30)
1967-69, UNESCO/FAO   Bilingual: E/F $12.00 the set
    The Ecological Study of the Mediterranean Zone: Vegetation Map of
the Mediterranean Zone; Explanatory Notes explains the vegetation
map, gives a general picture of the area in terms of the documentation, compares the vegetation map with the Bioclimatic Map of the
Mediterranean Zone (entry 219); bibliography. Prepared jointly by
UNESCO and FAO. Described in detail (with plan) in UNESCO map
catalogue (entry 1143).

223   METEOROLOGY AND THE MIGRATION OF DESERT LOCUSTS: Applications of Synoptic Meteorology in Locust Control. By R.C. Rainey.
117 pp., addenda and errata sheet and 2 amended captions; illus., maps.
(WMO. Technical Notes, 54; Anti-Locust Research Centre, London.
Anti-Locust Memoirs, 7)
1963, WMO/ALRC   WMO-No.138.TP.64 $9.00
    Deals with relevant biology and behavior of the desert locust; de-
scribes in detail the migration and breeding of the desert locust during one year of widespread infestation in relation to synoptic meteorology of the region and period and shows how locust movements
may be interpreted and forecast by use of current synoptic charts
and meso-scale meteorological observations. Extensive references.

224   METEOROLOGY AND THE DESERT LOCUST. Proceedings of the
WMO/FAO Seminar on Meteorology and the Desert Locust, Tehran, 25
November-11 December 1963. xxii, 310 pp., illus., maps. (Technical
Notes, 69)
1965, WMO   WMO-No.171.TP.85 $10.50
                Composite: E/F
    Report on seminar and lectures (35) on the application of synoptic
and meso-scale meteorology to the control of locusts, their biology,
their flight performance in flying swarms, the effects of convergence
and divergence of windfields on migration and distribution of locusts,
and the planning of air operations in locust control.

225   DESERT LOCUST PROJECT: Final Report. 142 pp. (FAO/SF: 34/DLC)
1968, UNDP/FAO $5.00
    Report on United Nations Development Program (Special Fund) Des-
ert Locust Project (1960-1966), executed by FAO, and carried out in
five main sections: ecological surveys; strengthening and coordination of research; reporting and forecasting; and training and operational research. Recommendations; assessment of the projects.

HYDROMETEOROLOGY AND HYDROLOGY

GENERAL, METHODOLOGY AND PRACTICE

226   Standards for Methods and Records of HYDROLOGIC MEASUREMENTS.
82 pp., illus. (ECAFE. Flood Control Series, 6; ST/ECAFE/SER.F/6)
1954, UN   Sales No. 54.II.F.3   Out of print
    Contains standard methods of measurement and standard forms for a
hydrologic yearbook recommended for adoption for general use in

the ECAFE region (pending adoption of standards for use on a global basis) by the Second Regional Conference on Water Resources Development, Tokyo, 1954.

227 Field Methods and Equipment in HYDROLOGY AND HYDROMETEOROLOGY. (Transactions of the Interregional Seminar on Field Methods and Equipment Used in Hydrology and Hydrometeorology held at Bangkok, Thailand, from 27 November to 11 December 1961.) 127 pp., illus. (ECAFE. Flood Control Series, 22; ST/ECAFE/SER.F/22)
1963, UN   Sales No. 63.II.F.4   Out of print
Contains proceedings, lectures (6) and papers (12) on measurement of water stage, sediment concentration and transport, storm precipitation, water discharge, evaporation from free-water and soil surfaces. The seminar was sponsored jointly by the United Nations (ECAFE) and WMO.

228 WEATHER AND WATER. 27 pp., illus.
1966, WMO   WMO-No.204.TP.107 $1.00
Briefly describes hydrometeorological activities, application of weather knowledge to water problems: hydrological forecasting, hydrologic structures, irrigation, land-water management, climatic trends, modification of the hydrologic cycle; the role of WMO; and programs of the International Hydrological Decade, 1965-1975. Most of the text was written by J.P. Bruce.

229 Instruments and Measurements in HYDROMETEOROLOGY. Lectures given at the Second Session of the Commission for Hydrometeorology, Warsaw—29 September-15 October 1964. 56 pp., illus. (Technical Notes, 76)
1966, WMO   WMO-No.191.TP.97 $2.50
Contains five lectures on automation of stream flow records; methods of measuring precipitation; instruments for limnometeorological research on the Great Lakes; some improvements by French hydrologists in hydrometeorological instruments; methods of evaporation measurement used in the U.S.S.R. for water balance studies of river catchments.

230 The Use and Interpretation of HYDROLOGIC DATA. (Proceedings of Nine Seminars conducted by an Advisory Group in Afghanistan, Ceylon, China (Taiwan), Hongkong, Iran, Malaysia, Pakistan, the Philippines and Thailand between 22 September 1965 and 26 July 1966.) 195 pp., illus. (ECAFE. Water Resources Series, 34; ST/ECAFE/SER.F/34)
1968, UN   Sales No. E.68.II.F.9 $3.00
Contains report on, and lectures (26) and selected assignments in analytical techniques in, the seminars conducted by the Advisory Group on the Use and Interpretation of Hydrologic Data.

231 HYDROLOGICAL REQUIREMENTS FOR WEATHER RADAR DATA. By A.F. Flanders. 16 pp. (Reports on WMO/IHD Projects, 9)
1969, WMO $2.00
Surveys briefly hydrological requirements for weather radar data based on the U.S. Weather Bureau's experience in its radar network. Traces the application of radar data to hydrological forecast problems.

232 The Use of ANALOG AND DIGITAL COMPUTERS IN HYDROLOGY. Proceedings of the Tucson Symposium. A Contribution to the Interna-

tional Hydrological Decade.... 2 vol. (755 pp.), illus. (Studies and Reports in Hydrology, 1)
1969, UNESCO/IASH   Composite: E/F $29.00 the set cl.
Papers (78) deal with the techniques of analogue computing, digital computing, digital data handling, hybrid computing; comparative studies of computing techniques; and mathematic computing techniques.

233   LAND SUBSIDENCE. Proceedings of the Tokyo Symposium, September 1969.... 2 vol. (xxix, 324 pp.; pp. 325-661); illus. (Studies and Reports in Hydrology, 8)
1970, UNESCO/IAHS   Composite: E/F $26.50 the set
Papers (65) on problems of land subsidence from various causes: especially from the lowering of water tables and the exploitation of oil fields (with many papers on situations in Japan and the U.S.A.), the dissolution of limestone deposits, the working of mines, or purely tectonic movements. Some studies concern the influence of soil components (sand, clay, loess, peat, etc.), others the reaction of hurricanes and tides. Some contributions are confined to the effects of subsidence upon surface and subsoil hydrology, irrespective of its causes.

234   Guide to HYDROMETEOROLOGICAL PRACTICES. 2nd ed. 330 pp., tables. Loose-leaf, with cover.
1970- , WMO   WMO-No.168.TP.82 $14.00
Provides information about practices, procedures, and instrumentation for implementation of work in the field of hydrometeorology. Chap. 1: General. Chap. 2: Instruments and methods of observation. Chap. 3: Design of networks. Chap. 4: Collection, processing, and publication of data. Annex A: Hydrological analysis. Annex B: Hydrological forecasting. Annex C: Application to water management. Amendments are included in Supplements, issued irregularly. First edition issued in 1965.

235   INTERNATIONAL LEGEND FOR HYDROGEOLOGICAL MAPS.... 101 pp.
1970, UNESCO   Quadrilingual: E/F/R/S $6.00
Legend for the preparation of hydrogeological maps in an internationally standardized form and resulting from decisions of two international organizations and their working groups——the International Association of Scientific Hydrology (IASH), forming part of the International Union of Geology and Geophysics, and the International Association of Hydrogeologists (IAH), forming part of the International Union of Geological Sciences. In eight parts: (1) foreword; (2) list (with addresses) of members and former members of the IASH Standing Committee and the IAH Working Group; (3) introduction; (4) recommended symbols; (5) select bibliography; (6) list of hydrogeological maps fully or largely based on the international legend; (7) model map; and (8) annex of recommended symbols for karst areas and permafrost features and additional symbols for arid zone and miscellaneous features.

236   NUCLEAR WELL LOGGING IN HYDROLOGY. Prepared by the Working Group on Nuclear Techniques in Hydrology of the International Hydrological Decade. 90 pp., fig. (Technical Reports Series, 126; STI/DOC/10/126)
1971, IAEA $3.00
Summarizes the status of nuclear well logging with respect to hydrological investigations, with conclusions and recommendations.

237   USE OF WEIRS AND FLUMES IN STREAM GAUGING. Report of a
Working Group of the Commission for Hydrology. v, 57 pp., tables.
(Technical Notes, 117)
1971, WMO   WMO-No.280 $4.00
    Provides technical information on the flow-measuring methods uti-
lizing artificial control sections of such a shape that head-discharge
relationships can be determined from measured water levels without
the necessity of calibration.

238   [Canceled.]

Africa

239   The ROLE OF HYDROLOGY AND HYDROMETEOROLOGY IN THE
ECONOMIC DEVELOPMENT OF AFRICA.... 2 vols.
1971- , WMO   WMO-No.301
Vol. 1: Proceedings, Reports and Recommendations.
In preparation
Vol. 2: Technical Papers presented to the ECA/WMO Conference,
Addis Ababa, 13-23 September 1971.... ix, 342 pp., fig.
1971, WMO/UN ECA   Composite: E/F $7.50
    Contains 43 papers in English or French with abstracts in both
languages.

EDUCATION AND TRAINING

240   TRAINING OF HYDROMETEOROLOGICAL PERSONNEL. Report of the
CHy Working Group on Training in Hydrometeorology. Prepared by
K.A. Hzmaljan and others. 41 pp., including annexes.
1967, WMO   WMO-No.219.TP.116 $2.50
    Contains detailed syllabi for the three main levels of hydrometeoro-
logical staff, i.e., the professional, technical assistant, and observer
levels.

241   TEXTBOOKS ON HYDROLOGY: Analyses and Synoptic Tables of Contents of Selected Books. 185 pp. (Technical Papers in Hydrology, 6)
1970, UNESCO $5.00
    Detailed analyses of 19 textbooks published in English, French,
Russian, Spanish, German, and Italian, on general hydrology (10
textbooks), on hydrology of surface waters (6), on hydrometry (1),
and on hydrological forecasting (2). Synoptic tables of contents of
works on general hydrology. The textbooks considered by a working
panel were of university level and were published after 1940.

HYDROLOGICAL NETWORKS

242   DESIGN OF HYDROLOGICAL NETWORKS. By Max A. Kohler. 16 pp.
(Technical Notes, 25)
1970, 3rd printing, WMO   WMO-No.82.TP.32 $3.50
    Treats problems in planning hydrological networks. Briefly de-
scribes application of hydrological data and examines certain techniques that can be used in the absence of sufficient data. Published
together with Techniques for Surveying Surface-water Resources
(Technical Notes, 26) (entry 280). First issued in 1958.

243   HYDROLOGIC NETWORKS AND METHODS. (Transactions of the Inter-
       regional Seminar on Hydrologic Networks and Methods, held at Bang-
       kok, Thailand, from 14 to 27 July 1959.) 157 pp., illus. (ECAFE.
       Flood Control Series, 15; ST/ECAFE/SER.F/15)
       1960, UN   Sales No. 60.II.F.2   Out of print
       Proceedings, lectures (3), and papers (15) deal with hydrologic data
       networks, calculation and forecasting streamflow, including methods
       used when adequate hydrometric readings are not available. Seminar
       was sponsored jointly by United Nations (ECAFE) and WMO.

244   HYDROLOGICAL BENCH MARKS. By W.B. Langbein. 8 pp. (Reports
       on WMO/IHD Projects, 8)
       1968, WMO $2.00
       Deals with analysis of the hydrological structure of a catchment,
       analysis of the statistical structure of a hydrological time series and
       the problem of changes in hydrological regimen. Deals also with se-
       lection and operation of hydrological bench marks as well as rela-
       tions to climatological reference stations, to representative catch-
       ments and vigil stations.

245   HYDROLOGICAL NETWORK DESIGN—NEEDS, PROBLEMS AND AP-
       PROACHES. Prepared by J.C. Rodda from an outline furnished by W.B.
       Langbein, and employing material provided by A.G. Kovzel, D.R. Dawdy,
       and K. Szesztay. 57 pp., illus. (Reports on WMO/IHD Projects, 12)
       1969, WMO $2.00
       Deals with definition and classification of hydrological networks,
       methods of design, and the implications of the development of new
       instruments. Gives examples of different approaches, taken from
       certain national networks, and treats the whole subject critically as
       a reminder that much has yet to be accomplished.

246   LIST OF INTERNATIONAL HYDROLOGICAL DECADE STATIONS OF
       THE WORLD. 148 pp. (Studies and Reports in Hydrology, 6)
       1969, UNESCO   Quadrilingual: E/F/R/S $12.00
       Provides information on data available from countries participating
       in IHD on types of network observations which have been selected for
       international exchange. Presents information on Decade stations in
       five tables: (I) River stations; (II) Lake stations; (III) Pan-evapora-
       tion stations; (IV) Lysimeter stations; (V) Ground water stations.
       Observations carried out cover water discharge of rivers, obser-
       vations on lakes, evaporation from the water surface, evaporation
       from the soil, and ground water regime.

247   DISCHARGE OF SELECTED RIVERS OF THE WORLD .... —vols.
       (Studies and Reports in Hydrology, 5, vol. 1-)
       1969- , UNESCO   Quadrilingual: E/F/R/S
       Vol. 1: General and Regime Characteristics of Stations Selected ....
       70 pp.
       1969 $6.00
       Introductory volume to an international compilation of data on the
       discharge of rivers throughout the world. Gives list of 1,000
       gauging stations selected, with both their general characteristics
       (Part 1)—name and number of identification, geographical loca-
       tion, period of operation—and their regime characteristics (Part
       2)—average monthly discharges for the longest period for which
       records are available.

Vol. 2: Monthly and Annual Discharges Recorded at Various Selected Stations (from Start of Observations up to 1964) .... 194 pp., tables. 1971 $16.00
Presents data for 137 stations chosen from those which are most characteristic and furnish the oldest and most valid data. Both volumes 1 and 2 contain permanent data, and so are published once only during the International Hydrological Decade (IHD, 1965-1974).
Vol. 3 [Part 1]: Mean Monthly and Extreme Discharges (1965-1969). 98 pp., tables.
1971 $8.00
Volume 3 will consist of separate parts, to be published periodically, giving the data (on monthly, annual, and extreme discharges) compiled each year, from 1965 onwards, by the stations selected for volume 1.

248   REPRESENTATIVE AND EXPERIMENTAL BASINS. An International Guide for Research and Practice. Ed. by C. Toebes and V. Ouryvaev. 348 pp., illus. (Studies and Reports in Hydrology, 4)
1970, UNESCO $13.00 cl.
Guide consists of six chapters: (1) Introduction describing the aims of the publication and the purpose of representative and experimental basins within the larger framework of hydrological research, (2) selection and organization of basin networks, (3) planning of observations according to the research objectives, (4) methods of observation and instrumentation, (5) data processing and publication, and (6) analysis techniques and interpretation of research results. References and bibliography.

249   Some Recommendations for the OPERATION OF REPRESENTATIVE AND EXPERIMENTAL BASINS AND THE ANALYSIS OF DATA. A report of the CHy Working Group on Representative and Experimental Basins. xiii, 33 pp. (Reports on WMO/IHD Projects, 15)
1971, WMO   WMO-No.302 $2.00
Recommendations for simplified procedures for separation of different forms of runoff and methods for determination of infiltration.

Africa

250   HYDROMETEOROLOGICAL INSTRUMENTS, OBSERVATIONS AND NETWORKS IN AFRICA. Proceedings of the Seminar on Hydrometeorological Instruments, Methods of Observation and Establishment of Hydrometeorological Networks in Africa, Addis Ababa, 2-20 October 1967. 231 pp., illus.
1969, WMO $4.50
Lectures (5) by consultants; and papers (6) by participants on work in Burundi, Ghana, Liberia, Madagascar, Mauritius and Egypt. The seminar was organized by the UN Economic Commission for Africa and WMO.

PRECIPITATION

251   PRECIPITATION MEASUREMENTS AT SEA. Review of the Present State of the Problem prepared by a Working Group of the Commission for Maritime Meteorology. xii, 18 pp. (Technical Notes, 47)
1970, reprint, WMO   WMO-No.124.TP.55 $4.50
After describing briefly the different methods available for measur-

ing precipitation aboard ship, the report discusses the difficulties encountered with particular stress on the disturbing effects of the ship and gauge on the air flow and of the movement of the gauge. Considers also the effect of sea spray. Results of comparative measurements are given. Conclusions and recommendations of the working group. Published together with Methods of Forecasting the State of the Sea on the Basis of Meteorological Data (Technical Notes, 46) (entry 142). First issued in 1962.

252   EVALUATION OF ATMOSPHERIC MOISTURE TRANSPORT FOR HYDROLOGICAL PURPOSES. By E. Palmén. 63 pp., illus. (Reports on WMO/IHD Projects, 1)
1967, WMO $1.50
Presents formulas for computation of the difference between evapotranspiration and precipitation from atmospheric moisture flux and its divergence; discusses the advantages of applying this aerological method for water-balance estimation.

253   METHODS OF ESTIMATING AREAL AVERAGE PRECIPITATION. By A.F. Rainbird. 45 pp., illus. (Reports on WMO/IHD Projects, 3)
1970, 2nd printing, WMO $3.00
Discusses a number of methods of estimating areal average precipitation from point precipitation data such as the arithmetic mean method, the polygon method, and the isohyetal method, and their merits and shortcomings. Deals with various aids to the interpretation and extension of precipitation data. First issued in 1967.

254   RADAR MEASUREMENT OF PRECIPITATION FOR HYDROLOGICAL PURPOSES. By E. Kessler and K.E. Wilk. 46 pp. and errata slip; illus. (Reports on WMO/IHD Projects, 5)
1968, WMO $1.50
Presents, in two parts, (1) a review of the status in the world of radar measurement for the assessment of rainfall over an area, and (2) a detailed description of the weather-radar data processing system at the U.S. National Severe Storms Laboratory.

255   PREPARATION OF CO-ORDINATED PRECIPITATION, RUNOFF AND EVAPORATION MAPS. By T.J. Nordenson. 20 pp., illus. (Reports on WMO/IHD Projects, 6)
1968, WMO $1.50
Describes methods for preparation of maps of average annual precipitation, runoff and evaporation. Emphasises coordinated development of these maps in order to obtain a consistency between them which will provide a nearly true water balance.

256   Manual for DEPTH-AREA-DURATION ANALYSIS OF STORM PRECIPITATION. xv, 114 pp., illus.
1969, WMO   WMO-No.237.TP.129 $7.00
Manual is based mainly on practices and procedures followed in the U.S.A., with information received from other countries incorporated into the text. In two parts, it describes (1) depth-area-duration analysis by standard methods, and (2) an experiment in depth-area-duration analysis by computer.

257   ARTIFICIAL MODIFICATION OF CLOUDS AND PRECIPITATION. By Morris Neiburger. 33 pp. (Technical Notes, 105)
1969, WMO   WMO-No.249.TP.137 $3.00
Describes in four chapters the scales and status of weather modification, the physical basis for cloud and precipitation modification,

the evaluation of attempts to increase precipitation, and the reduction of fog, hail, and lightning by cloud seeding. References. A revision of the out-of-print Artificial Control of Clouds and Hydrom-
eters, 1955 (Technical Notes, 13).

257   The PRECIPITATION MEASUREMENT PARADOX—the Instrument
bis   Accuracy Problem. By J.C. Rodda. xii, 42 pp., fig. (Reports on
        WMO/IHD Projects, 16)
        1971, WMO   WMO-No.316 $2.00
        Report on the problems of measuring precipitation at a point, with
        particular attention to the commonly used elevated type of raingauge
        compared with measurements made by instruments installed with
        their rims at ground level. It is hoped that the report may assist in
        the search for a means to provide more realistic assessments of
        precipitation amounts of rainfall and snowfall.

EVAPORATION
---

258   The Standardization of the MEASUREMENT OF EVAPORATION AS A
        CLIMATIC FACTOR. By G.W. Robertson. 10 pp. (Technical Notes,
        11)
        1970, 2nd printing, WMO   WMO-No.42.TP.16 $1.50
        Proposes a meteorological definition of evaporation (to be called
        "latent evaporation" or "L.E.") and considers evaporimeters suit-
        able for measuring L.E. Published together with The Forecasting
        from Weather Data of Potato Blight and Other Plant Diseases and
        Pests, by P.M. Austin Bourke (Technical Notes, 10). First issued in
        1955.

259   EVAPORATION REDUCTION; Physical and Chemical Principles and
        Review of Experiments. By J. Frenkiel. 79 pp., illus. (Arid Zone Re-
        search, 27)
        1965, UNESCO $2.00
        Reviews theoretical and practical aspects of evaporation reduction
        up to mid-1964. Extensive bibliography.

260   MEASUREMENT AND ESTIMATION OF EVAPORATION AND EVAPO-
        TRANSPIRATION. (Report of a Working Group on Evaporation Mea-
        surement of the Commission for Instruments and Methods of Observa-
        tion.) By M. Gangopadhyaya, Chairman, and others. 121 pp., illus.
        (Technical Notes, 83)
        1968, reprint, WMO   WMO-No.201.TP.105 $6.50
        Contains a comprehensive review of the evaporation instruments
        used in many countries of the world, international comparisons made
        between them under the auspices of WMO, and the many different
        methods at present in use for deriving evaporation losses. First is-
        sued in 1966.

261   PROBLEMS OF EVAPORATION ASSESSMENT IN THE WATER BAL-
        ANCE. By C.E. Hounam. 80 pp., tables. (Reports on WMO/IHD Pro-
        jects, 13)
        1971, WMO   Sales No. WMO-No.285 $2.00
        Describes various methods for measuring point evaporation which
        can be used to obtain the data needed for an assessment of the evap-
        oration from an area. Pays particular attention to the factors af-
        fecting the spatial variation in evaporation and to the networks of in-
        struments required.

USE OF NUCLEAR TECHNIQUES

262 APPLICATION OF ISOTOPE TECHNIQUES IN HYDROLOGY. A Comprehensive Report from the Panel on Application of Isotope Techniques in Hydrology, held in Vienna from 6-9 November 1961. 33 pp. (Technical Reports Series, 11; STI/DOC/10/11)
1962, IAEA $1.00
   Summarizes conclusions and recommendations as to the use of radioactive tracers for groundwater, the measurement of stream discharge, and their application to hydraulic problems, etc.

263 RADIOISOTOPES IN HYDROLOGY. Proceedings of the Symposium on the Application of Radioisotopes in Hydrology held by IAEA in Tokyo, 5-9 March 1963. 459 pp. and erratum sheet, illus. (Proceedings Series; STI/PUB/71)
1963, IAEA   Composite: E/F $9.00
   Includes papers (27) on water tracers; flow and course changes in rivers; flow and stratification and age of groundwater; and silt movements in rivers and harbors.

264 ISOTOPE TECHNIQUES FOR HYDROLOGY. Report of the Panel on Use of Isotopes in Hydrology held in Vienna, 17-21 December 1962. 36 pp. (Technical Reports Series, 23; STI/DOC/10/23)
1964, IAEA $1.00
   Discusses groundwater dating techniques and short-term groundwater tracing, and briefly describes field experiments and some specific groundwater problems. Considers also surface water applications.

265 GUIDE TO THE SAFE HANDLING OF RADIOISOTOPES IN HYDROLOGY. 38 pp. (Safety Series, 20; STI/PUB/131)
1966, IAEA $1.00
   Presents briefly the health and safety aspects involved in the use of radioisotopes in hydrology, once the radioactive tracer method has been chosen. Includes references and additional bibliography. Prepared by IAEA with the help of FAO and WHO.

266 TRITIUM AND OTHER ENVIRONMENTAL ISOTOPES IN THE HYDROLOGICAL CYCLE; Report of a Panel. 83 pp., illus. (Technical Reports Series, 73; STI/DOC/10/73
1967, IAEA $2.00
   Reviews the environmental isotopes (especially tritium), their role in hydrological investigations, their applications to oceanography and meteorology, and the advances in the techniques for the collection and analysis of isotopic samples; includes four papers on environmental tritium. Panel was convened in Vienna on 12-16 October 1964.

267 ISOTOPES IN HYDROLOGY. Proceedings in the Symposium on Isotopes in Hydrology held by IAEA in co-operation with the International Union of Geodesy and Geophysics in Vienna, 14-18 November 1966. 740 pp., illus. (Proceedings Series, STI/PUB/141)
1967, IAEA   Composite: E/F/R $15.00
   Contains papers (41) on hydrometeorology, stream flow measurements; sediment studies; geochronology and environmental studies; aquifer characteristics; unsaturated zone, seepage, tracer technology; surface water, limnology, glaciology.

268   Guidebook on NUCLEAR TECHNIQUES IN HYDROLOGY. Prepared by the Working Group on Nuclear Techniques in Hydrology of the International Hydrological Decade. 214 pp., illus. (Technical Reports Series, 91; STI/DOC/10/91)
1968, IAEA $5.50
    Provides general information on the principles and applications of present nuclear and isotopic methods in hydrological research. Contains an introduction and chapters on atmosphere, surface water (transient and in storage), subsurface water (unsaturated and saturated zones), and their interrelations. Text was adopted by the IHD Co-ordinating Council at its 4th session, May 1968.

269   ENVIRONMENTAL ISOTOPE DATA NO. 3. WORLD SURVEY OF ISOTOPE CONCENTRATION IN PRECIPITATION (1966-1967). Report from a Network organized by IAEA in co-operation with WMO and co-operating national laboratories. xviii, 402 pp., tables, maps. (Technical Reports Series, 129; STI/DOC/10/129)
1971, IAEA $8.00
    Data on the concentration of the environmental isotopes (tritium, deuterium and oxygen-18) in monthly samples of precipitation taken by a global network of 212 stations in 65 countries and territories in the period 1966-1967. Presents also selected meteorological data, such as amount of precipitation, vapor pressure and temperature. Data before 1965 which were not appropriate and available at the time of the second survey have also been included as late reports in the latter part of this volume.— The first survey, covering the period 1953-1963, was issued in 1969 (xviii, 421 pp. and corrigendum sheet; Technical Reports Series, 96; STI/DOC/10/96; $8.00). The second survey, covering the period 1964-1965, was issued in 1970 (xvii, 402 pp.; Technical Reports Series, 117; STI/DOC/10/117; $8.00).

270   ISOTOPE HYDROLOGY 1970. Proceedings of a Symposium on Use of Isotopes in Hydrology, held by IAEA in co-operation with UNESCO, in Vienna, 9-13 March 1970. 918 pp., illus. (Proceedings Series; STI/PUB/255)
1970, IAEA   Composite: E/F/R/S $24.00
    Papers (53) on the development of isotope hydrology through the co-operation of isotope specialists and hydrologists. Sessions covered environmental isotopes in the water cycle (12 papers); problems in carbon-14 dating of ground water (5); field studies using environmental isotope techniques (8); tracer techniques in surface water studies (7); sediment studies (3); borehole techniques and nuclear instruments (6); leakage from reservoirs and canals (4); computational studies (3); and new techniques in tracing and dating (5).

GLACIOLOGY, SNOW AND ICE, POLAR REGIONS

271   FLUCTUATIONS OF GLACIERS, 1959-1965. By Peter Kasser. [130] pp., maps.
1967, UNESCO/IASH $6.00
    Study of the measurement of glacier variations for the period 1959-1965, with addenda from earlier years. Continues, on an extended

basis, the international reports made for the period 1894-1958. Contains a separate folding map of the Aletsch Glaciers. (See also entry 273)

272   SATELLITE APPLICATIONS TO SNOW HYDROLOGY 1968. By Robert W. Popham. 10 pp. (Reports on WMO/IHD Projects, 7)
1968, WMO $2.00
Summarizes satellite snow-surveillance activities; reviews existing and future satellite instrumentation applicable to snow hydrology; and suggests areas where further efforts could yield significant results.

273   VARIATIONS OF EXISTING GLACIERS. A Guide to International Practices for Their Measurement. 19 pp. (Technical Papers in Hydrology, 3)
1969, UNESCO/IASH $1.25
Guide prepared with the cooperation of the International Union of Geodesy and Geophysics and the IASH International Commission of Snow and Ice (ICSI). Should be studied in connection with Peter Kasser, Fluctuations of Glaciers, 1959-1965 (entry 271).

274   PERENNIAL ICE AND SNOW MASSES. A Guide for Compilation and Assemblage of Data for a World Inventory. 59 pp., illus. (Technical Papers in Hydrology, 1)
1970, UNESCO/IASH $4.00
Guide to the provision of information on ice and snow with a view to mapping permanent ice and snow masses and the compilation and assemblage of data for publication in order to obtain the elements necessary for the establishment of the regional distribution of permanent snow and ice in various territories and the degree of accuracy in each area. Prepared by a working group of the IASH International Commission of Snow and Ice (ICSI).

275   SEASONAL SNOW COVER. A Guide for Measurement, Compilation and Assemblage of Data. 38 pp., illus. (Technical Papers in Hydrology, 2)
1970, UNESCO/IASH/WMO $3.00
Guide proposing methods for international standardization of data collection concerning seasonal ice and snow masses, including mapping. Prepared by the IASH International Commission of Ice and Snow (ICSI).

276   COMBINED HEAT, ICE AND WATER BALANCES AT SELECTED GLACIER BASINS. A Guide for Compilation and Assemblage of Data for Glacier Mass Balance Measurements. 20 pp., illus. (Technical Papers in Hydrology, 5)
1970, UNESCO/IASH $2.00
Deals with ice and water balances and provides a basis for international cooperation in attaining standardized measurements of glacier mass balances. Prepared by a working group of the IASH International Commission of Snow and Ice (ICSI).

277   POLAR METEOROLOGY. Proceedings of the WMO/SCAR/ICPM Symposium on Polar Meteorology, Geneva, 5-9 September 1966. 540 pp., illus.
1967, WMO   WMO-No.211.TP.111 $27.00
Papers (31) on atmospheric constituents: water vapor and ozone; boundary layer; physical processes radiation; heat balance; free atmosphere studies; circulation, local, tropopause, stratosphere.

278   ANTARCTIC GLACIOLOGY IN THE INTERNATIONAL HYDROLOGICAL
       DECADE. 16 pp. (Technical Papers in Hydrology, 4)
       1969, UNESCO/IASH $1.25
          Hydrological appraisal of glaciological research in Antarctica pre-
          pared by the Secretary of the Scientific Committee on Antarctic
          Research (SCAR) reviewed and approved by the IHD Working Group
          on World Water Balance.

WATER RESOURCES

APPRAISAL AND DEVELOPMENT

279   Proceedings of the United Nations Scientific Conference on the CON-
       SERVATION AND UTILIZATION OF RESOURCES, 17 August-6 Septem-
       ber 1949, Lake Success, New York. 8 vol. (E/CONF.7/7, vol. 1-8)
       1950-51, UN
          Vol. 4: Water Resources. 466 pp., illus. (E/CONF.7/7, vol. 4)
          Sales No. 50.II.B.5   Out of print
          Papers (82) on appraisal of water resources, water supply and
          pollution problems, river-basin development, drainage-basin
          management, water-control structures, flood control and naviga-
          tion, irrigation and drainage, and hydro power and recreational
          uses of water (including protection of fish and wildlife).

280   TECHNIQUES FOR SURVEYING SURFACE-WATER RESOURCES. By
       Ray K. Linsley. 41 pp. (Technical Notes, 26)
       1970, 3rd printing, WMO   WMO-No.82.TP.32 $3.50
          Considers techniques for surveying surface water resources.
          Examines methods to be utilized in the absence of adequate hydro-
          logical data. Describes simple techniques for observing hydrologi-
          cal phenomena. First issued in 1958. Published together with De-
          sign of Hydrological Networks (Technical Notes, 25) (entry 242).

281   LARGE-SCALE GROUND-WATER DEVELOPMENT. 84 pp. (E/3424-
       ST/ECA/65)
       1960, UN   Sales No. 60.II.B.3   Out of print
          Basic considerations relating to groundwater development, economic
          and financial aspects, organization, administration, and legislation.

282   Manual of STANDARDS AND CRITERIA FOR PLANNING WATER RE-
       SOURCE PROJECTS. 53 pp. (ECAFE. Water Resource Series, 26;
       ST/ECAFE/SER.F/26)
       1964, UN   Sales No. 64.II.F.12   Out of print
          Guidelines for investigations, standards and criteria in formulation
          and appraisal of plans.

283   Methods of HYDROLOGICAL FORECASTING FOR THE UTILIZATION
       OF WATER RESOURCES. (Transactions of the Inter-Regional Seminar
       on Methods of Hydrological Forecasting for the Utilization of Water Re-
       sources, held at Bangkok, Thailand, from 4 to 17 August 1964.) 167 pp.,
       illus. (ECAFE. Water Resources Series, 27; ST/ECAFE/SER.F/27)
       1965, UN   Sales No. 65.II.F.5 $2.00
          Lectures and papers (18) on hydrological forecasting. Seminar was
          sponsored jointly by United Nations (ECAFE) and WMO.

284   WORLD WEATHER WATCH AND ITS IMPLICATIONS IN HYDROLOGY
       AND WATER RESOURCES MANAGEMENT. By J.P. Bruce and J.
       Němec. 10 pp. (Reports on WMO/IHD Projects, 4)
       1970, 2nd printing, WMO $3.00
       Describes basic elements of the World Weather Watch together with
       hydrological and water resources management aspects of these
       elements. Discusses significance to the hydrological community of
       the data which will be obtained through WWW and interrelationship of
       this program with other programs in hydrology and water resources
       management. Presents future plans for augmenting WWW activities
       in these fields. First issued in 1967.

285   HYDROLOGY OF FRACTURED ROCKS. Proceedings of the Dubrovnik
       Symposium, October 1965, organized by Unesco in the Framework of
       the International Hydrological Decade with the support of IASH, FAO
       and IAH.... 2 vol. (420; 269 pp.); illus., maps.
       1967, UNESCO/IASH   Composite: E/F $22.00 the set
       Papers (80) cover ground waters in karstic regions: flow character-
       istics, salt encroachment, geological characteristics, natural and
       artificial recharge, karstic hydrology in volcanic terrains; influence
       of exploitation on runoff; erosion. The two volumes, with a special
       cover, also constitute Publications No. 73 and 74 of the International
       Association of Scientific Hydrology.

286   Methods and Techniques of GROUND-WATER INVESTIGATION AND
       DEVELOPMENT. (Transactions of the Second Ground-Water [i.e. Re-
       gional] Seminar on Methods and Techniques of Ground-Water Investiga-
       tion and Development, held at Tehran, Iran, from 16 October to 5 No-
       vember 1966). 206 pp., illus. (ECAFE. Water Resources Series, 33;
       ST/ECAFE.F/33)
       1968, UN   Sales No. E.68.II.F.6 $3.50
       Lecture (13) and papers (20). (For the first regional seminar—1962
       —see entry 365.)

287   WATER AND MAN: a World View. By Raymond L. Nace. 46 pp., illus.
       (Unesco and Its Programme)
       1969, UNESCO
       Presents the historical and scientific background of the world water
       shortage and the machinery of international cooperation set in
       motion by the International Hydrological Decade (1965-1974).

288   "Man in Quest of Water." By Raymond L. Nace. Unesco Courier, vol.
       23, no. 6, June 1970, pp. 4-13, 32; illus.
       1970, UNESCO $0.50
       Brief account of the historical and scientific background of the
       search for water resources and description of some areas of inter-
       national cooperation during the International Hydrological Decade.

289   HYDROLOGY OF DELTAS; Proceedings of the Bucharest Symposium,
       6-14 May 1969.... 2 vols.; 491 pp., illus., fig., tables. (Studies and
       Reports in Hydrology, 9, vol. 1-2)
       1970, UNESCO/IASH   Composite: E/F $26.50 the set
       Papers (52) on hydrology of deltas.

290   SCIENTIFIC FRAMEWORK OF WORLD WATER BALANCE. 27 pp.
       (Technical Papers in Hydrology, 7)
       1971, UNESCO $2.50
       This paper seeks to define the world water balance, to specify the
       purposes of study, to summarize the current state of knowledge, to

review available or proposed means for accomplishment of objectives, and to suggest how necessary work and study may be carried out. References. Prepared by the Panel on Scientific Framework of the International Hydrological Decade's Working Group on the World Water Balance.

291   Triennial Report on WATER RESOURCES DEVELOPMENT 1968-1970. vii, 202 pp. (ST/ECA/143)
1971, UN   Sales No. E.71.II.A.15) $4.00
Sixth review (the previous reports being biennial) of the activities of the United Nations system in the field of water resources development prepared by the Water Resources Development Centre, Department of Economic and Social Affairs, United Nations Secretariat. Annexes include information on conferences, working parties and seminars; research and studies; publications; technical and financial assistance to regional and national projects.

Arid Zones

292   Reviews of Research on ARID ZONE HYDROLOGY. 212 pp., maps. (Arid Zone Programme, 1)
1953, UNESCO   Out of print
Contains regional reports (8) and homoclimatic maps; and a paper by Peveril Meigs on world distribution of arid and semiarid homoclimates (with four folding maps), as a background paper for the Ankara Symposium of 1952 (entry 293).

293   Proceedings of the ANKARA SYMPOSIUM ON ARID ZONE HYDROLOGY, jointly organized by the Government of Turkey and Unesco. 268 pp., illus., 2 fold. charts. (Arid Zone Programme, 2)
1953, UNESCO   Out of print
Papers (29) deal with underground water, its physical and chemical properties, its statics and dynamics; the hydrological balance and the influence of the use of underground water upon it; prospection and drilling for underground water; the relation between the hydrology of underground water and other sciences. (See also entry 292.)

294   CLIMATOLOGY; Reviews of Research. 190 pp., illus. (Arid Zone Research, 10)
1958, UNESCO   Out of print
Review papers (8) on evaporation and water balance; climatic factors in arid zone plant and animal ecology; radiation and the thermal balance; modification of microclimates; the chemical climate and saline soils; climatological observational requirements. Papers for the Canberra Symposium, 1956 (entry 295).

295   CLIMATOLOGY AND MICROCLIMATOLOGY. Proceedings of the Canberra Symposium.... 355 pp., illus. (Arid Zone Research, 11)
1958, UNESCO   Composite: E/F   Out of print
Papers (50) of a 1956 symposium cover aspects of evaporation and water balance; radiation and thermal balance; interrelationships of climatic elements and flora and fauna; microclimate of man and domestic animals; modification of microclimate; salting and chemistry of rain water; climatological observational requirements. (See also entry 294.)

296   ARID ZONE HYDROLOGY: Recent Developments. By H. Schoeller. 125
       pp. (Arid Zone Research, 12)
       1959, UNESCO   Out of print
          Reviews research since 1952 in hydrology in general and hydrology
          in particular in such branches as utilization of ground water, geo-
          chemistry of ground water, utilization of radioactive tracers. Ex-
          tensive bibliography.

297   A HISTORY OF LAND USE IN ARID REGIONS. Ed. by L. Dudley Stamp.
       388 pp., maps. (Arid Zone Research, 17)
       1965, 2nd printing, UNESCO   Out of print
          Synthesizes various factors—geological, climatic, biological as well
          as human—which have determined the history of land use in arid re-
          tions in the main geographical areas of the world. Special chapters
          deal with climate and its variations and public health hazards in land
          use. First issued in 1961.

298   The PROBLEMS OF THE ARID ZONE. Proceedings of the Paris Sym-
       posium. 481 pp., illus. (Arid Zone Research, 18)
       1962, UNESCO $15.00 cl.
          Papers (27) cover state of scientific knowledge of the arid zone;
          nomadism in relation to grazing resources; alternative uses of
          limited water supplies; and public awareness and the educational
          problem. There are papers on surface water, ground water hydrol-
          ogy, hydrometeorology, saline water conversion, and irrigation,
          among other topics.

299   AGRICULTURAL PLANNING AND VILLAGE COMMUNITY IN ISRAEL.
       E. by Joseph Ben-David. 159 pp. (Arid Zone Research, 23)
       1964, UNESCO $2.50
          Describes the planning and social processes which went into agricul-
          tural development and settlement in Israel with special reference to
          the arid southern part of the country—the Negev—and to the prob-
          lems of creating new communities. One chapter is on land and
          water.

300   LAND USE IN SEMI-ARID MEDITERRANEAN CLIMATES. Unesco/In-
       ternational Geographical Union Symposium, Irklion (Greece), 19-26
       September 1962.... Ed. by D.H.K. Amiran. 170 pp., illus. (Arid Zone
       Research, 26)
       1964, UNESCO   Composite: E/F $6.50
          Contains papers (23) on the classification and mapping of geomor-
          phological and land use aspects of Mediterranean-type arid regions,
          regional studies, and land use and urbanization.

301   INCIDENCE AND SPREAD OF CONTINENTAL DROUGHT. By V.P.
       Subrahmanyam. 51 pp. (Reports on WMO/IHD Projects, 2)
       1967, WMO $1.50
          Discusses current status of research on problems associated with
          droughts in general and on continental drought in particular.

302   PHYSICAL PRINCIPLES OF WATER PERCOLATION AND SEEPAGE.
       By J. Bear, D. Zaslavsky, S. Irmay (ed.). 465 pp., illus. (Arid Zone
       Research, 29)
       1968, UNESCO $20.00 cl.; $17.00 pa.
          Presents systematically results of studies on loss of water through
          percolation and seepage. Should serve as a guide for engineers, hy-
          drologists, and scientists. Prepared by a team of staff members of
          the Faculty of Civil Engineering, Technion-Israel Institute of Tech-
          nology, at the request of UNESCO.

303   "The Sahara: an Ever-present Challenge." By R. St. Barbe Baker.
       Unasylva, vol. 23, no. 2, 1969, pp. 3-10, illus.
       1969, FAO $0.65
       Describes land reclamation in the Sahara through dune control and
       reafforestation activities. An Article by the FAO Secretariat (pp.
       11-14) summarizes opinions and recommendations of 1967-1969
       meetings of foresters.

Africa

304   "The Search for Groundwater in the Crystalline Regions of Africa."
       By Robert Dijon. Natural Resources Forum, vol. 1, no. 1 (ST/ECA/
       142), 1971, pp. 32-37.
       1971, UN   Sales No. E/F/S.71.II.A.13 $2.00
       Provides a general view of the techniques used in exploration for
       ground water resources in crystalline rock regions of Africa, after
       describing the main factors of ground water occurrence in such a
       geological and climatic environment.

Asia and the Far East

305   Proceedings of the Regional Technical Conference on WATER RE-
       SOURCES DEVELOPMENT IN ASIA AND THE FAR EAST. 451 pp.,
       illus. (ECAFE. Flood Control Series, 9; ST/ECAFE/SER.F/9)
       1956, UN   Sales No. 56.II.F.3   Out of print
       Papers (85) of the Second Regional Conference, Tokyo, May 1954,
       deal with economic and social aspects of multipurpose river basin
       development; hydrologic problems; selection of types of hydraulic
       structures; planning of sediment control works; organization for
       water resources development. Three papers for this conference
       were published in No. 7 of this series (entry 361A), and one paper,
       revised, as No. 6 and 10 (entries 226 and 1099). The First Regional
       Conference, New Delhi, 1951, dealt primarily with flood control (en-
       try 377). The name of these conferences became stabilized with the
       Fifth Regional Conference, Bangkok, 1962 (entry 308).

306   Proceedings of the Third Regional Technical Conference on WATER
       RESOURCES DEVELOPMENT IN ASIA AND THE FAR EAST. 173 pp.,
       illus. (ECAFE. Flood Control Series, 13; ST/ECAFE/SER.F/13)
       1958, UN   Sales No. 59.II.F.2   Out of print
       Papers (14) of the conference, held in Manila, December 1957, con-
       cern current programs for water resources development, basic hy-
       drological data with special emphasis on deficiencies, manual labor
       vs. machines in earthwork, government agency vs. private contrac-
       tor in construction.

307   Proceedings of the Fourth Regional Technical Conference on WATER
       RESOURCES DEVELOPMENT IN ASIA AND THE FAR EAST. (Held in
       Colombo, Ceylon. 5 to 13 December 1960.) 160 pp., illus. (ECAFE.
       Flood Control Series, 19; ST/ECAFE/SER.F/19)
       1962, UN   Sales No. 62.II.F.2   Out of print
       Papers (4) and abstracts (48) review the developments in the region
       during 1951-1960 and deal with the organization of river valley proj-
       ects, the development of ground water resources, and flood problems
       in deltaic areas.

308   Proceedings of the Fifth Regional Conference on WATER RESOURCES
DEVELOPMENT IN ASIA AND THE FAR EAST (held at Bangkok, Thailand, 20 to 26 November 1962). 208 pp., illus. (ECAFE. Water Resources Series, 23; ST/ECAFE/SER.F/23)
1963, UN   Sales No. 63.II.F.7 $2.50
   Papers (31) include a review of developments in the ECAFE area
   during 1960-1962 and treat of the formulation of water resource development plans, the control of water pollution, and flood control in
   the ECAFE region; with some papers on experience in the U.S.S.R.,
   United Kingdom, and U.S.A.

309   Proceedings of the Sixth Regional Conference on WATER RESOURCES
DEVELOPMENT IN ASIA AND THE FAR EAST (held at Bangkok, Thailand, 12 to 19 November 1964). 400 pp. and erratum sheet; illus.
(ECAFE. Water Resources Series, 28; ST/ECAFE/SER.F/28)
1966, UN   Sales No. 66.II.F.2 $4.50
   Report and papers (53) include a review of developments in the
   ECAFE area during 1962-1964, and deal with national policies, conservation and use of water as related to watershed management, and
   the development of international rivers. (See also entry 366.)

310   Proceedings of the Seventh Regional Conference on WATER RE-
SOURCES DEVELOPMENT IN ASIA AND THE FAR EAST (held at Canberra, Australia, 19 to 26 September 1966). 262 pp., illus. (ECAFE.
Water Resources Series, 32; ST/ECAFE/SER.F/32)
1968, UN   Sales No. E.68.II.F.5 $3.50
   Papers (64) include a review of water resources development in the
   ECAFE region during 1964-1966, and deal with measures to accel-
   erate planned results from development projects, coordination of
   development and the formulation of a national master water plan,
   manpower and training, allocation of costs in multipurpose projects,
   organization of applied hydraulic research and experimentation.

311   PLANNING WATER RESOURCES DEVELOPMENT. Report and Background Papers of the Working Group of Experts on Water Resources
Planning convened at Bangkok, Thailand, from 29 August to 9 September
1968. 135 pp., illus. (ECAFE. Water Resources Series, 37; ST/
ECAFE/SER.F/37)
1969, UN   Sales No. E.69.II.F.13 $2.50
   Report of the Working Group; and papers (8) on measurement standards and techniques for project evaluation (cost/benefit analysis,
   etc.); social and noneconomic considerations; coordination of activities of participating agencies.

312   Proceedings of the Eighth Session of the Regional Conference on
WATER RESOURCES DEVELOPMENT IN ASIA AND THE FAR EAST
(held in Bangkok, Thailand, 18 to 25 November 1968). 293 pp., maps.
(ECAFE. Water Resources Series, 38; ST/ECAFE/SER.F/38)
1968 [i.e. 1970], UN   Sales No. E.70.II.F.13 $4.00
   Papers include a review of water resources development in the
   ECAFE region during 1966-1968 and deal with policies and methods
   pertaining to the financing of water resources projects and the re-
   payment of project costs; efficient management, operation, and
   maintenance of water resources projects; and the integration of
   water resources plans with national economic development plans.

Europe

313   The HYDRO- ELECTRIC POTENTIAL OF EUROPE'S WATER RE-
        SOURCES. 2 vol. (ST/ECE/EP/39, vol. 1-2)
        1968, UN   Sales No. E.68.II.E/Mim.31, vol. 1-2 $3.50 the set
            Vol. 1: Methods of Analysis and Their Application. [110] pp., in-
            cluding annexes; fold. maps.
            Vol. 2: The Present State of Assessment in Europe. [137] pp., in-
            cluding annexes; illus., maps.
        Analyzes certain basic aspects of Europe's water resources considered
        as a means for the production of primary energy. Vol. 1 discusses
        methods applied to determine gross and net surface or river hydro po-
        tentials in an average year or within-year period. Also contains results
        of a survey of gross surface potential completed in respect of virtually
        the whole of Europe, including the European part of the U.S.S.R., in the
        form of an international map prepared at the scale of 1:2,500,000. Vol.
        2 contains a concise appraisal of the changes in the knowledge of hydro-
        electric resources since 1950 and of the degree of exploitation of these
        resources that has occurred over the same period, and is continuing.

314   TRENDS IN WATER RESOURCES USE AND DEVELOPMENT IN THE
        ECE REGION. A Survey prepared under the auspices of the Committee
        on Water Problems of the United Nations Economic Commission for
        Europe. ii, [154] pp., including appendices and annexes. (ST/ECE/
        WATER/1)
        1970, UN   Sales No. E.70.II.E/Mim.28 $1.70
            Surveys trends in water availability and needs and attempts at
            balancing, water quality, and water management.

315   "Water Resources in the Ukrainian S.S.R., Present and Future." By
        S.M. Perekhrest. Nature and Resources, vol. 7, no. 3, September 1971,
        pp. 9-14.
        1971, UNESCO   Free on request
            Surveys the distribution of water resources in the Ukrainian S.S.R.,
            noting the uneven territorial distribution of flows and the consider-
            able fluctuations of annual and seasonal flows, the effects of drainage
            and reclamation of bogs and marshes on the water balance and water
            reserves, and the program of flow control essential for the develop-
            ment of irrigation and supply of water to all consumers.

Latin America

316   LOS RECURSOS HIDRÁULICOS DE AMÉRICA LATINA.
        1960- , UN   In Spanish only
        a. No. 1: Chile. 190 pp. (E/CN.12/501)
            1960, UN   Sales No. 60.II.G.4   Out of print
        b. No. 2: Venezuela. 127 pp. (E/CN.12/593/Rev.1)
            1963, UN   Sales No. 63.II.G.6   Out of print
        c. No. 3: Bolivia y Colombia. 177 pp. (E/CN.12/695)
            1964, UN   Sales No. 64.II.G.11   Out of print

317   "ECLA Activities in connexion with the Development of Latin Amer-
        ica's Water Resources." Economic Bulletin for Latin America, vol.
        16, no. 1, 1971, pp. 134-152.
        1971, UN   Sales No. E.71.II.G.5 $3.00
            Appraises Latin American planning in the use of water resources
            and reviews the work of the Joint Survey Group on Water Resources
            under the ECLA Natural Resources and Energy Program since 1957.

The first version of this paper was presented at the Regional Technical Conference on the Role of Meteorological Services in Economic Development in Latin America, organized by WMO with the cooperation of ECLA, Santiago, Chile, December 1970.

WATER LAWS

Asia and the Far East

318 WATER LEGISLATION IN ASIA AND THE FAR EAST. (ECAFE. Water Resources Series, 31- ; ST/ECAFE/SER.F/31-)
1967- , UN
a. Part 1: Afghanistan, Brunei, Burma, Republic of China, Hong Kong, Iran, Japan, New Zealand, Philippines, and Thailand. 183 pp. (ECAFE. Water Resources Series, 31; ST/ECAFE/SER.F/31)
1967, UN   Sales No. 67.II.F.11 $3.00
b. Part 2(A): Water Legislation in Australia (South Australia and Victoria as Examples), Cambodia, Ceylon, India, Republic of Korea, Laos, Singapore, Republic of Viet-Nam, and Western Samoa.—Part 2(B): Proceedings of the Working Group of Experts on Water Codes. xv, 277 pp., fold. table. (ECAFE. Water Resources Series, 35; ST/ECAFE/SER.F/35)
1968, UN   Sales No. E.69.II.F.6 $3.50
Comparative review of water legislation in the ECAFE countries, covering ownership of waters, right to use, order of priorities, quality and pollution control, water rights administration, etc. The report and documents of the Working Group (1967) note the essential considerations for drafting water codes and legislation and for implementing and enforcing water laws.

Europe

319 GROUNDWATER LEGISLATION IN EUROPE. 175 pp. (FAO Legislative Series, 5)
1964, FAO $3.00
Analyzes legislation in 23 countries, including Cyprus, Israel, and Turkey, with respect to ownership rights and rights of use; groundwater exploration, withdrawal, and use; works capable of changing the behavior of ground waters; pollution; implementation of legislation and jurisdiction.

320 WATER LAWS IN ITALY. By Dante A. Caponera. 28 pp. (FAO Agricultural Development Papers, 22)
1953, FAO   Out of print
Discusses Italy's simple and complete system of water laws based on a background of Roman law and customs.

Latin America

321 Las LEYES DE AGUAS EN SUDAMÉRICA. (Estudio Comparado de su Régimen Económico y Administrativo) Por Guillermo J. Cana y F.F. Vargas Galíndez. 242 pp., illus., maps. (FAO Cuadernos de Fomento Agropecuario, 56)
1956, FAO   In Spanish only   Out of print
Comparative survey of Latin American water laws until 1953. List of references and bibliography.

Moslem Countries

322   WATER LAWS IN MOSLEM COUNTRIES. By Dante A. Caponera. 202
pp. (FAO Agricultural Development Papers, 43)
1954, FAO   Out of print
    Compilation and discussion of water laws and customs. Describes
canon, customary, and codified law. Selected references.

North America

323   WATER LAWS IN THE UNITED STATES OF AMERICA relating to
Water Rights, Irrigation, Conservation, Drainage, and Flood and Over-
flow Protection for Agricultural Lands. By Milo B. Williams. 161 pp.
(FAO Agricultural Development Papers, 2)
1950, FAO   Out of print

COMMUNITY AND INDUSTRIAL WATER SUPPLY

324   WATER SUPPLY FOR RURAL AREAS AND SMALL COMMUNITIES.
By Edmund G. Wagner and J.N. Lanoix. 340 pp., illus. (WHO Monograph Series, 42)
1959, 2nd printing, WHO $6.75 cl.
    Deals with the development of a water-supply program, the installa-
tion of various types of water-supply systems, and their management.

325   URBAN WATER SUPPLY CONDITIONS AND NEEDS IN SEVENTY-
FIVE DEVELOPING COUNTRIES. By Bernd H. Dieterich and John M.
Henderson. 92 pp. (Public Health Papers, 23)
1963, WHO $1.00
    Study of present water supplies and future needs in representative
developing countries. It reveals the present unsatisfactory state of
urban water supply.

326   OPERATION AND CONTROL OF WATER TREATMENT PROCESSES.
By Charles R. Cox. Prepared in consultation with 24 specialists in
various countries. 390 pp., illus. (WHO Monograph Series, 49)
1964, WHO $7.25 cl.
    Guide for the operation and control of water treatment processes
designed to produce safe water for domestic purposes—but does not
deal with treatment of water for industrial uses. Guide is primarily
intended to serve the needs of plant superintendents, operators, and
laboratory personnel. A supplement (pp. 277-373) covers laboratory
procedures.

327   COMMUNITY WATER SUPPLY; Report of a WHO Expert Committee.
21 pp. (WHO Technical Report Series, 420)
1969, WHO $0.60
    Reviews results of WHO-assisted programs for improving commu-
nity water supply in developing countries, with recommendations.

Latin America

328   TARIFAS DE AGUA. Informe del Seminario sobre Tarifas de Agua
       (Montevideo, Uruguay, 25 de Septiembre-1 de Octubre de 1960). 80 pp.
       (Publicaciones Científicas, 54)
       1961, PAHO   In Spanish only   Out of print
       Reviews situation in Latin America concerning rates charged for
       supply of water, considers financing of water supply systems, and
       makes recommendations.

329   SEMINARIO SOBRE DISEÑO DE ABASTECIMIENTOS DE AGUA, Buenos
       Aires, Argentina, 20-29 de septiembre de 1962. 211 pp. (Publicaciones
       Científicas, 95)
       1964, PAHO   In Spanish only   Out of print
       Papers and lectures deal with planning of water supply systems in
       the Americas.

330   UTILIZACIÓN DE TUBERÍAS PLÁSTICAS EN ABASTECIMIENTOS DE
       AGUA POTABLE. Trabajos presentados por el Prof. Charles A.
       Farish en el Simposio celebrado en Caracas, Venezuela, del 21 de
       octubre al 1 de noviembre de 1963. 170 pp. (Publicaciones Científicas,
       113)
       1965, PAHO   In Spanish only   Out of print
       Contains technical papers on use of plastic tubing and pipes in drink-
       ing water supply systems.

331   BOMBAS DE AGUA. Apuntes del Curso Intensivo. 269 pp., illus.
       (Publicaciones Científicas, 145)
       1966, PAHO   In Spanish only $3.00
       Contains technical papers on construction, testing, operation, and
       maintenance of pumps for municipal water supply systems.

332   "Organization of Water Supply Agencies in the Americas." By Oscar
       Terrevazzi. Boletín de la Oficina Sanitaria Panamericana, English
       edition, Selections from 1966, pp. 60-81.
       1967, PAHO
       Deals with water supply and sewerage agencies in the Americas,
       local and national bodies, laws and recently enacted legislation,
       and recently established funds. First published in Spanish in
       Boletín de la Oficina Sanitaria Panamericana, vol. 60, no. 1, enero
       de 1966, pp. 1-27.

DRINKING WATER AND FLUORIDATION

333   EXPERT COMMITTEE ON WATER FLUORIDATION. First Report. 25
       pp. (WHO Technical Report Series, 146)
       1958, WHO $0.30
       Deals with prevention of dental caries by fluoridation, based on ex-
       perience of a number of countries over previous years. A compre-
       hensive investigation into the biological effects of fluoridation shows
       the safety of this chemical in appropriate concentrations and its
       favorable effect on the teeth. Some technological considerations of
       water fluoridation, as well as other methods for the administration
       of fluorine, conclude the report.

334   EXPERT COMMITTEE ON HYGIENE AND SANITATION IN AVIATION.
First Report. 62 pp. (WHO Technical Report Series, 174)
1959, WHO $0.60
Discusses water supply and distribution; food and milk sanitation;
sewage and waste disposal; insect and rodent control; and refuse
disposal. Also considers the relationships which should exist at the
national or local level between the airport and constituted health
authorities, the organization of personnel for airport sanitation, and
the role of WHO.

335   Guide to HYGIENE AND SANITATION IN AVIATION. 51 pp., illus.
1960, WHO   Out of print
Shows how crews and passengers can be protected against infection
throughout the world. Specific recommendations are formulated for
standards of drinking water, food hygiene, waste disposal, and vector
control. Prepared by the WHO Expert Committee on Hygiene and
Sanitation in Aviation.

336   INTERNATIONAL STANDARDS FOR DRINKING-WATER. 3rd ed. 70
pp.
1971, WHO $3.00
States the minimum requirements as to chemical and bacterial
quality that suppliers of water for domestic use can reasonably be
expected to observe.

337   Guide to SHIP SANITATION. By Vincent B. Lamoureux. 119 pp.
1967, WHO $4.00
Attempts a synthesis of the best national practices, based on replies
to a WHO questionnaire to all its Member States. Provides practical
recommendations for such measures as the protection of potable
water from contamination, the preservation of food quality, the safe
disposal of wastes, and the ratproofing and deratting of ships. Supplements the requirements of the International Sanitary Regulations.

338   "Fluoridation and Dental Health." WHO Chronicle, vol. 23, no. 11,
November 1969, pp. 505-512.
1969, WHO $0.60
Article indicates that reports from countries with experience show
that water fluoridation, where it is practicable, remains the safest
and most effective means of preventing dental caries. Gives statistics of use and includes text of resolution of the 22nd World Health
Assembly, 23 July 1969, arising from the report on the subject by
the Director-General of WHO.

339   FLUORIDES AND HUMAN HEALTH. Prepared in consultation with 93
dental and medical specialists in various countries. 364 pp., (WHO
Monograph Series, 59)
1970, WHO $10.00 cl.
In nine chapters: Introduction; supply of fluorine to man (ingestion
from water, foods, and drugs, inhalation from dust or vapors); absorption of fluorides; distribution of fluorides in the body; excretion
of fluorides; physiological effects of small doses; toxic effects of
large doses; fluorides in general health (studies in U.S.A., India,
Japan; industrial fluoride hazards); fluorides and dental health.

USE OF SALINE WATER

340   UTILIZATION OF SALINE WATER; Reviews of Research. 2nd ed.
       102 pp. (Arid Zone Programme, 4)
       1956, UNESCO   Out of print
       Covers biological and agricultural problems presented by plants
       tolerant of saline or brackish water and the use of such water for ir-
       rigation; plant growth under saline conditions; and use of sea water
       (desalination).

341   SALINITY PROBLEMS IN THE ARID ZONES. Proceedings of the
       Teheran Symposium.... 395 pp., illus. (Arid Zone Research, 14)
       1961, UNESCO   Composite: E/F   Out of print
       Papers of a symposium of 1958 on: hydrology in relation to salinity
       (13); on the physiology of plants and animals in relation to consump-
       tion of saline water (12); use of brackish water in irrigation and
       saline soils (16); demineralization of saline water(11).

342   Proceedings of the United Nations Interregional Seminar on ORE CON-
       CENTRATION IN WATER-SHORT AREAS, New York, United States of
       America, 14-25 February 1966. 342 pp., illus. (ST/TAO/SER.C/95)
       1968, UN   Sales No. E.68.II.B.4   Out of print
       Technical papers (15) on methods of dry concentration and methods
       using limited or saline water for processing ores, covering well es-
       tablished (e.g., thermal concentration, magnetic separation) and
       newly developed techniques (e.g., electrostatic separation, heavy-
       liquid separation, and sorting by electronic selection).

DESALINATION OF WATER

343   DESALINATION OF WATER USING CONVENTIONAL AND NUCLEAR
       ENERGY. A Report on the Present Status of Desalination and the Pos-
       sible Role Nuclear Energy May Play in This Field. 54 pp., illus.
       (Technical Reports Series, 24; STI/DOC/10/24)
       1964, IAEA $1.00
       Reviews generally desalination of water on a large scale and is in-
       tended to aid preliminary evaluation of projects which may be con-
       sidered, with attention to demand for water, desalination processes
       in use, existing plants, design of large plants and economic consid-
       erations. Selected bibliography.

344   WATER DESALINATION IN DEVELOPING COUNTRIES. 325 pp.,
       maps. (ST/ECA/82)
       1964, UN   Sales No. 64.II.B.5 $4.00
       Surveys economic and technical aspects of saltwater conversion up
       to the end of 1962, covering 43 reports on the countries and territo-
       ries surveyed. (See also entry 346)

345   Proceedings of the Third INTERNATIONAL CONFERENCE ON THE
       PEACEFUL USES OF ATOMIC ENERGY, held in Geneva, 31 August-
       9 September 1964. Multilingual ed. 16 vol. (A/CONF.28/1, vol. 1-16)
          Vol. 6: Nuclear Reactors—II. Fast Reactors and Advanced Con-
          cepts. xv, 528 pp., illus. (A/CONF/28/1, vol. 6)
          1965, UN   Sales No. 65.IX.6 Composite: E/F/R/S $12.50
          Includes three papers on use of reactors in desalination of water
          in the U.S.A., Israel, and Tunisia.

346 WATER DESALINATION: PROPOSALS FOR A COSTING PROCEDURE
     AND RELATED TECHNICAL AND ECONOMIC CONSIDERATIONS. 56
     pp. (ST/ECA/86)
     1965, UN   Sales No. 65.II.B.5 $0.75
     Report of a study group convened in October 1964. Embodies crite-
     ria for application and guidelines for costing with respect to de-
     salination. Considers both single-purpose and combined (desalina-
     tion and power) plants. A follow-up to Water Desalination in Devel-
     oping Countries, 1964 (entry 344).

347 NUCLEAR ENERGY FOR WATER DESALINATION. Report of a Panel
     on the Use of Nuclear Energy for Water Desalination held in Vienna,
     5-9 April 1965. 133 pp., illus. (Technical Reports Series, 51; STI/
     DOC/10/51)
     1966, IAEA $3.00
     Summary report and recommendations of the fifth panel convened by
     IAEA on this subject, with texts or abstracts of 18 reference docu-
     ments. Annexes contain resumes of the fourth (1964) and third
     (1963) panels.

348 COSTING METHODS FOR NUCLEAR DESALINATION. 39 pp., illus.
     (Technical Reports Series, 69; STI/DOC/10/69)
     1966, IAEA $1.00
     Reviews basic principles for costing desalination plants (single-
     purpose and dual-purpose) and of various methods proposed for al-
     locating costs in dual-purpose plants, with indications of possible
     limitations of each method.

349 Proceedings of the Interregional Seminar on the ECONOMIC APPLICA-
     TION OF WATER DESALINATION, New York, 22 September-2 October
     1965. 367 pp., illus. (ST/TAO/SER.C/90)
     1967, UN   Sales No. 66.II.B.30 $4.50
     In three parts: (1) highlights of the seminar and its recommenda-
     tions; (2) texts of 19 lectures; (3) reports on the operational experi-
     ence of desalination plants in the Bahamas, Kuwait, and the U.S.
     Virgin Islands. The seminar dealt with possible alternatives to de-
     salination, economic application of desalinated water, and factors
     influencing selection of the appropriate size and type of desalination
     plants required in water-short areas.

350 Guide to the COSTING OF WATER FROM NUCLEAR DESALINATION
     PLANTS. Work Done on IAEA Research Contract No. 484. 83 pp.,
     illus. (Technical Reports Series, 80; STI/DOC/10/80)
     1967, IAEA $1.00
     Report prepared by Applied Research and Engineering Limited
     (AREL). Investigates relationship of the cost of steam from a nu-
     clear heat source and the cost of the desalted water and (dual-
     purpose) electricity produced. With the data presented in this publi-
     cation, a community can decide whether or not it is worth while to
     undertake for its own particular needs a relatively costly detailed
     study on the costing of water from nuclear desalination plants.

351 COST OF WATER FROM A SINGLE-PURPOSE MULTI-STAGE FLASH
     PLANT WITH VAPOUR COMPRESSION. Report prepared for the IAEA
     by Applied Research and Engineering Limited, Washington, Co. Dur-
     ham, United Kingdom. 73 pp., illus. (Technical Reports Series, 93;
     STI/DOC/10/93)
     1968, IAEA $2.50
     Presents cost of desalted water from a single-purpose MSF/Recom-

pression plant and compares cost of water from a dual-purpose plant. Also shows cost of water from a plant producing both water and power separately from a common reactor.

352 The DESIGN OF WATER SUPPLY SYSTEMS BASED ON DESALINA-TION. Selection of Plant Sizes and Associated Storage Facilities to Meet Variations in Demand and Plant Outages. 64 pp., illus. (ST/ECA/106)
1968, UN   Sales No. E.68.II.B.20   Out of print
Summary report of detailed discussions of a panel of experts, 1966, as well as additional findings subsequently contributed by the Secretariat.

353 VALUE TO AGRICULTURE OF HIGH-QUALITY WATER FROM NU-CLEAR DESALINATION. Report of a Panel on the Value of High-quality Water from Nuclear Desalting in Agriculture, organized by IAEA and held in Vienna, 30 October-3 November 1967. 278 pp., illus. (Panel Proceedings Series; STI/PUB/210)
1969, IAEA   Composite: E/F $7.00
Papers (22) and discussions of economic and agricultural aspects such as value to agriculture of high-quality water from nuclear desalination, cost of irrigation water, water quality and yield depression, tolerance of certain crops to salinity, and agronomic aspects of irrigation with brackish water. A conversion table is given in an annex.

354 NUCLEAR DESALINATION. Proceedings of a Symposium on Nuclear Desalination held by IAEA in Madrid, 18-22 November 1968. 941 pp., illus.
1969, IAEA/Elsevier   Composite: E/F/S $38.50
Papers (62) and discussions covering rational programs, experience with large-scale desalination plants, case studies, nuclear reactors, desalination and power-plant cycles, current and projected desalination techniques, economics of nuclear desalination, agro-industrial complexes, and implementation of large desalination plant studies: research and development. First supplement to Desalination; the International Journal on the Science and Technology of Water De-salting (Amsterdam, Elsevier).

355 First United Nations DESALINATION PLANT OPERATION SURVEY. A Technical and Economic Analysis of the Performance of Desalination Plants in Operation. 122 pp. (ST/ECA/112)
1969, UN   Sales No. E.69.II.B.17 $2.00
Surveys all plants in commercial operation for which it was possible to obtain records of operation throughout 1965 (87 plants in 21 countries). Excludes plants with a capacity of less than 10,000 gallons per day, specialized boiler feed-water plants and experimental plants.

356 SOLAR DISTILLATION as a Means of Meeting Small-scale Water De-mands. 86 pp., illus. (ST/ECA/121)
1970, UN   Sales No. E.70.II.B.1 $2.00
Report of a panel of experts, meeting in October 1968, on conditions under which solar distillation may prove an economic solution to the problems of fresh water shortage in small communities.

357   NUCLEAR ENERGY COSTS AND ECONOMIC DEVELOPMENT. Pro-
       ceedings of a Symposium on Nuclear Energy Costs and Economic De-
       velopment held by IAEA in Istanbul, 20-24 October 1969. 746 pp., il-
       lus. (Proceedings Series; STI/PUB/239)
       1970, IAEA   Composite: E/F/R/S $20.00
       Papers (54) on general criteria for the selection of nuclear stations,
       present capital costs experience, forecasting of energy requirements
       and nuclear prospects, power cost estimates for conventional and
       nuclear plants, economic comparison of various power generation
       strategies, nuclear power and economic development, nuclear desal-
       ination and its economics (5 papers), and energy centers and agro-
       industrial complexes (5 papers and a round-table discussion).

RIVER BASIN DEVELOPMENT

358   RIVER TRAINING AND BANK PROTECTION. 100 pp., illus. (ECAFE.
       Flood Control Series, 4; ST/ECAFE/SER.F/4)
       1953, UN   Sales No. 53.II.F.6   Out of print
       General study made in cooperation with various technical organiza-
       tions, with emphasis on the training of rivers to change the configu-
       ration of their beds rather than for flood control or navigation pur-
       poses.

359   LEGISLATIVE TEXTS AND TREATY PROVISIONS CONCERNING THE
       UTILIZATION OF INTERNATIONAL RIVERS FOR OTHER PURPOSES
       THAN NAVIGATION. 934 pp.
       1963, UN   Sales No. E/F.63.V.6   Out of print

360   INTEGRATED RIVER BASIN DEVELOPMENT; Report of a Panel of
       Experts. [2nd ed.] xvi, 80 pp., 2 fold. maps. (E/3066/Rev.1)
       1970, UN   Sales No. E.70.II.A.4 $1.50
       The new Chairman of the Panel of Experts, Professor Gilbert F.
       White, has reviewed the 1958 report and has contributed a preface to
       the second edition summarizing the major technological and manage-
       ment developments in the field. Changes in and additions to the an-
       nexes attached to the report have been made as a consequence of his
       review.

Asia and the Far East

361   MULTIPLE-PURPOSE RIVER BASIN DEVELOPMENT.
       1955- , UN

361A   Part 1: Manual of River Basin Planning. 83 pp., illus. (ECAFE. Flood
       Control Series, 7; ST/ECAFE/SER.F/7)
       1955, UN   Sales No. 55.II.F.1   Out of print
       Covers general aspects of planning for multipurpose river basin de-
       velopment with general reference to the ECAFE region: water re-
       sources development, watershed management, fisheries, health
       problems; cost/benefit analyses; coordination with other develop-
       ments.

361B   Part 2A: Water Resource Development in Ceylon, China: Taiwan, Japan
       and the Philippines. 122 pp., illus. (ECAFE. Flood Series, 8; ST/
       ECAFE/SER.F/8)
       1956, UN   Sales No. 56.II.F.2   Out of print

361C   Part 2B: Water Resource Development in Burma, India and Pakistan.
        135 pp., illus. (ECAFE. Flood Control Series, 11; ST/ECAFE/SER.F/
        11)
        1956, UN   Sales No. 56.II.F.8   Out of print

361D   Part 2C: Water Resources Development in British Borneo, Federation
        of Malaya, Indonesia and Thailand. 135 pp., illus. (ECAFE. Flood
        Control Series, 14; ST/ECAFE/SER.F/14)
        1959, UN   Sales No. 59.II.F.5   Out of print

361E   Part 2D: Water Resources Development in Afghanistan, Iran, Republic
        of Korea and Nepal. 76 pp., illus. (ECAFE. Flood Control Series, 18;
        ST/ECAFE/SER.F/18)
        1961, UN   Sales No. 61.II.F.8   Out of print

361F   Part 2E: Water Resources Development in Australia, New Zealand and
        Western Samoa. 116 pp., illus., 5 fold maps. (ECAFE. Water Re-
        sources Series, 36; ST/ECAFE/SER.F/36)
        1969, UN   Sales No. E.69.II.F.7 $2.50
        Country surveys reviewing the stage of development attained in re-
        lation to flood control, irrigation and drainage, hydroelectric power,
        navigation, water supply, watershed management and multipurpose
        development. Needs and possibilities for further development are
        touched upon along with various problems that have to be faced.

362   Development of Water Resources in the LOWER MEKONG BASIN. 75
        pp., illus. (ECAFE. Flood Control Series, 12; ST/ECAFE/SER.F/12)
        1957, UN   Sales No. 57.II.F.8 $1.50
        Examines the potential water resources of this international river
        and its current and contemplated utilization as well as the need for
        further development. Reviews certain promising projects for fur-
        ther detailed examination.

363   A Case Study of the DAMODAR VALLEY CORPORATION AND ITS
        PROJECTS. 100 pp., illus. (ECAFE. Flood Control Series, 16; ST/
        ECAFE/SER.F/16)
        1960, UN   Sales No. 60.II.F.7   Out of print
        Study of operations (1948-1960) of the Damodar Valley Corporation
        (DVC) In India in development of the Damodar Valley (States of
        Behar and West Bengal), involving flood and drought control, water-
        shed management, soil conservation, fisheries, irrigation, naviga-
        tion, and a power program. Describes organization and administra-
        tion of DVC.

364   A Case Study of Comprehensive Development of the KITAKAMI RIVER
        BASIN. 51 pp., illus. (ECAFE. Flood Control Series, 20; ST/ECAFE/
        SER.F/20)
        1962, UN   Sales No. 62.II.F.7   Out of print
        Study of the comprehensive development of a river basin in north-
        eastern Honshu, Japan, under a comprehensive plan executed sepa-
        rately by the ministries and prefectures concerned. Briefly com-
        pares the Kitakami Special Area with the Tennessee Valley Author-
        ity and the Damodar Valley Corporation.

365   The Development of GROUNDWATER RESOURCES WITH SPECIAL
        REFERENCE TO DELTAIC AREAS. (Transactions of the Regional
        Seminar on the Development of Groundwater Resources with special

reference to Deltaic Areas, held at Bangkok, Thailand, from 24 April to 8 May 1962.) 244 pp., illus. (ECAFE. Water Resources Series, 24; ST/ECAFE/SER.F/24)
1964, UN   Sales No. 64.II.F.5   Out of print
Report, lectures (20) and papers (21) concern groundwater investigation and development in deltaic areas, including well drilling and maintenance, determination of recharge rates and hydrological mapping. (For the second regional seminar, see entry 286.)

366   A Compendium of MAJOR INTERNATIONAL RIVERS IN THE ECAFE REGION. 74 pp., illus. (ECAFE. Water Resources Series, 29; ST/ECAFE/SER.F/29)
1966, UN   Sales No. 66.II.F.8 $1.50
Describes and provides basic data for twelve of the largest international rivers in the ECAFE region: Mekong, Red, Brahmaputra, Meghna, Ganges, Kosi, Gandak, Gogra, Indus, Sutlej, Kabul, and Helmand. Paper was submitted to the Sixth Regional Conference on Water Resources Development in Asia and the East, 1964 (entry 309).

Humid Tropics

367   Scientific Problems of the HUMID TROPICAL ZONE: DELTAS AND THEIR IMPLICATIONS. Proceedings of the Dacca Symposium jointly organized by the Government of Pakistan and Unesco, 24 February to 2 March 1964.... 422 pp., illus. (Humid Tropics Research)
1966, UNESCO   Composite: E/F $14.00 cl.
Papers (53) cover hydrography and hydrology; geomorphology, sedimentation, and pedology; vegetation; biology; human influence; descriptive studies of the Irawaddy and Amazon deltas; classification of deltas.

FLOOD CONTROL

GENERAL

368   ASSESSMENT OF THE MAGNITUDE AND FREQUENCY OF FLOOD FLOWS. (Transactions of the Inter-regional Seminar on the Assessment of the Magnitude and Frequency of Flood Flows, held at Bangkok, Thailand, from 19 to 26 April 1966.) 206 pp., illus. (ECAFE. Water Resources Series, 30; ST/ECAFE/SER.F/30)
1967, UN/WMO   UN Sales No. 67.II.F.7 $3.00
Lectures and papers (12) on the hydrometeorological approach to the assessment of the frequency and magnitude of floods and statistical methods of flood-frequency analysis.

369   MEASUREMENT OF PEAK DISCHARGE BY INDIRECT METHODS. Prepared by M.A. Benson. 161 pp., illus. (Technical Notes, 90)
1968, WMO   WMO-No.225.TP.119 $3.50
Describes in detail procedures involved in applying hydraulic theory to the post facto measurement of peak flood discharge in streams. While it reflects primarily the practices and procedures followed in the U.S.A., it takes into account comments of experts from other

countries. Gives detailed descriptions of procedures for collecting field data after floods to determine the location and elevation of high-water marks and the characteristics of the channel, and for computing the discharge by such indirect methods as slope-area, contracted opening, flow-over-dam, and flow-through-culvert.

370   FLOODS AND THEIR COMPUTATION. Proceedings of the Leningrad Symposium, August 1967. 2 vols. (985 pp.); illus. (Studies and Reports in Hydrology, 3)
1969, UNESCO/IASH/WMO   Composite: E/F $31.00 the set cl.
Papers (105) covering as the main subjects: general problems, theory of flood formation, and methods of flood computation; theory of formation and methods of computation of snowmelt floods; influence of basin characteristics on the elements of flood hydrographs.

371   HYDROLOGICAL FORECASTING. Proceedings of the WMO/Unesco Symposium on Hydrological Forecasting (Surfers' Paradise, Queensland, Australia, 1967). xvi, 325 pp., illus. (Technical Notes, 92)
1969, WMO/UNESCO   WMO-No.228.TP.122 $21.00
Papers (28) deal with forecasting of precipitation, data acquisition and instrumentation, river and flood forecasting techniques, operational aspects of forecasting. The principal theme is the forecasting of rainfall floods, especially for short time intervals, and attention is focused on those aspects of hydrological forecasting of regional importance.

372   ESTIMATION OF MAXIMUM FLOODS. Report of a Working Group of the Commission for Hydrometeorology. 288 pp., illus. (Technical Notes, 98)
1969, WMO   WMO-No.233.TP.126 $12.50
Describes methods for estimating the extremes of rainfall and snowmelt on the basis of physical analysis, and methods for converting these into estimates of extreme flood flows; provides background on statistical analysis and outline of techniques used in various countries in flood-frequency analysis; describes use of meteorological data in estimating flood frequencies.

373   FLOOD STUDIES: AN INTERNATIONAL GUIDE FOR COLLECTION AND PROCESSING OF DATA. Ed. by F. Snyder, A. Sokolov, and K. Szesztay. 49 pp. (Technical Papers in Hydrology, 8)
1971, UNESCO $4.00
Technical report in three chapters: (1) purposes and scope of the report and general treatment of basin characteristics, streamflow and publication of data; (2) procedure for collecting and processing data on rainfall floods: and (3) procedure for collecting and processing data on snowmelt floods. Text was adopted by the Working Group on Floods and Their Computation, established by the Coordinating Council of the International Hydrological Decade (IHD, 1965-74).

374   "Flood Damage Prevention Policies." By Gilbert F. White. Natural Resources Forum, vol. 1, no. 1 (ST/ECA/142), 1971, pp. 39-44.
1971, UN   Sales No. E/F/S.71.II.A.13 $2.00
Discusses the major types of public policies intended to guide the adjustments that can be made to floods, with some recommendations to developing countries concerning the delimitation of flood hazards.

Asia and the Far East

375   FLOOD DAMAGE AND FLOOD CONTROL ACTIVITIES IN ASIA AND
THE FAR EAST. 82 pp., illus., fold. maps. (ECAFE. Flood Control
Series, 1; ST/ECAFE/SER.F/1)
1951, UN   Sales No. 51.II.F.2   Out of print
Covers hydrology of the area (summary); the development of flood
control works; and flood damage and flood control activities of 19
rivers in the region.

376   METHODS AND PROBLEMS OF FLOOD CONTROL IN ASIA AND THE
FAR EAST. 45 pp., illus. (ECAFE. Flood Control Series, 2; ST/
ECAFE/SER.F/2)
1951, UN   Sales No. 51.II.F.5 $1.75
Reviews flood control methods in use in ECAFE region. A revised
background paper of the First Regional Conference, New Delhi, 1951
(entry 377).

377   Proceedings of the REGIONAL TECHNICAL CONFERENCE ON FLOOD
CONTROL IN ASIA AND THE FAR EAST. 320 pp., illus. (ECAFE.
Flood Control Series, 3; ST/ECAFE/SER.F/3)
1952, UN   Sales No. 53.II.F.1 $3.00
Papers (32) of the First Regional Conference on Water Resources
Development in Asia and the Far East, New Delhi, January 1951,
deal with flood control methods in use in the ECAFE region. (See
also entries 376, 380.)

378   Proceedings of the REGIONAL SYMPOSIUM ON FLOOD CONTROL,
RECLAMATION, UTILIZATION AND DEVELOPMENT OF DELTAIC
AREAS (held at Bangkok, Thailand, 2 to 9 July 1963). 224 pp., illus.
(ECAFE. Water Resources Series, 25; ST/ECAFE/SER.F/25)
1964, UN   Sales No. 64.II.F.6 $3.00
Contains report of the symposium, report of the ECAFE Mission on
Deltaic Areas (1962/63) and 11 papers.

379   FORECASTING OF HEAVY RAINS AND FLOODS. Proceedings of a
Joint Training Seminar held by Regional Associations II and V of WMO
at Kuala Lumpur, Malaysia, from 11 to 23 November 1968. Sponsored
by WMO. 293 pp., illus.
1970, WMO $7.00
Lectures (15) by consultants and papers (7) presented by participants
on methods of forecasting heavy rains and floods, particularly in
connection with the meteorological and hydrological conditions (typhoons and monsoons) of the two WMO regions, Asia and the Southwest Pacific.

DAMS AND RESERVOIRS, EARTHMOVING

380   The SEDIMENT PROBLEM. 92 pp., illus. (ECAFE. Flood Control
Series, 5; ST/ECAFE/SER.F/5)
1953, UN   Sales No. 53.II.F.7   Out of print
Deals with soil erosion, transportation of sediment, silting, and
scouring of canals, silting of reservoirs, action of sediment on the
regime of rivers, sampling and analysis of sediment. Revision of a
background paper of the Regional Technical Conference on Flood
Control in Asia and the Far East, New Delhi, 1951 (entry 377).

381   Proceedings of the REGIONAL SYMPOSIUM ON DAMS AND RESER-
       VOIRS (held at Tokyo, Japan, 18 to 23 September 1961). 238 pp., illus.
       (ECAFE. Flood Control Series, 21; ST/ECAFE/SER.F/21)
       1962, UN   Sales No. 62.II.F.11 $3.00
       Report and papers (19) deal with the factors affecting the choice of a
       site for a dam, the type of dam for a selected site, and the coordina-
       tion of reservoir storage requirements for multipurpose reservoirs.
       Examples of dams and reservoirs in the ECAFE region, Italy, and
       the U.S.A.

382   RECOMMENDATIONS CONCERNING RESERVOIRS. 35 pp.
       1967, UNESCO $1.50
       Provides general information on preparing regulations concerning
       dams and reservoirs, their supervision, and the revision of existing
       regulations. Recommendations are in the form of a set of principles
       and abstracts of existing regulations, arranged by chapters, com-
       piled from legislation of ten countries. The recommendations were
       prepared by an international working group of experts.

383   EARTHMOVING BY MANUAL LABOUR AND MACHINES. (Report and
       Discussion Papers of the Working Party on Earthmoving Operations)
       114 pp., illus. (ECAFE. Flood Control Series, 17; ST/ECAFE/SER.
       F/17)
       1961, UN   Sales No. 61.II.F.4   Out of print
       Subjects of papers include the measure of capital intensity in engi-
       neering construction, work study, incentive schemes, and earth-
       moving operations in the U.S.S.R. The Working Party met in 1959
       in New Delhi.

384   MAN-MADE LAKES: Planning and Development. Ed. by Karl L. Lagler.
       71 pp., illus.
       1969, FAO $1.50
       Study of development and management of man-made lakes or reser-
       voirs; an appendix outlines briefly four projects of the United Na-
       tions Development Program (Special Fund) concerning man-made
       lakes: Lake Kainji (Nigeria), Lake Kariba (Zambia and Rhodesia),
       Lake Nasser (Egypt) and Lake Volta (Ghana).

IRRIGATION AND DRAINAGE

385   Some Aspects of SURFACE WATER DEVELOPMENT IN ARID RE-
       GIONS. By A. De Vajda. 45 pp., illus. (FAO Agricultural Develop-
       ment Papers, 21)
       1952, FAO   Out of print
       Outlines the techniques of utilization of natural flow without storage
       by retarding surface runoff by drainage and by the storage of flood-
       water.

386   COMMUNITY ORGANIZATION FOR IRRIGATION IN THE UNITED
       STATES. By F. Adams. 39 pp. (FAO Agricultural Development Pa-
       pers, 19)
       1953, FAO   Out of print
       Outline and discussion of the forms of organization which have been
       developed in the western states of the U.S.A. to deal with irrigation.

387   WATER-LIFTING DEVICES FOR IRRIGATION. By A. Molenaar. 76 pp., illus. (FAO Agricultural Development Papers, 60)
1956, FAO   Out of print
Describes and illustrates the principal types of devices used to raise water for irrigation in rice-growing countries, with information on construction, operation, power, and costs.

388   IRRIGATION BY SPRINKLING. By A. Molenaar. 90 pp., illus. (FAO Agricultural Development Papers, 65)
1962, 2nd printing, FAO   Out of print
Explains the design and use of sprinkling systems as an aid in effecting better use and control of irrigation water. First issued in 1960.

389   METHODS AND MACHINES FOR TILE AND OTHER TUBE DRAINAGE. By G.H. Theobald. 104 pp., illus. (FAO Agricultural Development Papers, 78)
1963, FAO $3.00
Describes in detail the practical considerations in drainage work and can serve therefore as a useful handbook for teaching drainage methods by hand and by machine.

390   APPLIED IRRIGATION RESEARCH. By Albert W. Marsh. 54 pp., illus.
1967, FAO
Methodology of irrigation treatments according to water requirements, giving measurement formulas. A main document (reference no. 00531-67-MR).

391   SURFACE IRRIGATION. By L.J. Booher. 161 pp. and corrigendum sheet; illus.
1967, FAO
Analyzes methods for obtaining adequate and efficient irrigation through the uniform distribution of the water supply to the field. A main document (reference no. 01060-67-MR).

392   SOIL-MOISTURE AND IRRIGATION STUDIES. Proceedings of a Panel on the Use of Isotope and Radiation Techniques in Soil-moisture and Irrigation Studies organized by the Joint FAO/IAEA Division of Atomic Energy in Food and Agriculture and held in Vienna, 14-18 March 1966. 109 pp., illus. (Panel Proceedings Series; STI/PUB/133)
1967, IAEA $2.50
Contains an introductory paper on the need of arid and semiarid regions for water-use efficiency studies; five papers on radiation equipment; one paper on radioisotopes and labeled salts; two papers on water-use efficiency; the group experiment; and recommendations of the panel.

393   ISOTOPE AND RADIATION TECHNIQUES IN SOIL PHYSICS AND IRRIGATION STUDIES. Proceedings of a Symposium on the Use of Isotope and Radiation Techniques in Soil Physics and Irrigation Studies organized jointly by FAO and IAEA and held in Istanbul, 12-16 June 1967. 446 pp., illus. (Proceedings Series; STI/PUB/158)
1967, IAEA   Composite: E/F/R/S $9.00
Papers (32) cover radiation equipment; soil-moisture studies; water-movement studies; soil-water-plant relationships; desalted water for agriculture.

394   RICE IRRIGATION IN JAPAN. By H. Fukuda and H. Tsutsui. 766 pp.,
illus., map.
1968, FAO
A main document (reference no. 03794-68-MR).

395   SPRINKLER IRRIGATION. By A.F. Pillsbury. 179 pp. (FAO Agricultural Development Papers, 88)
1968, FAO $3.50
Reviews use of sprinkler irrigation systems. Discusses in detail
design criteria of which the engineer should be aware in planning
such systems.

396   SUCCESSFUL IRRIGATION: Planning, Development, Management. By
Robert M. Hagan, Clyde E. Houston, Stephen V. Allison. 53 pp., illus.
1968, FAO $1.00
Booklet intended to assist policy makers, administrators, and planners to discharge more effectively their roles in the development of
irrigated agriculture.

397   "Irrigation Development at the Farm Level." By B.D. van't Woudt.
Economic Bulletin for Asia and the Far East, vol. 19, no. 3 (E/CN.11/
849), December 1968, pp. 32-46.
1968, UN   Sales No. E.68.II.F.16 $2.00

SCHISTOSOMIASIS (BILHARZIASIS) CONTROL

398   Study Group on the ECOLOGY OF INTERMEDIATE SNAIL HOSTS OF
BILHARZIASIS. Report. 38 pp. (WHO Technical Report Series, 120)
1957, WHO   Out of print
Covers distribution of intermediate snail hosts in relation to hydrogeology; factors conditioning the habitat and breeding conditions and
influencing the life cycle; application of knowledge of ecology and
physiology to control methods.

399   Second AFRICAN CONFERENCE ON BILHARZIASIS (WHO/CCTA),
Lourenço Marques, Mozambique, 30 March-8 April 1960. Report. 37
pp. (WHO Technical Report Series, 204)
1960, WHO $0.30
Deals with water management, land use, and agricultural practices
in relation to the transmission and spread of schistosomiasis, in addition to assessment of the medical and public health importance of
schistosomiasis and snail control by chemical means. Report of
the first conference, 1956 (Technical Report Series, 139, 1957) is
out of print.

400   "Bilharziasis." Bulletin of the World Health Organization, vol. 27, no.
1, 1962, pp. 1-187, illus.
1962, WHO $3.25
In addition to articles on snails, snail counts, and molluscicides,
contains articles (3) on schistosomiasis control and water resources
development; on pump schemes and efficacy of chemical barriers;
and water flow velocities in irrigation canals and their mathematical
analysis.

401 "Bilharziasis and Malaria." Bulletin of the World Health Organization, vol. 35, no. 3, 1966, pp. 281-458, illus.
1966, WHO $3.25
Contains articles dealing with epidemiology of schistosomiasis in Egypt (including sources of water supply), molluscicides, etc.

402 SNAIL CONTROL IN THE PREVENTION OF BILHARZIASIS. By various contributors. 247 pp., illus. (WHO Monograph Series, 50)
1965, WHO $6.00
Examines the measures of a snail control program applicable to natural habitats, to irrigation schemes, and to other man-made habitats; discusses the chemicals used as molluscicides and the use of the latter, as well as methods of control not using molluscicides; and such related topics as the life history and bionomics of the snail host, the training of personnel, and the factors deciding a control program.

403 WHO Expert Committee on BILHARZIASIS. Third Report. 56 pp. (WHO Technical Report Series, 299)
1965, WHO $1.00
Covers the health and economic importance of schistosomiasis (including relationship with water resource development schemes), epidemiology, pathology, control (including provision of safe water supply and sanitation), and research needs. First and second reports were published in 1953 and 1961 (WHO Technical Reports Series, 65 and 214).

404 EPIDEMIOLOGY AND CONTROL OF SCHISTOSOMIASIS. Report of a WHO Scientific Group. 35 pp. (WHO Technical Report Series, 372)
1967, WHO $1.00
Deals with parasitology, epidemiology, health and economic importance, control, techniques of control; recommendations.

405 HEALTH IMPLICATIONS OF WATER-RELATED PARASITIC DISEASES IN WATER DEVELOPMENT SCHEMES. 22 pp.
1967, FAO/WHO Free on request to WHO
Brochure designed to brief experts engaged in water-development schemes on the possible effects on water-borne parasitic diseases of dam construction, irrigation schemes, fish ponds, and related population movements. The potential health dangers are noted from malaria, schistosomiasis (bilharziasis), filariasis, onchoceriasis (river blindness), and distomatosis (fascioliasis).

SOIL SCIENCE (PEDOLOGY)

GENERAL, SOIL CONSERVATION

406 SOIL CONSERVATION; an International Study. 189 pp., illus. (FAO Agricultural Studies, 4)
1952, 2nd printing, FAO $2.00
Brief account of soil deterioration and of some of the measures being taken in various countries to correct the misuse of land. Emphasizes particularly soil erosion and the conservation of soil through control of erosion. First issued in 1948.

407   Proceedings of the United Nations Scientific Conference on the CON-
        SERVATION AND UTILIZATION OF NATURAL RESOURCES, 17
        August-6 September 1949, Lake Success, New York. 8 vol. (E/CONF.
        7/7, vol. 1-8)
        1950-51, UN
            Vol. 6: Land Resources. 629 pp., illus. (E/CONF.7/7, vol. 6)
            Sales No. 50.II.B.7   Out of print
                Papers (44) on soil conservation; aids to farming; improving soil
                productivity; plant breeding; protection of crops and grasslands;
                storage and preservation of agricultural products; livestock
                breeding; crop policy and the feeding of livestock; livestock dis-
                eases and pests; grazing lands; more effective use of new agri-
                cultural lands.

408   TROPICAL SOILS AND VEGETATION. Proceedings of the Abidjan
        Symposium jointly organized by Unesco and the Commission for Tech-
        nical Co-operation in Africa South of the Sahara, 20-24 October 1959....
        115 pp., illus. (Humid Tropics Research)
        1961, UNESCO   Composite: E/F   Out of print
            Contains 12 papers and the recommendations of the symposium.

409   SOIL EROSION BY WIND AND MEASURES FOR ITS CONTROL ON
        AGRICULTURAL LANDS. 88 pp., illus. (FAO Agricultural Develop-
        ment Papers, 71)
        1960, FAO   Out of print
            Describes control measures in North America and Australia, as well
            as those adaptable to developing countries. Devotes particular at-
            tention to the part that improved tillage and cultural practices play
            in control of erosion by wind in dry farming areas. Describes and
            illustrates equipment for control of soil erosion. Bibliography.

410   SOIL EROSION BY WATER: Some Measures for Its Control on Culti-
        vated Lands. 284 pp., illus. (FAO Agricultural Development Papers,
        81)
        1965, FAO $4.00
            Discusses some measures of water erosion control in temperate and
            tropical conditions that are feasible from an economic, social, and
            agricultural production point of view. Bibliography.

411   DARK CLAY SOILS OF TROPICAL AND SUBTROPICAL REGIONS. Ed.
        by R. Dudal. 161 pp., illus. (FAO Agricultural Development Papers,
        83)
        1965, FAO $3.00
            Summary of knowledge of the nature and distribution of dark clay
            soils and of the means of improving their management and increas-
            ing their output, intended for land planners, extension officers, ad-
            ministrators, farm management specialists, and all others involved
            in the application of soil studies.

412   Review of Research on LATERITES. By R. Maignien. 148 pp., illus.
        (Natural Resources Research, 4)
        1966, UNESCO $5.00
            Background, definition and scope of the problem; morphological and
            analytical characteristics of laterites; global distribution and rela-
            tion with environmental factors; origin of laterites; classification;
            utilization: fertility, source of ores, use in civil engineering, stra-
            tigraphy, hydrology.

SOIL MOISTURE

413   PRACTICAL SOIL MOISTURE PROBLEMS IN AGRICULTURE. Report
       of the Working Group on Practical Soil Moisture Problems in Agricul-
       ture of the Commission for Agricultural Meteorology. Prepared by G.
       Stanhill and others. 69 pp. (Technical Notes, 97)
       1968, WMO   WMO-No.235.TP.128 $3.50
       Contains an annotated short list (16 items) of literature reviewing
       the moisture balance of the soil; general description of four tech-
       niques of soil moisture content in order of preference (with a more
       detailed description of the first choice, the direct gravimetric
       method, in an appendix); guide to meteorological aids to the efficient
       use of water for irrigation (with details based on replies to a ques-
       tionnaire in an appendix). Bibliography, and recommendations for
       further research.

414   WATER IN THE UNSATURATED ZONE. Proceedings of the Wagenin-
       gen Symposium. Ed. by P.E. Rijtema and H. Wassink. 2 vol. (995
       pp.), illus. (Studies and Reports in Hydrology, 2)
       1969, UNESCO/IASH   Composite: E/F $31.00 the set cl.
       Papers on determination of soil moisture, capillary conductivity and
       diffusitivity; relation between soil characteristics and soil proper-
       ties; infiltration and redistribution of soil moisture following precip-
       itation; mathematics of unsaturated flow; energy relations in water
       transport and supply of water by capillary rise; evaporation from
       bare soils and accumulation of salts; extraction of soil moisture by
       plants; influence of temperature on the transport of water; rate of
       recharge of ground water; effect of the capillary fringe on ground-
       water flow; storage capacity.

415   NEUTRON MOISTURE GAUGES. A Guide-book on Theory and Practice.
       95 pp., illus. (Technical Reports Series, 112; STI/DOC/10/112)
       1970, IAEA $3.00
       Guide to the design and application of neutron moisture gauges in
       measurements on soils in civil engineering, agriculture, and hydrol-
       ogy, prepared by a panel of experts. Appendix: definitions of terms
       relating to neutron soil moisture gauges.

416   DIRECT METHODS OF SOIL MOISTURE ESTIMATION FOR WATER
       BALANCE PURPOSES. By M. Kutilek. viii, 58 pp., fig. (Reports on
       WMO/IHD Projects, 14)
       1971, WMO   WMO-No.286 $2.00
       Report on the status of the art of measurement and estimation of
       soil moisture over large areas for water balance purposes. Bibliog-
       raphy.

SOIL-PLANT RELATIONS

417   RADIOISOTOPES IN SOIL-PLANT NUTRITION STUDIES. Proceedings
       of the Symposium on the Use of Radioisotopes in Soil-Plant Nutrition
       Studies jointly organized by IAEA and FAO and held in Bombay, 26
       February-2 March 1962. 461 pp. and corrigendum slip; illus. (Pro-
       ceedings Series; STI/PUB/55)
       1962, IAEA   Composite: E/F/R/S   Out of print
       Papers (32) on soil chemistry and physics, ion uptake and transloca-
       tion, biological measurement of soil characteristics, and fertilizer
       usage.

418   LABORATORY TRAINING MANUAL ON THE USE OF ISOTOPES AND
        RADIATION IN SOIL-PLANT RELATIONS RESEARCH. 166 pp., illus.
        (Technical Reports Series, 29; STI/PUB/15/29)
        1964, IAEA   Out of print
        In two parts: a basic part which contains lectures and laboratory
        exercises on the properties and use of radioisotopes and radiation;
        and a second part containing a series of detailed lectures and labo-
        ratory exercises on the application of radioisotopes and radiation in
        the field of soil-plant relationships.

419   PLANT NUTRIENT SUPPLY AND MOVEMENT. Report of a Panel on
        the Use of Radioisotopes and Radiation in Plant Nutrient Supply and
        Movement in Soil Systems, convened by IAEA and FAO and held in
        Vienna, 9-13 November 1964. 160 pp., illus. (Technical Reports Se-
        ries, 48; STI/DOC/10/48)
        1965, IAEA $3.50
        Papers (17), summary, and recommendations of the panel present
        the results, obtained by using isotopes, of some basic research on
        soils.

420   ISOTOPES AND RADIATION IN SOIL-PLANT NUTRITION STUDIES.
        Proceedings of the Symposium on the Use of Isotopes and Radiation in
        Soil-Plant Nutrition Studies jointly organized by IAEA and FAO and
        held in Ankara, 28 June-2 July 1965. 609 pp., illus. (Proceedings
        Series; STI/PUB/108)
        1965, IAEA   Composite: E/F/R/S $12.00
        Papers (45) on soil chemistry, soil physics, ion uptake and translo-
        cation, and fertilizer usage.

421   LIMITING STEPS IN ION UPTAKE BY PLANTS FROM SOIL. Report of
        a Panel on Limiting Steps in Ion Uptake by Plants from Soil, convened
        by the Joint FAO/IAEA Division of Atomic Energy in Agriculture and
        held in Vienna, 22-26 November 1965. 154 pp. and corrigenda sheet;
        illus. (Technical Reports Series, 65; STI/DOC/10/65)
        1966, IAEA $3.50
        Papers (16) include, among other subjects, kinetics of phosphate re-
        lease from soil, distribution of ions at surfaces and implications for
        diffusion, role of mass movement and ionic diffusion in ionic migra-
        tion through soils, types of limitations to ion uptake by plants, spec-
        ulations on the oxidation-reduction status of the rhizosphere of rice
        shoots in submerged soils, and a comprehensive paper prepared by
        the panel.

422   ISOTOPES IN PLANT NUTRITION AND PHYSIOLOGY. Proceedings of
        the Symposium on the Use of Isotopes in Plant Nutrition and Physiology
        jointly organized by IAEA and FAO, and held in Vienna, 5-9 September
        1966. 596 pp., illus. (Proceedings Series; STI/PUB/137)
        1967, IAEA   Composite: E/F/R/S $12.50
        Papers (47) on soil chemistry and fertility; experimental techniques;
        assimilation and metabolism; transport phenomena; ion absorption,
        accumulation and transport; genetical aspects of plant nutrition; in-
        cluding papers (5) of a seminar on the nature, genetics, and screen-
        ing of nutritional genotants and mutants in higher plants.

423   ISOTOPE STUDIES ON THE NITROGEN CHAIN. Proceedings of the
        Symposium on the Use of Isotopes in Studies of Nitrogen Metabolism in
        the Soil-Plant-Animal System jointly organized by IAEA and FAO in

co-operation with the Joint Commission on Applied Radioactivity (ICSU)
and held in Vienna, 28 August-1 September 1967. 343 pp., illus. (Proceedings Series; STI/PUB/161)
1968, IAEA   Composite: E/F/R $7.00
   Papers (23) on soil science and plant physiology; animal physiology
   and nutrition; techniques; plant breeding and protein quality.

424   ISOTOPES AND RADIATION IN SOIL ORGANIC-MATTER STUDIES.
        Proceedings of the Symposium on the Use of Isotopes and Radiation in
        Soil Organic-matter Studies jointly organized by IAEA and FAO in co-
        operation with the International Soil Science Society and held in Vienna,
        15-19 July 1968. 584 pp., illus. (Proceedings Series; STI/PUB/190)
        1968, IAEA   Composite: E/F/R/S $13.50
           Papers (48) on laboratory techniques for studying soil organic mat-
           ter; properties of soil organic compounds; studies of nitrogen trans-
           formation in soils using 15N; organic-matter synthesis and decom-
           position in soils; and organic-matter effects on nutrient availability
           in soils.

425   SOIL BIOLOGY; Reviews of Research. 240 pp., errata slip; illus. (Nat-
        ural Resources Research, 9)
        1969, UNESCO $11.00 cl.
           Reviews (5) of results of research in soil biology in various parts of
           the world: methodological problems; biological fixation of atmo-
           spheric nitrogen by free-living bacteria; ecological associations
           among soil microorganisms; biology and soil fertility; and microbial
           degradation and biological effects of pesticides in soil. Bibliogra-
           phies.

426   Methods of Study in SOIL ECOLOGY. Proceedings of the Paris Sym-
        posium organized by Unesco and the International Biological Pro-
        gramme.... Ed. by J. Phillipson. 303 pp., illus. (Ecology and Con-
        servation, 2)
        1970, UNESCO   Composite: E/F $17.00 cl.
           Papers on general problems in soil ecology (3 papers); structural
           aspects of soil ecosystems (3); functional aspects of soil ecosystems
           (3); methods for the study of production by soil microorganisms (7),
           by macrophytes (4), by soil mesofauna (6), by arthropods of the mac-
           rofauna (4), by root-fluid feeders and nematodes (3), and by earth-
           worms, enchytraeids and molluscs (4); and one paper each on soil
           respiration and on problems in soil ecology requiring most urgent
           attention.

427   NITROGEN-15 IN SOIL-PLANT STUDIES. Proceedings of a Research
        Co-ordination Meeting on Recent Developments in the Use of Nitrogen-
        15 in Soil-Plant Studies, organized by the Joint FAO/IAEA Division of
        Atomic Energy in Food and Agriculture and held in Sofia, 1-5 December
        1969. 255 pp., fig., tables. (Panel Proceedings Series; STI/PUB/278)
        1971, IAEA $7.00
           Papers (17) concerning recent advances in the use of nitrogen-15 in
           soil-plant studies, and panel recommendations on the use of the
           isotope technique in studying the efficiency of fertilizer nitrogen
           conversion into protein. A special section was devoted to optical
           spectroscopy techniques for nitrogen-15 assay.

FERTILIZERS

428   CROP PRODUCTION LEVELS AND FERTILIZER USE. By Moyle S.
        Williams and John W. Couston. 48 pp., illus. (Fertilizer Program,
        Freedom from Hunger Campaign)
        1962, FAO $0.50
            Confirms the close relationship between fertilizer use, on a national
            scale, and average crop yields.

429   "Fertilizer Use: Spearhead of Agricultural Development." The State of
        Food and Agriculture 1963, pp. 135-175, 218-227.
        1963, FAO $3.00
            A special chapter and annex tables. Reviews use of fertilizer in this
            century, its effects on crop production and in forestry and fisheries,
            physical factors affecting response to fertilizer, and economic and
            social factors affecting its use by the farmer. (See entry 1174 for
            The State of Food and Agriculture.)

430   FERTILIZERS AND THEIR USE; a Pocket Guide for Extension Officers.
        2nd ed., rev. x, 54 pp., illus. (Fertilizer Program, Freedom from
        Hunger Campaign)
        1970, FAO $1.75 cl.
            First edition issued in 1965. (For related filmstrip, see entry 1079.)

431   Report of the Interregional Seminar on the PRODUCTION OF FERTI-
        LIZERS, Kiev, Ukrainian Soviet Socialist Republic, 24 August to 11
        September 1965. 51 pp. and corrigenda sheet. (ST/TAO/SER.C/78
        and Corr.1)
        1966, UN   Sales No. 66.II.B.7 $1.50
            Contains a brief account of the papers and discussions of the
            seminar and its conclusions and recommendations. (For seminar
            papers, see entry 434.)

432   STATISTICS OF CROP RESPONSES TO FERTILIZERS. 112 pp.
        1966, FAO $2.00
            Manual on design of fertilizer experiments and reporting and analy-
            sis of resulting quantitative data, as applied to rice production. One
            aim is to improve and standardize statistical information on crop
            responses to fertilizers as the basis for sounder estimates of na-
            tional fertilizer requirements and of the possibilities of increased
            agricultural production.

433   FERTILIZER MANUAL. xvii, 218 pp., illus. (ST/CID/15)
        1967, UN   Sales No. 67.II.B.1 $3.50
            Contents include: world survey—people, food, and fertilizer; role of
            fertilizer in agricultural production; criteria for production versus
            importation of fertilizer; demand, marketing, distribution, and
            pricing of fertilizer; production of ammonia, ammonia salts, nitric
            acid, nitrates, urea, phosphate, and potash fertilizers; location of
            plants; investment and production costs.

434   FERTILIZER PRODUCTION, TECHNOLOGY AND USE. Papers pre-
        sented at the United Nations Interregional Seminar on the Production of
        Fertilizers, Kiev, Ukrainian Soviet Socialist Republic, 24 August-11
        September 1965. 404 pp., illus. (ID/2)
        1968, UN   Sales No. E.68.II.B.1 $5.00
            Papers (54) on production—world, regional, and country reports;
            technology: processes, maintenance, and safety measures; planning,
            economics, marketing, and use in agriculture. (For seminar report,
            see entry 431.)

435 Handbook of Utilization of AQUATIC PLANTS. A Compilation of the
World's Publications. Ed. by E.C.S. Little. 128 pp., illus. (PL:CP/20)
1968, FAO $1.50
Reprints articles on utilization of the water hyacinth and other
aquatic weeds as composts, fodder and source of potash, and for
other purposes. A main document (reference no. 03000-68-MI).

436 EFFICIENT USE OF FERTILIZERS. Revised and enlarged ed. Ed. by
Vladimir Ignatieff and Harold J. Page. xxi, 367 pp., fold. map. (FAO
Agricultural Studies, 43)
1968, 5th printing, FAO $4.00
Describes the role of fertilizers; plant nutrients; preparation and
use of organic manures; commercial fertilizers and soil amendments; factors affecting the use of fertilizers and manures; application of fertilizers; plant-nutrient relationships to soil regions; soil
and nutrient needs of specific crops; economic aspects of the use of
fertilizers; the farmers and the agricultural services. First edition
issued in 1949; revised and enlarged edition first issued in 1958.

437 CHEMICAL FERTILIZER PROJECTS: Their Creation, Evaluation and
Establishment. By Christopher J. Pratt. 52 pp. (UNIDO. Fertilizer
Industry Series Monographs, 1; ID/SER.F/1)
1968, UN   Sales No. E.68.II.B.17 $0.75
Guidelines for evaluating potential chemical fertilizer projects and
selecting the most suitable in terms of national interests, technical
feasibility, and maximum economic benefit.

438 Report of the Meeting of the Ad Hoc Expert Group on FERTILIZER
PRODUCTION IN SIX SELECTED COUNTRIES WITH GOOD NATURAL
GAS RESOURCES, United Nations Headquarters, New York, 9-16
December 1966. 68 pp., tables. (UNIDO. Fertilizer Reports; ID/5)
1969, UN   Sales No. E.69.II.B.5 $0.75
Report, with conclusions and recommendations, of the expert group
on its examination of the possibilities of large-scale production of
fertilizers using natural gas in six selected countries—Iran, Kuwait,
Libya, Nigeria, Saudi Arabia, and Venezuela.

439 Guide to BUILDING AN AMMONIA FERTILIZER COMPLEX. By J.A.
Finneran and P.J. Masur. 24 pp. (UNIDO. Fertilizer Industry Series
Monographs, 2; ID/SER.F/2)
1969, UN   Sales No. E.69.II.B.10 $0.75

440 The REDUCTION OF SULPHUR NEEDS IN FERTILIZER MANUFAC-
TURE. By Christopher J. Pratt. 61 pp., illus. (UNIDO. Fertilizer Industry Series Monographs, 3; ID/SER.F/3)
1969, UN   Sales No. E.69.II.B.26 $0.75

441 New Process for the PRODUCTION OF PHOSPHATE FERTILIZERS
USING HYDROCHLORIC ACID. By Y. Araten and R. Brosh. 31 pp.,
illus. (UNIDO. Fertilizer Industry Series Monographs, 5; ID/SER.F/5)
1969, UN   Sales No. E.69.II.B.23 $0.75

442 Industrialization of Developing Countries: Problems and Prospects.
FERTILIZER INDUSTRY. Based on the Proceedings of the International Symposium on Industrial Development (Athens, November-De-
cember 1967). 76 pp. (UNIDO Monographs on Industrial Development,
6; ID/40/6)
1969, UN   Sales No. E.69.II.B.39, vol. 6 $0.50
In nine chapters: (1) trends in the world fertilizer industry; (2) fac-

tors that inhibit the development of a fertilizer industry; (3-5) nitrogen, phosphate, and potash fertilizers; (6) economic issues to be faced in developing a fertilizer industry; (7) financing of fertilizer projects; (8) discussions and recommendations of the symposium; and (9) United Nations action and bilateral assistance to promote the fertilizer industry in developing countries.

443   DIRECTORY OF FERTILIZER PRODUCTION FACILITIES. Parts I-(ID/44, vol.1-)
1970, UN
Part I: Africa. 271 pp., tables, map. (ID/44, vol. 1)
1970   Sales No. E.70.II.B.28 $3.00
Contains data on present and projected demand and production of fertilizer; existing fertilizer production facilities; and projects being implemented or in the planning state. Also contains information on availability and production of fertilizer raw materials and fuels and other relevant data briefly illustrating the state of development of the national economies and of the agricultural and manufacturing sectors. Contains a summary of regional data and of country data grouped by subregions. Further volumes are in course of preparation for Asia, Latin America, and the Middle East.

Asia and the Far East

444   MINERAL RAW MATERIAL RESOURCES FOR THE FERTILIZER INDUSTRY IN ASIA AND THE FAR EAST. 75 pp. (ECAFE. Mineral Resources Development Series, 28; E/CN.11/798)
1967, UN   Sales No. 68.II.F.3 $1.50
Summarizes consumption, production and present and future sources of supply of fertilizers for the ECAFE region; surveys, country by country, known occurrences of mineral raw materials.

445   Proceedings of the SEMINAR ON SOURCES OF MINERAL RAW MATERIALS FOR THE FERTILIZER INDUSTRY IN ASIA AND THE FAR EAST. xvi, 392 pp., illus., maps. (ECAFE. Mineral Resources Development Series, 32; E/CN.11/837)
1968, UN   Sales No. E.69.II.F.2 $5.00
Report and documentation of a seminar held in Bangkok in December 1967, and jointly sponsored by ECAFE and FAO. Deals with present demand and projected requirements for fertilizers in the ECAFE region in 1970, 1975, and 1980; availability of primary fertilizer materials in the area; and deposits of phosphate rocks, potash, sulfur, and other mineral raw materials.

446   The AMMONIUM CHLORIDE AND SODA ASH DUAL MANUFACTURING PROCESS IN JAPAN. By Shozaburo Seki. 33 pp., illus. (UNIDO. Fertilizer Industry Series Monographs, 4; ID/SER.F/4)
1969, UN   Sales No. E.69.II.B.20 $0.75

Latin America

447   FERTILIZER DEMAND AND SUPPLY PROJECTIONS TO 1980 FOR SOUTH AMERICA, MEXICO AND CENTRAL AMERICA. 80 pp., tables. (UNIDO. Fertilizer Industry Series, Monograph No. 6; ID/SER.F/6)
1971, UN   Sales No. E.71.II.F.9 $1.50
Describes the agricultural sectors of the major countries of Latin

America, with special reference to past fertilizer demand and supply and projections to 1980. Data are based on information available at the end of 1968. Prepared by Christopher J. Pratt, consultant to UNIDO.

PLANT RESOURCES

PLANT ECOLOGY

448   GRASSES IN AGRICULTURE. By R.O. Whyte, T.R.G. Moir and J.P.
       Cooper. 417 pp., illus. (FAO Agricultural Studies, 42)
       1968, 4th printing, FAO $4.00
       Reviews adaptation, management, improvement, and utilization of
       cultivated grasses in dryland and irrigated agriculture throughout
       the world, in three parts: (1) agronomy of grasses, (2) biology of
       grasses, and (3) genera and species of grasses. Bibliography.
       First issued in 1959.

449   Functioning of TERRESTRIAL ECOSYSTEMS AT THE PRIMARY
       PRODUCTION LEVEL. Proceedings of the Copenhagen Symposium....
       Ed. by F.E. Eckardt. 516 pp., illus. (Natural Resources Research, 5)
       1968, UNESCO   Composite: E/F $19.00 cl.
       Contains the proceedings of the symposium held 24-30 July 1965 and
       attended by specialists in the fields of micrometeorology, plant
       physiology, and soil science. In three parts: a series of introductory
       lectures (10); short papers (31) on recent techniques and procedures
       facilitating the study of the production of the plant cover; a general
       discussion mainly concerned with the problem of choice, delimita-
       tion and equipment of experimental sites (4 papers).

Africa

450   CROP ECOLOGIC SURVEY IN WEST AFRICA (Liberia, Ivory Coast,
       Ghana, Togo, Dahomey, Nigeria). By J. Papadakis. 2 vol.
       1965-66, FAO
       Vol. 1: Test. 1966. 103 pp. (PL/FFC/2) $2.00
       Vol. 2: Atlas. 1965. 44 pp. Bilingual: E/F. $5.00
       Presents results of a survey carried out in 1963. Volume I covers all
       environmental conditions; describes the ecological regions and the
       crops that can be grown in each; discusses separately the production of
       each crop; and states the implications of the survey concerning agri-
       cultural development and agronomic research; bibliography. Volume II
       (bilingual) comprises 24 maps and three tables.

451   The GRASS COVER OF AFRICA. By J.M. Rattray. 168 pp., map.
       (FAO Agricultural Studies, 49)
       1968, 2nd printing, FAO $2.50
       Summarizes the extent of knowledge of the grass cover of Africa and
       is useful as a guide not only in the field of range or grassland man-
       agement but also in the wider reaches of land use generally. Bibli-
       ography. A ten-color 1:10,000,000 map of Africa's grass cover is
       included in a separate folder. First issued in 1960.

Arid Zones

452   PLANT ECOLOGY. Proceedings of the Montpellier Symposium....
124 pp., illus. (Arid Zone Research, 5)
1955, UNESCO   Composite: E/F   Out of print
Papers of the symposium of October 1953 on structural and physiological features of vegetation (5 papers), climatic, ecoclimatic, and
hydrological effects on vegetation (5), soil and vegetation (6), other
factors (3). (See also entry 453.)

453   PLANT ECOLOGY. Reviews of Research.... 377 pp., illus., maps.
(Arid Zone Research, 6)
1955, UNESCO   Composite: E/F   Out of print
Ten reports on research on the plant ecology of specific regions,
with particular attention to wild species which might be suitable for
agricultural crops. Background material for the Montpellier symposium (entry 452). Bibliographies.

454   PLANT-WATER RELATIONSHIPS IN ARID AND SEMI-ARID CONDI-
TIONS. Reviews of Research. 225 pp., illus. (Arid Zone Research, 15)
1960, UNESCO $7.25 cl.
Review papers (9) deal with income and loss of water; soil-water relations; physiological and morphological changes in plants due to
water deficiency; adaptation to drought; methods of research on
water relations; management of native vegetation; principles of dry
land crop management; fallow as a management technique; and principles of irrigated cropping. Prepared for the Madrid Symposium of
1959 (entry 455).

455   PLANT-WATER RELATIONSHIPS IN ARID AND SEMI-ARID CONDI-
TIONS. Proceedings of the Madrid Symposium.... 352 pp., illus.
(Arid Zone Research, 16)
1962, UNESCO   Composite: E/F/S   Out of print
Papers (39) cover methodology of water relation studies of plants;
water resources for plants; water balance of plants under arid and
semiarid conditions; drought and heat resistance of plants; practical
applications to agronomy. (See also entry 454.)

456   MEDICAL PLANTS OF THE ARID ZONES. 96 pp. (Arid Zone Research, 13)
1960, UNESCO $3.00
Covers both botanical and pharmacological aspects. Contains a list
of arid-zone medicinal species.

457   METHODOLOGY OF PLANT ECO-PHYSIOLOGY. Proceedings of the
Montpellier Symposium.... Ed. by F.E. Eckardt. 531 pp., illus. (Arid
Zone Research, 25)
1965, UNESCO   Composite: E/F $20.00 cl.
Papers (58) of this 1962 symposium are grouped into three sections:
environmental factors; physiology of plants considered individually;
physiology of plant cover. Among the topics covered are humidity,
vapor pressure, evaporation and transpiration, soil-water potential
and movement, precipitation, water tension in plants, drought resistance.

Humid Tropics

458   Study of TROPICAL VEGETATION. Proceedings of the Kandy Symposium jointly organized by the Government of Ceylon and Unesco, Queen's Hotel, Kandy, Ceylon, 19-21 March 1956.... 226 pp., illus. (Humid Tropics Research)
1958, UNESCO   Composite: E/F   Out of print
Contains papers (31) on tropical ecology and vegetation and the recommendations of the symposium.

FOREST RESOURCES

459   Proceedings of the United Nations Scientific Conference on the CONSERVATION AND UTILIZATION OF NATURAL RESOURCES, 17 August-6 September 1949, Lake Success, New York. 8 vol. (E/CONF. 7/7, vol. 1-8)
1950-51, UN
Vol. 5: Forest Resources. 325 pp., illus. (E/CONF.7/7, vol. 5)
Sales No. 50.II.B.6   Out of print
Papers (77) on forest inventories; protection of forests; forest management; protective functions of the forests (including effect on climate and water regime, soils, wildlife); administration of forests; logging and sawmill techniques; preservation and chemical utilization of wood.

460   EUCALYPTS FOR PLANTING. By André Métro. 403 pp., illus. (FAO Forestry and Forest Products Studies, 11)
1965, 3rd printing, FAO $3.50
Part I gives a brief account of the geography and eucalypt forest types of Australia and of the experience with eucalypts as exotics in other countries. Part II covers the adaptability of eucalypts, planting techniques, pathology, stand treatment, and utilization. Part III covers the systematics of the eucalypts and contains monographs (pp. 172-374) on each of the species most used for planting, giving for each its botanical description, a description of the natural stands, and a summary of observations on that species in artificial plantations. Bibliography; index of species. Forms part of the World Forest Tree Planting Manual. First issued in 1955.

461   POPLARS IN FORESTRY AND LAND USE. 511 pp., illus. (FAO Forestry and Forest Products Studies, 12)
1965, 2nd printing, FAO $3.50
Deals with classification, identification, and distribution of poplar types; cultivation; animal pests and diseases; genetics and breeding; characteristics of poplar wood, its exploitation and utilization. Bibliography. Prepared by the Standing Executive Committee of the International Poplar Commission. Section of the World Forest Tree Planting Manual series. First issued in 1958.

462   CHOICE OF TREE SPECIES. By Sir Harry Champion and N.V. Brasnett. 307 pp., illus. (FAO Forestry Development Papers, 13)
1958, FAO   Out of print
Deals with factors which should guide the choice of tree species for planting on various sites. Part 1 covers general matters. Part 2 covers ecological principles. Part 3 deals with four selected species: European or Norway spruce, Monterey pine, teak, and blue gum. Bibliography. A section of the World Forest Tree Planting Manual series.

463   PINUS RADIATA. By C.W. Scott. 328 pp., illus. (FAO Forestry and
Forest Products Studies, 14)
1960, FAO   Out of print
Monograph on <u>Pinus radiata</u>: its botany, ecology, and growth in North
America; the tree as an exotic in other countries; planting techniques, thinning, and pruning; natural regeneration; growth and yield;
pathology and protection; wood technology, utilization, harvesting,
marketing, and trade. Selected references. A section of the World
Forest Tree Planting Manual series.

464   FOREST INFLUENCES. 307 pp., illus. (FAO Forestry and Forest
Products Studies, 15)
1962, FAO   Out of print
Study which aims to clarify the position of forest influences within a
program of sound land use policy, with particular stress on the fact
that forest lands not only protect the soil and water regime but also
produce clear water.

465   WORLD FOREST INVENTORY 1963 .... 113 pp., illus.
1966, FAO   Trilingual: E/F/S $2.50
Fourth quinquennial survey of the world's forest reserves. Tables
cover: I. Land categories; II. Forest land: ownership, management
status, silvicultural treatment, permanent forest reserves; III.
Density and composition of all forests; IV. Growing stock and annual
increment in all forests; V. Forest in use; VI. Removals of industrial wood and fuelwood. The three previous inventories—of 1948
(published in 1948), 1953 (1955) and 1958 (1960)—are out of print.

466   WOOD: WORLD TRENDS AND PROSPECTS. 131 pp., maps. (Freedom
from Hunger Campaign. Basic Studies, 16)
1967, FAO $2.50
Survey of the situation, problems, and prospects for wood and the
wood-using industries and their products over the period 1950-1975,
based on national and regional appraisals.

467   FORESTS, FOOD AND PEOPLE. By Henry Beresford-Peirse. 72 pp.,
illus. (Freedom from Hunger Campaign. Basic Studies, 20)
1968, FAO $1.00
Introductory study of the world of wood designed as useful reading
for both students and general readers. Provides an understanding of
the close relationship between forests and farming, dealing with the
indispensability of forestry products (in unprocessed form as molded
boards, and as food and chemicals) and the importance of forests for
wildlife. Discusses also the usefulness of forest areas for tourism
and leisure time of man.

468   PLANNING A FOREST INVENTORY. By B. Husch. x, 121 pp. and corrigenda slip. (FAO Forestry and Forest Products Studies, 17)
1971, FAO $2.50
Guide to the planning and execution of a forest inventory, represent-
ing the views of the FAO Forestry Department which have been
formulated after extensive consultation with governmental and commercial authorities. Revision of an earlier (1950) study, <u>Planning a
National Forest Inventory,</u> by J.D.B. Harrison (FAO Forestry and
Forest Products Studies, 1).

Africa

469   TREE PLANTING PRACTICES IN TROPICAL AFRICA. By M.S. Parry.
302 pp., illus. (FAO Forestry Development Papers, 8)
1956, FAO   Out of print
Describes characteristics of species most commonly planted in
tropical Africa and discusses environment, thinning and pruning,
nursing techniques, planting problems, and protection against insect and animal damage. References. A section of the World
Forest Tree Planting Manual series.

470   TIMBER TRENDS AND PROSPECTS IN AFRICA. A Study prepared
jointly by the secretariats of the FAO and the United Nations Economic
Commission for Africa. 90 pp.
1967, FAO/UN $2.50
In six chapters: forest resources; primary wood-using industries;
past consumption of, future requirements for, and trade in, wood
products; prospects and problems.

Arid Zones

471   TREE PLANTING PRACTICES FOR ARID ZONES. By A.Y. Goor. 234
pp., illus., 2 fold. maps. (FAO Forestry Development Papers, 16)
1963, FAO $3.00
After brief introduction to the climate, soil, and ecology of vegeta-
tion in arid zones, deals with the handling of seed, nurseries, af-
forestation, planting, the care of plantations, and special plantations.
Selected references. A revised and expanded edition of the author's
Tree Planting Practices for Arid Areas (FAO Forestry Development
Papers, 6), 1955. A section of the World Forest Tree Planting
Manual series.

Asia and the Far East

472   TREE PLANTING PRACTICES IN TEMPERATE ASIA: JAPAN. 156
pp., illus. (FAO Forestry Development Papers, 10)
1956, FAO   Out of print
Deals with current planting practice in Japan and concerns seed,
nurseries, nursery stock, planting methods, and the tending of forests, after giving a general outline of the climate, soil, and distribution of forests in Japan. Section of the World Forest Tree Planting
Manual series.

473   TREE PLANTING PRACTICES IN TEMPERATE ASIA: Burma, India,
Pakistan. By P.N. Suri and S.K. Seth. 150 pp., illus. (FAO Forestry
Development Papers, 14)
1959, FAO $1.50
Deals with seed, nursery, and planting practices in the southern
countries of parts of Temperate Asia and the characteristics of the
species most frequently planted in the region. Section of the World
Forest Tree Planting Manual series.

474   TIMBER TRENDS AND PROSPECTS IN THE ASIA-PACIFIC REGION.
A Study prepared jointly by the secretariats of the FAO and the United
Nations Economic Commission for Asia and the Far East. 224 pp. (E/
CN.11/533)
1961, FAO/UN   Out of print
Deals with present consumption and future demand; present re-
sources and future supply possibilities; development of the forest
estate; forest industries and trade.

Europe

475   EUROPEAN TIMBER TRENDS AND PROSPECTS. A New Appraisal
1950-1975, prepared jointly by the secretariats of the FAO and the
United Nations Economic Commission for Europe. xix, 233 pp.,
diagrams.
1964, UN/FAO   UN Sales No. 64.II.E.4   Out of print
Deals with consumption; primary processing industries; production;
trade; prospects. Chapter 16 deals with forest resources of 1950
and 1960 and the felling and removal of roundwood. Supersedes a
previous study published in 1953.

Latin America

476   PRÁCTICAS DE PLANTACIÓN FORESTAL EN AMÉRICA LATINA.
Por Carlos M. Flinta. 499 pp., illus. (FAO: Cuadernos de Fomento
Forestal, 15)
1966, 2nd printing, FAO   In Spanish only $5.00
After describing the ecological and phytogeographical setting, deals
with methods of handling seed, nurseries and plantations; describes
characteristics of the species most frequently planted in the region.
Bibliography. Section of the World Forest Tree Planting Manual
series. First issued in 1960.

477   LATIN AMERICAN TIMBER TRENDS AND PROSPECTS. 117 pp. and
erratum sheet. (E/CN.12/624-FAO/LAFC 62/5)
1963, UN/FAO   UN Sales No. 63.II.G.1 $1.50
Covers forest resources, production, consumption and future requirements, trade, and problems and opportunities.

478   SURVEY OF PINE FORESTS, HONDURAS: Final Report. Report prepared for the Government of Honduras by FAO acting as executing
agency for the United Nations Development Program. 80 pp., illus., 3
fold. maps in pocket. (FAO/SF: 26-HON 50)
1968, UNDP/FAO $6.50 cl.
Summarizes the results of a United Nations Development Program
(Special Fund) project in Honduras organized to provide a nucleus of
trained Honduran forest-inventory personnel and to collect the basic
information needed for promoting forestry and forest industries development in the pine forest areas of the country.

479   Manual para la Identificación de Campo de los Principales ARBOLES
TROPICALES DE MÉXICO. Por T.D. Pennington y José Sarukhán.
Ilustraciones de Silvia Garcia, Rosa María C. de Dies, Julia Laken y
Rosemary Wise. 413 pp., fe de erratas; illus.
1968, FAO/Instituto Nacional de Investigaciones (Mexico, D.F.) $17.50
     Forestales
Five chapters: Introduction; forest vegetation of the Mexican humid
tropical zone; glossary of terms used; keys to the species; descrip-
tions of the species. Bibliography, and list of specimens observed.

Tropics

480   TROPICAL SILVICULTURE. 3 vol. (FAO Forestry and Forest Products Studies, 13)
1957-58, FAO   Out of print
Vol. 1. By I.T. Haig, M.A. Huberman and U Aung Din. 190 pp., illus., maps.
Vol. 2. 415 pp., illus. Composite: E/F
Vol. 3. 101 pp. Composite: E/F

Vol. 1 sums up the status of tropical silviculture with particular regard
to natural regeneration as a guide to future efforts in developing optimum silvicultural practices, and is based on a series of general invitation papers on the topic presented at the Tropical Section of the Fourth
World Forestry Congress, Dehra Dun, India, December 1954, together
with a broad survey of other pertinent literature. Selected invitation
papers are included in Vol. 2 and 3.

ANIMAL RESOURCES

GENERAL

481   ANIMAL DISEASE AND HUMAN HEALTH. By James H. Steele. 50 pp.,
        illus. (Freedom from Hunger Campaign. Basic Studies, 3)
        1966, 3rd printing, FAO $0.50
        Describes role of veterinary medicine in controlling and eradicating
        animal and poultry diseases. First issued in 1962.

482   METEOROLOGICAL OBSERVATIONS IN ANIMAL EXPERIMENTS. Ed.
        by C.V. Smith. 37 pp. (Technical Notes, 107)
        1970, WMO   WMO-No.253.TP.140 $2.50
        Survey of meteorological measurements relevant to animal experi-
        ments, both outdoors and indoors, and observations relative to ani-
        mal diseases. Bibliography.

483   OBSERVATIONS ON THE GOAT. By M.H. French. 204 pp., tables.
        (FAO Agricultural Studies, 80)
        1970, FAO $4.50
        Comprehensive survey of the goat and goat raising and the use of its
        milk, meat, skin, and hair. Dispels a number of myths and empha-
        sizes factors in favor of the goat. Bibliography.

484   WEATHER AND ANIMAL DISEASES. By L.P. Smith. x, 49 pp. (Tech-
        nical Notes, 113)
        1970, WMO   WMO-No.268.TP.252 $3.50
        Reviews recent progress in the application of standard meteorologi-
        cal observations to the problems of forecasting the incidence and
        intensity of animal diseases, and surveys the principles and tech-
        niques involved and details of the methods that have been published
        in full elsewhere. Bibliography.

485   "Wildlife Conservation and Management." By Thane Riney. Unasylva,
        vol. 24, no. 4, 1970, pp. 15-17.
        1970, FAO $0.65
        Surveys FAO experience in forest and wildlife conservation and
        management and considers the effects upon the forestry profession
        of trends toward nontimber aspects of forest management, notably
        the search for new uses of both plants and animals in the attempt to
        stabilize ecosystems of marginal lands, the effect of forest manage-
        ment practices upon animal habits, and the growing ease of intro-
        ducing animal wildlife from one geographical region to another.
        Notes also the economic value of wildlife as a local source of pro-
        tein, in terms of exports of live wild animals, birds and fish, of
        hides and skins and tinned meat, and the value of parks and reserves
        in terms of tourism; forms of exploitation of wildlife resources
        which need regulation.

Africa

486   The CONSERVATION OF WILD LIFE AND NATURAL HABITATS IN
      CENTRAL AND EAST AFRICA. Report on a Mission Accomplished
      for Unesco, July-September 1960. By Julian Huxley. 113 pp., illus.,
      maps.
      1961, UNESCO   Out of print

487   Conservation and Management of AFRICAN WILDLIFE. By T. Riney.
      Sponsored by FAO and the International Union for the Conservation of
      Nature and Natural Resources. 35 pp. (African Special Project Stage
      III Reports)
      1967, FAO $1.50
      Describes the method of the project, during which 19 countries were
      visited, and presents its conclusions as to wildlife resources: avail-
      ability, contribution to national economies, problems, potentialities,
      legal and administrative aspects of development, management of
      game and its habitat, training of personnel, and international impli-
      cations.

      a. English-speaking Country Reports. By T. Riney and P. Hill. 145
         pp., maps.
         1967, FAO
            Covers Botswana, Ethiopia, Kenya, Malawi, Nigeria, Sierra
            Leone, Somalia, Tanzania, Uganda, and Zambia.
      b. Rapports sur les pays francophones. Par T. Riney et P. Hill. 135
         pp., maps.
         1967, FAO   In French only
            Covers Burundi, Cameroon, Central African Republic, Chad,
            Dahomey, Mali, Senegal, and Upper Volta.

Arid Zones

488   SHEEP BREEDS OF THE MEDITERRANEAN. By I.L. Mason. [296]
      pp. including plates; maps.
      1967, FAO/Commonwealth Agricultural Bureaux $13.00
      Describes the sheep population of the Mediterranean countries, in-
      cluding Portugal. Each breed is classified according to tail type
      and fleece type and described succinctly, with information on distri-
      bution, nomenclature, origin, management, and breeding cycle. The
      sheep industry in each country is outlined briefly and numbers of the
      local breeds are indicated. Breeds and varieties are shown in 157
      photographs. Extensive bibliography.

ANIMAL NUTRITION
___

489   USE OF RADIOISOTOPES IN ANIMAL BIOLOGY AND THE MEDICAL
      SCIENCES. Proceedings of a Conference held in Mexico City, 21 No-
      vember-1 December 1961. 2 vol., illus. ([Proceedings Series: EP/4])
      1962, IAEA/Academic Press, Inc.   Composite: E/F/S Vol. 1: $16.00 cl.
                                              Vol. 2: Out of print
      Vol. 1 contains papers (35) on general physiology, haematology, and
      glandular functions; vol. 2, papers (17) on mineral metabolism, lac-
      tation and metabolism; and eight clinical studies.

490 RADIOISOTOPES IN ANIMAL NUTRITION AND PHYSIOLOGY. Proceedings of the Symposium on the Use of Radioisotopes in Animal Nutrition and Physiology jointly organized by IAEA and FAO and held in Prague, 23-27 November 1964. 874 pp., illus. (Proceedings Series; STI/PUB/90)
1965, IAEA   Composite: E/F/R $9.00
Papers (49) on biochemistry and physiology of milk secretion; trace elements including magnesium; influence of environmental factors on animal production and reproduction.

491 POULTRY FEEDING IN TROPICAL AND SUBTROPICAL COUNTRIES. 96 pp., illus. (FAO Agricultural Development Papers, 82)
1967, 2nd printing, FAO   Out of print
Provides expert information on the principles and practices to be followed in feeding poultry for the most successful results as an aid to developing tropical and subtropical countries in modernizing and intensifying poultry production in order to raise their standards of nutrition. Summarizes the use made of locally produced feedstuffs in 39 countries in Africa, Asia, Latin America, and the Southwest Pacific. References. First issued in 1965.

492 The MINERAL NUTRITION OF LIVESTOCK. By E.J. Underwood. xvii, 237 pp., illus.
1966, FAO/Commonwealth Agricultural Bureaux $8.00
Gives a brief history of the recognition of imbalanced intakes of essential minerals, describes the symptoms and metabolic disturbances associated with excesses or deficiencies in the diet, reviews the methods of diagnosis, and summarizes the requirements of the various classes of livestock. Published by arrangement with FAO.

493 NONPROTEIN NITROGEN IN THE NUTRITION OF REMINANTS. By J.K. Loosli and I.W. McDonald. 94 pp. (FAO Agricultural Studies, 75)
1968, FAO $2.00
Report on the use of urea and other nonprotein nitrogen (NPN) containing compounds in feeding ruminants. Summarizes research findings and serves as a guide for feeding cattle, sheep, and other ruminants, particularly in countries where little or no research has yet been done. Bibliography.

494 TRACE MINERAL STUDIES WITH ISOTOPES IN DOMESTIC ANIMALS. Proceedings of a Panel on the Use of Isotopes in Studies of Mineral Metabolism and Disease in Domestic Animals organized by the Joint FAO/IAEA Division of Atomic Energy in Food and Agriculture, held in Vienna, 28-31 October 1968. 151 pp., illus. (Panel Proceedings Series; STI/PUB/218)
1969, IAEA $4.00
Contents include papers (12) on such topics as placental transfer of iron, copper metabolism in cows, manganese metabolism in poultry and in ruminants, competition between trace elements, and neutron activation analysis for the determination of selenium in biological materials; and the recommendations and conclusions of the panel.

495 ISOTOPE TECHNIQUES FOR STUDYING ANIMAL PROTEIN PRODUCTION FROM NON-PROTEIN NITROGEN. Report of a Consultants' Group on the Application of Isotope Techniques to the Study of Animal Protein Production from Non-protein Nitrogen, organized by the Joint

FAO/IAEA Division of Atomic Energy in Food and Agriculture and held in Vienna, 25-28 August 1969. 29 pp. (Technical Reports Series, 111; STI/DOC/10/111)
1970, IAEA $2.00
Outlines isotope research methods for investigating nonprotein nitrogen (NPN) supplementation of feedstuffs. Protein-deficient cereals and other high-carbohydrate feeds may become available in developing countries to support ruminant production of meat and milk and will require supplementation with NPN. The economic feasibility of NPN application must be carefully evaluated.

496   MINERAL STUDIES WITH ISOTOPES IN DOMESTIC ANIMALS. Proceedings of a Panel on the Use of Nuclear Techniques in Studies of Mineral Metabolism and Disease in Domestic Animals, organized by the Joint FAO/IAEA Division of Atomic Energy in Food and Agriculture, held in Vienna, 28 September-2 October 1970. 203 pp., fig., tables. (Panel Proceedings Series; STI/PUB/293)
1971, IAEA $6.00
Papers (19) and recommendations resulting from discussions.

FISH AND AQUATIC MAMMALS

497   Proceedings of the United Nations Scientific Conference on the CONSERVATION AND UTILIZATION OF RESOURCES, 17 August-6 September 1949, Lake Success, New York. 8 vol. (E/CONF.7/7, vol. 1-8)
1950-51, UN
Vol. 7: Wildlife and Fish Resources. 259 pp. and erratum slip; illus. (E/CONF.7/7, vol. 7)
Sales No. 50.II.B.8   Out of print
Contains papers on: changes in abundance of fish population (7); developing fishery resources (8); fisheries statistics and technological development; fishing, handling and preserving fish, processing fish products (14); management and cultivation of freshwater fish (12); research in the conservation and utilization of living marine resources: fish and algae (6); game and fur conservation (5); management of wildlife resources (15).

498A   Report of the International Technical Conference on the CONSERVATION OF THE LIVING RESOURCES OF THE SEA, 18 April to 10 May 1955, Rome. 17 pp. and corrigendum sheet; illus. (A/CONF.10/6 and Corr.1)
1955, UN   Sales No. 55.II.B.2   Out of print

498B   Papers presented at the International Technical Conference on the CONSERVATION OF THE LIVING RESOURCES OF THE SEA, Rome, 18 April to 10 May 1955. 371 pp. and corrigendum sheet. (A/CONF.10/7 and Corr.1)
1956, UN   Sales No. 56.II.B.1   Out of print
Report of the Conference and papers (23) on conservation of stocks of fish and sea mammals.

499   The Application of Science to INLAND FISHERIES; a Review of the Biological Basis of Freshwater Fisheries. By E.D. Le Cren. (FAO Fisheries Studies, 8)
1958, FAO   Out of print
Considers the biological basis of fisheries and the limnological and ecological processes leading to the harvest of fish. Prospective integration of fish farming with land farming is studied. References.

500   FISHERIES IN THE FOOD ECONOMY. 79 pp. (Freedom from Hunger
Campaign. Basic Studies, 19)
1968, FAO $1.25
    Examines the place of fish in food consumption, trends in fish pro-
duction and utilization, production and trade in fish commodities,
and the factors influencing future supply and demand. Notes the
work being done to increase supply availability and to promote consumption.

501   The STATE OF WORLD FISHERIES. 49 pp., illus. (World Food Problems, 7)
1968, FAO $0.80
    States the case for management of the high sea fisheries in order to
prevent overfishing and to allow stocks to continue to sustain high
yields of an important source of protein. Reprint of a special chapter in The State of Food and Agriculture 1967 (entry 1174).

Africa

502   FISH CULTURE IN CENTRAL EAST AFRICA. By A. Maar, M.A.E.
Mortimer and I. Van der Lingen. 160 pp., illus.
1966, FAO $1.50
    Manual of fish culture practice in ponds and dams, and related
studies. Deals also with transport and preservation of fish. A joint
FAO/WHO/UNICEF publication intended to encourage the utilization
of fish as a protein-rich food.

FOOD PRODUCTION AND SUPPLY

GENERAL

503   ASPECTS OF ECONOMIC DEVELOPMENT. 84 pp. (Freedom from
Hunger Campaign. Basic Studies, 8)
1962, UN   Sales No. 62.I.27   Out of print
    Outlines the main aspects of the economic development process, in-
cluding the production of more food, the improvement of health, the
spread of education, and the promotion of social welfare and human
rights.

504   INCREASING FOOD PRODUCTION THROUGH EDUCATION, RE-
SEARCH, AND EXTENSION. By C.W. Chang. 78 pp., illus. (Freedom
from Hunger Campaign. Basic Studies, 9)
1965, 2nd printing, FAO $1.00
    Shows how agricultural education, research, and extension services,
when adequately supported and coordinated, can benefit farm populations and increase agricultural production. First issued in 1962.

505   POSSIBILITIES OF INCREASING WORLD FOOD PRODUCTION. By
Walter H. Pawley. 231 pp., illus. (Freedom from Hunger Campaign.
Basic Studies, 19)
1967, 2nd printing, FAO $2.50
    Examines the technical possibilities of improving food production
and nutrition so that the population which is already foreseeable may

be adequately fed. These possibilities include extending areas under cultivation, increasing crop yields, the output of livestock products, fish and fishery products, and the development of forest resources. The relationship between forestry and food production is examined. First issued in 1963.

506   EDUCATION AND AGRICULTURAL DEVELOPMENT. 62 pp. (Freedom from Hunger Campaign. Basic Studies, 15)
1963, UNESCO $0.75
Planning of education and training for agricultural development at the primary and secondary levels and in institutions for technical, vocational, and higher education.

507   AGRICULTURE AND INDUSTRIALIZATION. 129 pp. (Freedom from Hunger Campaign. Basic Studies, 17)
1967, FAO $1.50
Sets out the main aspects of the mutually dependent relationship between agriculture and industry, emphasizing the two branches of industry most closely related to agriculture—the industries using agricultural, fishery, and forest products as raw materials, and the industries serving agriculture by the provision of equipment and other requisites for production.

508   TOWARD A STRATEGY FOR AGRICULTURAL DEVELOPMENT. FAO's Five Areas of Concentration. 66 pp. (Freedom from Hunger Campaign. Basic Studies, 21)
1969, FAO $1.00
Five papers embody the results of careful analysis by interdivisional working parties of the problems involved in the five areas selected where FAO's operational strategy could be concentrated to achieve the practical results in agricultural development with the limited resources at the disposal of the Organization. These areas are: high-yielding varieties of basic food crops; filling the protein gap; war on waste; mobilization of human resources for rural development; and earning and saving foreign exchange.

509   NUCLEAR TECHNIQUES FOR INCREASED FOOD PRODUCTION. Prepared under the auspices of the Joint FAO/IAEA Division of Atomic Energy in Food and Agriculture, Vienna. 76 pp., illus. (Freedom from Hunger Campaign. Basic Studies, 22)
1969, FAO/IAEA $1.25
Describes how radiation aids food producers in the induction of mutations, control of insects, making of radiation-attenuated animal vaccines, and irradiation of food; how isotopes are used in agricultural research on plant-soil relationships, fertilizer studies, animal and fish biology, and in food protection; and how nuclear techniques are employed in irrigation to trace underground water movement, in soil moisture studies, in desalting, and use of water in agroindustrial projects.

510   SMALLER FARMLANDS CAN YIELD MORE. Raising Agricultural Productivity by Technological Change. 73 pp. (World Food Problems, 8)
1969, FAO $1.25
Discusses the elements that contribute to increased agricultural productivity: the role of science and technology, and the organizational and institutional requirements for promoting technological improvement. Reprint of special chapter of The State of Food and Agriculture 1968 (entry 1174).

511 FIVE KEYS TO DEVELOPMENT. 44 pp., illus.
1970, FAO
Emphasizes the need to bring to bear in five crucial areas a multi-
tude of connected skills under sound administrative direction in
order to deal with the problem of increasing food production in the
developing countries: the reduction of waste, the closing of the protein gap, the promotion of high-yielding cereal varieties, earning and
saving of foreign exchange in developing countries, and better utilization of human resources.

512 FOOD LOSSES: the Tragedy and Some Solutions. 38 pp., illus.
1970, FAO
Considers briefly some of the causes of food losses: ineffective use
of land, erosion, wasteful irrigation, shifting cultivation, pests and
diseases in crops, and crop storage defects.

512 AGRICULTURAL DEVELOPMENT: A Review of FAO's Field Activities.
bis xii, 194 pp., illus. (Freedom from Hunger Campaign. Basic Studies,
23)
1970, FAO $2.50
In two parts: (1) a general review of development aid in agriculture
and FAO's contribution thereto through FAO/UNDP (Technical Assistance), Freedom from Hunger Campaign, the FAO/UNICEF, and
FAO/UNDP (Special Fund) programs; and (2) a description of specific projects under these programs. First issued in June 1969 as
document C 69/18, Review of FAO's Field Activities, for the fifteenth session of the FAO Conference.

STATISTICS
***

513 FOOD BALANCE SHEETS, 1964-66 AVERAGE.... xxxvii, 766 pp.
1971, FAO   Trilingual: E/F/S $6.00
Fourth issue of three-year averages. Contains tables of worldwide
coverage for food production, changes in stocks, and foreign trade
by country and food commodity; utilization of food supply during
this period; consumption per head, including specific daily intake of
calories and nutrients. Derived from the annual national food balance sheets received by FAO. Previous documents cover 1957-1959
average (1963), 1954-1956 (1958), 1936-1952 (1954), 1934-1948 (1949)
and 1960-1962 (1966).

514 FOOD COMPOSITION TABLES FOR INTERNATIONAL USE. By
Charlotte Chatfield. 56 pp. (FAO Nutritional Studies, 3)
1949, FAO   Out of print
Prepared to facilitate the drawing-up of food balance sheets and to
introduce a greater degree of uniformity in the completion of nutritional statistics.

515 FOOD COMPOSITION TABLES—MINERALS AND VITAMINS—FOR
INTERNATIONAL USE. By Charlotte Chatfield. 117 pp. and erratum
slip. (FAO Nutritional Studies, 11)
1964, 3rd printing, FAO   Out of print
First issued in 1954.

Monthly Bulletin of Agricultural Economics and Statistics. <u>See entry</u> <u>1187</u>.

Production Yearbook. <u>See entry 1180</u>.

Trade Yearbook. <u>See entry 1185</u>.

Yearbook of Fisheries Statistics. <u>See entry 1173</u>.

SURVEYS

516   Third WORLD FOOD SURVEY. 102 pp. (Freedom from Hunger Campaign. Basic Studies, 11)
       1964, 2nd printing, FAO $1.50
       Presents a comprehensive picture of present and past world food
       situation. Survey is based on food balance sheet data for over 80
       countries covering some 90 percent of the world's population. It
       also draws on the food consumption and dietary surveys conducted
       in various parts of the world and introduces new statistical tech-
       niques in the study of food supplies and needs. First issued in 1963.
       The first (1946) and second (1952) world food surveys are out of
       print.

The State of Food and Agriculture. <u>See entry 1174</u>.

PROJECTIONS

517   AGRICULTURAL COMMODITY PROJECTIONS, 1970-1980. 2 vol.
       (329; lxi, 407 pp.) (CCP 71/20)
       1971, FAO $5.00 per volume
       Vol. 1: Part I: General Outlook; Part II: Projections by Commodi-
       ties.
       Vol. 2: Part I: General Methodology; Part II: Statistical Appendix.
       Third FAO projections study. The first part ("General Outlook") is in
       four chapters: (1) scope, assumptions, and methodology; (2) major
       findings of the projections study; (3) food demand and nutrition; and (4)
       international trade. In this study there has been applied to national
       situations the revised energy and protein requirements established re-
       cently (March-April 1971) by the FAO/WHO Ad Hoc Committee of
       Experts on Energy and Protein, whose report will be published.

CROP STORAGE

518   PRESERVATION OF GRAINS IN STORAGE. 174 pp. (FAO Agricul-
       tural Studies, 2)
       1952, 2nd printing, FAO   Out of print
       Papers presented at the International Meeting on Infestation of Food-
       stuffs, 5-12 August 1947. First issued in 1948.

519   STORING AND DRYING GRAIN, IN CANADA, IN THE UNITED STATES,
        IN THE UNITED KINGDOM. By L.E. Kirk, Lee Ling, and T.A. Oxley.
        42 pp., illus. (FAO Agricultural Studies, 6)
        1955, 3rd printing, FAO $0.50
        Describes briefly existing methods of handling, drying, and storing
        grain. First issued in 1948 and reprinted in 1949.

520   TERMITES IN THE HUMID TROPICS. Proceedings of the New Delhi
        Symposium jointly organized by Unesco and the Zoological Survey of
        India on 4-12 October 1960. 259 pp., illus. (Humid Tropics Research)
        1962, UNESCO $7.00
        Papers (41) and discussions on systematics and morphology; physi-
        ology and development; general biology; ecology; intestinal cellu-
        lose-digesting symbionts; termite control and termite-proof con-
        struction (including protection of agricultural products).

521   THE EFFECT OF WEATHER AND CLIMATE UPON THE KEEPING
        QUALITY OF FRUIT. Part I: General. Report of a Working Group of
        the Commission for Agricultural Meteorology.—Part II: The Banana
        Plant. By G.C. Green.—Part III: The Pineapple Plant. By G.C. Green.
        xxii, 180 pp. (Technical Notes, 53)
        1963, WMO   WMO-No.137.TP.63 $3.00
        The report covers effects of factors other than weather and climate
        upon the storage life of fruit; the effects of weather and climate upon
        the keeping quality of deciduous fruits, tropical and subtropical
        fruits, and citrus; and gives practical guidance to the grower on the
        storage or disposal of this crop.

522   BIGGER CROPS—AND BETTER STORAGE: the Role of Storage in
        World Food Supplies. 49 pp. (World Food Problems, 9)
        1969, FAO $1.00
        Notes the contribution of improved storage to expansion of world
        food supplies. Surveys storage problems in developing countries for
        both durable and perishable commodities and factors to be consid-
        ered in determining storage policy.

523   METEOROLOGY AND GRAIN STORAGE. By C.V. Smith. xvi, 65 pp.,
        illus. (Technical Notes, 101)
        1969, WMO   WMO-No.243.TP.133 $3.50
        Reviews and summarizes available knowledge and experience on the
        storage of cereals and other small seed crops. Examines the bio-
        logical characteristics of grain and of potentially damaging associ-
        ated organisms, and the physical characteristics of grain in bulk.
        Considers the preparation of food grains for safe storage, drying,
        ventilation, refrigeration, and monitoring of temperature and mois-
        ture content. Suggests ways in which the agricultural meteorologist
        may be of use. Bibliography.

524   "Waste and the Protein Gap—They Can Both Be Reduced." By H.A.B.
        Parpia. Ceres, vol. 2, no. 5, September/October 1969, pp. 20-24.
        1969, FAO $0.50
        Heavy losses which occur at different stages of food handling, proc-
        essing, distribution, and consumption appear to be the main causes
        of the difference between the average availability of protein in India
        and its actual consumption, which is nearly 20.5 grams per day.

525   HANDLING AND STORAGE OF FOOD GRAINS IN TROPICAL AND SUB-
       TROPICAL AREAS. By D.W. Hall. xiv, 350 pp., illus. (FAO Agricul-
       tural Development Papers, 90)
       1970, FAO $6.00
       Manual intended primarily for agricultural and public health offi-
       cers, for those responsible for designing storage and marketing
       facilities, and for agricultural colleges. Covers the causes of grain
       loss, deterioration and contamination, methods of drying and stor-
       age, the design of storage facilities, and also methods to control
       fungi, insects, and rodents. Extensive references.

FOOD AND POPULATION

526   SO BOLD AN AIM: Ten Years of International Co-operation toward
       Freedom from Want. By P. Lamartine Yates. 174 pp.
       1955, FAO $1.50 cl.
       Describes the ideas and events which led to the establishment of
       FAO and those which have since shaped the activities of the Or-
       ganization, particularly with respect to the impact upon food and
       agriculture of the world population problem.

527   MILLIONS STILL GO HUNGRY. 102 pp.
       1962, 3rd printing, FAO   Out of print
       A critical reexamination of some of the problems faced by FAO. Il-
       lustrated by detailed accounts of field projects. First issued in
       1957.

528   MAN AND HUNGER. Rev. ed. 77 pp., illus. (World Food Problems, 2)
       1962, 2nd printing, FAO   Out of print
       Discusses the problem of rising population and the food supply,
       means of raising levels of nutrition, and the work of FAO in this
       field. This edition originally issued in 1961; previous editions in
       1959 and 1957.

529   "The Share of Agriculture in a Growing Population." By F. Dovring.
       Monthly Bulletin of Agricultural Economics and Statistics, vol. 8, no.
       8/9, August/September 1959, pp. 1-11.
       1959, FAO $0.50
       Challenges the assumption in development economics that a primary
       requisite for developing a less developed country is the reduction of
       its agricultural population.

530   POPULATION AND FOOD SUPPLY. McDougall Memorial Lecture
       1959. By Arnold Toynbee. 20 pp.
       1959, FAO   Out of print
       A tribute to the work of Frank Lidgett McDougall of Australia in the
       founding of FAO and to the work of the organization to reduce world
       problems of food production and distribution and improvement of
       nutrition.

531   AGRICULTURE IN THE WORLD ECONOMY. 2nd rev. ed. 65 pp., il-
       lus., and corrigenda sheet.
       1962, FAO   Out of print
       Agriculture's development, resource use, production trends, and
       proportion of world income as seen against the background of a
       growing population and inadequate living standards in many parts of
       the world.

532   POPULATION AND FOOD SUPPLY. 49 pp., illus. (Freedom from
       Hunger Campaign. Basic Studies, 7)
       1962, UN   Sales No. 62.I.22 $0.50
       Report prepared by Robert C. Cook, President of the Population
       Reference Bureau, Inc., on the relation of food supply to popula-
       tion growth and policy.

533   SIX BILLIONS TO FEED. 41 pp., illus. (World Food Problems, 4)
       1963, 2nd printing, FAO   Out of print
       Presents in nontechnical terms the results of a statistical appraisal
       of hunger in the world, describing the methods used and the broad
       conclusions reached as to the needs to increase food supplies if by
       the year 2000 the world population is to have enough to eat. First
       issued in 1962.

534   INDICATIVE WORLD PLAN FOR AGRICULTURAL DEVELOPMENT TO
       1975 AND 1985. Provisional Regional Studies, No. 1-4.
       1966-69, FAO
       a. Near East. (Subregional Study No. 1)
          Vol. 1: Text. 1966. 177 pp.
          Vol. 2: Explanatory Notes and Statistical Tables. 1966. 243 pp.
          Covers 12 countries: Afghanistan, Federation of South Arabia, Iran,
          Iraq, Jordan, Kuwait, Lebanon, Saudi Arabia, Sudan, Syria, U.A.R.,
          Yemen.
       b. South America. (Provisional Regional Study No. 2)
          Vol. 1: Text. 1968. 286 pp. (IWP/68/RS.2)
          Vol. 2: Explanatory Notes and Statistical Tables. 1968. xxv, 229 pp.
          Supplement 1: Trends and Prospects for Major Export Commodities.
          1969. 39 pp. (IWP/68/RS.2, Supplement 1)
          Covers ten countries: Argentina, Bolivia, Brazil, Chile, Colombia,
          Ecuador, Paraguay, Peru, Uruguay, Venezuela.
       c. Africa, South of the Sahara. (Provisional Regional Study No. 3)
          Vol. 1: Text. 1968. 398 pp. (IWP/68/RS.3)
          Vol. 2: Explanatory Notes and Statistical Tables. 1968. xxvii, 608
          pp. Supplement 1: Trends and Prospects for Major Export Com-
          modities. 1969. 22 pp. (IWP/68/RS.3, Supplement 1)
          Covers 24 countries: Cameroon, Central African Republic, Chad,
          Congo (Brazzaville), Gabon; Democratic Republic of the Congo;
          Ethiopia, Kenya, Madagascar, Malawi, Tanzania, Uganda, Zambia;
          Gambia, Mali, Mauritania, Niger, Senegal, Upper Volta; Dahomey,
          Ghana, Ivory Coast, Togo; Nigeria.
       d. Asia and the Far East. (Provisional Regional Study No. 4)
          Vol. 1: Text. 1968. 471 pp. (IWP/68/RS.4)
          Vol. 2: Explanatory Notes and Statistical Tables. 1968. xxii, 189
          pp. (IWP/68/RS.4)
          Covers eight countries: Ceylon, China (Taiwan), India, Republic of
          Korea, West Malaysia, Pakistan, the Philippines, Thailand.

535   "Population Growth and Agricultural Development." By K.C. Aber-
       crombie. Monthly Bulletin of Agricultural Economics and Statistics,
       vol. 18, no. 4, April 1969, pp. 1-9.
       1969, FAO $0.50
       An attempt to put into better perspective the problems arising from
       the complex relationship between population growth and agricultural
       development and the effects of rapid population growth on employ-
       ment and levels of living.

536   Provisional INDICATIVE WORLD PLAN FOR AGRICULTURAL DE-
VELOPMENT. A Synthesis and Analysis of Factors Relevant to World,
Regional and National Agricultural Development. 2 vols.
1970, FAO $8.50 the set

     a. Vol. 1 (pp. 1-327): Part I. The Setting: Chap. 1 The Challenge of
1985; Part II. Programmes and Policies for Production in the Developing Countries: Chap. 2. Development of Land and Water Resources; Chap. 3. Crop Production; Chap. 4. Material Inputs Required to Support Crop Production Objectives; Chap. 5. Targets,
Policies and Inputs for Livestock Production; Chap. 6. The Prospects for World Fishery Development by 1975 and 1985; Chap. 7.
Targets, Policies and Inputs in Forestry.

     b. Vol.2 (pp. 331-672): Part III. Mobilizing the Human Resources:
Chap. 8. Economic Incentive Policies; Chap. 9. Agricultural Credit
and Marketing; Chap. 10. The Basic Organizational Structure;
Chap. 11. Land Reforms as an Instrument of Progress; Chap. 12.
Requirements and Policies for Professional and Technical Manpower
for Agricultural Development; Part IV. Agriculture in the Devel-
oping Countries within a World Framework: Chap. 13. Consumption
and Nutrition Perspectives and Food Policies; Chap. 14. Commodity
Balances and Trade Policy Issues; Chap. 15. Some Conclusions and
Broad Lines of Strategy.

     These two volumes of the provisional IWP were first issued as a work-
ing paper for the fifteenth session of the FAO Conference in 1969 (document C 69/4, vol. 1-2 and Corr.1).

537   Provisional INDICATIVE WORLD PLAN FOR AGRICULTURAL DE-
VELOPMENT. Summary and Main Conclusions. 72 pp.
1970, FAO $1.50

     First issued as a working paper for the fifteenth session of the FAO
Conference in 1969 (document C69/4, vol. 3 and Corr. 1).

538   A STRATEGY FOR PLENTY: THE INDICATIVE WORLD PLAN FOR
AGRICULTURAL DEVELOPMENT. 63 pp., illus. (World Food Problems, 11)
1970, FAO $1.50

     Presents to the public the main issues and conclusions of the provi-
sional IWP and the broad strategy for national and international action it suggests to resolve the problems confronting world agriculture during the 1970s and early 1980s.

538   WORLD AGRICULTURE: the Last Quarter Century. v, 42 pp., tables.
bis   (World Food Problems, 13)
1970, FAO $1.50

     Reviews the progress in agriculture since the end of the second
world war, especially in the developing countries. Reprint of a special chapter in The State of Food and Agriculture 1970 (entry 1174).

539   "Changing Views on the Man-Food Relationship." By K.C. Abercrombie. Ceres, vol. 4, no. 2, March/April 1971, pp. 22-26.
1971, FAO $1.00

     Notes the changes during the course of history in the views of man-
kind concerning survival in the face of the demands of growing population upon food supply.

540   "In the Year 2070." By Walter H. Pawley. Ceres, vol. 4, no. 4, July/
August 1971, pp. 22-27.
1971, FAO $1.00

     Looks ahead to consider the consequences of continued population growth at current rates in relation to living space, food supply, education, unemployment and underemployment, and other economic, social, and political effects. If two major technical breakthroughs were to occur—techniques for continuous cultivation of the soil in the humid tropics after removal of the tree cover, and desalination of sea water at sufficiently low cost and its transportation to irrigate arid areas through cheap enough power for pumping—there could be a great expansion of the cultivable area of the globe, but enormous shifts of population would be required. Considers that the time is quickly running out for further hesitation in creating a sense of urgency leading to the political will to action in both the developed and developing countries.

Africa

541 ''Agricultural Production, Population Growth, and Food Consumption per Capita.'' Economic Bulletin for Africa, vol. 9, no. 2 (E/CN.14/470), 1969, pp. 1-9; and ''Agricultural Development and Population Growth.'' Ibid., pp. 41-43.
1969, UN   Sales No. E.69.II.K.10 $1.50
Chap. 1 and 7, respectively, of Part I: Agriculture in the East African Subregion.

FOOD POLICY

542 NATIONAL FOOD RESERVE POLICIES IN UNDERDEVELOPED COUNTRIES. 70 pp. (FAO Commodity Policy Studies, 11)
1958, FAO $1.00
Suggests possible ways of promoting the use of surplus foodstuffs to build adequate reserves in vulnerable countries, dealing with such matters as crop fluctuations, marketed surpluses, imports, long-term remedies and emergency measures. Results of two field inquiries in India and Pakistan and a summary of program and stock policies in five Central American countries are included.

543 MARKETING—ITS ROLE IN INCREASING PRODUCTIVITY. By J.C. Abbott and others. 116 pp., illus. (Freedom from Hunger Campaign. Basic Studies, 4)
1967, 3rd printing, FAO $1.50
Shows how improvements in the marketing of agricultural products can result in the more effective distribution of food at lower costs to the consumer and in stimulation of production for an expanded market. First issued in 1962.

544 Report of the WORLD FOOD CONGRESS, Washington, D.C., 4 to 18 June 1963. 2 vol.
1963-65, FAO
[Vol. 1: Proceedings, Declaration and Resolution.] 160 pp. (C 63/13)
1963, FAO $3.00
Vol. 2: [Major Addresses and Speeches.] 115 pp.
1965, FAO $3.50
The congress was assembled under the Freedom from Hunger Campaign to take the measure of the problems of hunger and malnutrition

and to explore the means for their solution. It urges all governments of developing countries to act for a planned and integrated use of their resources and to adapt their institutions to the requirements of economic and social development; it further urges that international cooperation be strengthened to assist national efforts to develop, including equitable commodity agreements, increased technical assistance, and expanded and improved sharing of food surpluses.

545   Report of the SECOND WORLD FOOD CONGRESS, The Hague, Netherlands, 16-30 June 1970. 2 vols.
        1970, FAO
            Vol. 1: Story of the Congress; Final Declaration; Major Speeches.
            viii, 141 pp., illus. $2.50
            Vol. 2: Proceedings; Panels: Record of Proceedings; Organization;
            List of Participants. v, 199 pp., illus. $3.00
        The program of the Congress covered the main findings of the Indicative World Plan for Agricultural Development (IWP), related them to the FAO strategy of five areas of concentration, and considered how the necessary support for proposed action priorities could be rapidly and effectively organized. There were recommendations from eight commissions; and five panel discussions—on the IWP, the role and attitudes of youth in development, the perspectives of international development, population growth in relation to economic development, and the conservation of man's environment.

546   NATIONAL DEVELOPMENT EFFORTS. 67 pp. (Freedom from Hunger Campaign. Basic Studies, 13)
        1963, UN   Sales No. 63.I.10 $0.75
        Report on the economic and social development of the advanced and the developing countries, stressing the importance of strategic planning and of industrial and agricultural production.

547   HUNGER AND SOCIAL POLICY. 66 pp. (Freedom from Hunger Campaign. Basic Studies, 14)
        1966, ILO $0.50
        Describes the efforts of ILO through its international labor conventions and recommendations to improve national policies in regard to employment, wages, vocational training, housing, small industries and cooperation, and living and working conditions. Describes also its operational program in such fields as rural development, manpower assessment, vocational training and productivity, and labor studies, etc. These activities are aimed at building up reservoirs of skills and aptitudes adequate to the tasks of economic development.

548   AGRARIAN RECONSTRUCTION. By Erich H. Jacoby. 82 pp. (Freedom from Hunger Campaign. Basic Studies, 18)
        1968, FAO $1.25
        Deals with problems concerning agrarian reconstruction: land redistribution, tenure legislation, ceiling and floor operations, land consolidation, settlement. Gives particular attention to cooperative farming operations, the adjustment of indigenous tenure systems of specific importance to agricultural development in Africa, and to the administrative aspects of agrarian reconstruction programs.

549   "The Nutritional Approach in Food Policy Planning." By J. Périssé.
        Nutrition Newsletter, vol. 6, no. 1, January/March 1968, pp. 30-45.
        1968, FAO   Free on request to FAO

550   "Nutrition—a New Operational Technique in Planning." By Patrick J.
       Francois. Nutrition Newsletter, vol. 6, no. 1, January/March 1968,
       pp. 46-62.
       1968, FAO   Free on request to FAO

551   Manual on FOOD AND NUTRITION POLICY. Prepared with the as-
       sistance of B.F. Johnston and J.P. Greaves. 95 pp. (FAO Nutritional
       Studies, 22)
       1969, FAO $2.00
       Aims to define food and nutrition policy in the context of economic
       and social development, to describe the cooperation needed and the
       programs necessary to put this policy into effect. Bibliography.

552   MARKETING; a Dynamic Force in Agricultural Development. 40 pp.
       (World Food Problems, 10)
       1970, FAO $1.25
       Discusses the general relationship of agricultural marketing and
       economic development, and examines in some depth two aspects of
       marketing improvement which are of particular importance to most
       developing countries: the relation of marketing to the planning and
       implementation of agricultural price policies, and the relative role
       of private enterprise and public bodies in marketing. This is a re-
       print of a special chapter in The State of Food and Agriculture 1969
       (entry 1174).

Africa

553   Report of the Fourth INTER-AFRICAN CONFERENCE ON FOOD AND
       NUTRITION, Douala, Cameroun, 4-13 September 1961. Organized
       jointly by CCTA, FAO and WHO on the Topic "National Food and Nutri-
       tion Policies." 54 pp. (FAO Nutrition Meetings Report Series, 33)
       1963, FAO $1.00
       Contains recommendations of the conference and, among the an-
       nexes, the report of the Subcommittee on Techniques of Food Con-
       sumption Surveys, and Statutes for a Joint FAO/WHO/CCTA Re-
       gional Food and Nutrition Commission for Africa.

Asia and the Far East

554   Report of the Technical Meeting on NUTRITION IN FOOD POLICY AND
       PLANNING IN ASIA AND THE FAR EAST, Bangkok, Thailand, 6-13
       June 1960. 50 pp. (FAO Nutrition Meetings Report Series, 28)
       1961, FAO $0.50
       Appraises the food and nutrition situation in the region, and contains
       recommendations on establishment of food consumption targets, nu-
       tritional improvement through food policies and plans, and organiza-
       tion of national nutrition services and coordination of activities.

Latin America

555   Elements of a FOOD AND NUTRITION POLICY IN LATIN AMERICA.
       Report of a Technical Group Meeting, Washington, D.C., 19-23 May
       1969. 27 pp. (Scientific Publications, 194)
       1970, PAHO $0.50
       Summarizes the deliberations of the meeting in hopes that it will
       provide motivation and guidance for the respective national authori-

ties, particularly the planning units that are responsible for preparing the national policy for socioeconomic development, of which food and nutrition policy should be an integral part; includes conclusions and recommendations.

Food Laws and Regulations

General. See entry 1195.

Food additives. See entries 655-662, 1194, 1195.

Food hygiene. See entries 596, 604, 605, 1195.

Food irradiation. See entries 596, 604, 605, 1195.

Pesticide residues. See entries 1023, 1194-1196.

FOOD AID

556   FUNCTIONS OF A WORLD FOOD RESERVE: Scope and Limitations.
77 pp. (FAO Commodity Policy Studies, 10)
1961, 2nd printing, FAO $1.00
In two parts: (1) A general analysis of the possible functions and objectives of an international food reserve; (2) a chronological account of relevant international consultations and action, 1943-1955. An appendix contains the Principles of Surplus Disposal Recommended by FAO, 1955. First issued in 1956. Circulated to the twenty-second session of the Economic and Social Council by document E/2855, 16 April 1956.

557   DEVELOPMENT THROUGH FOOD: a Strategy for Surplus Utilization.
Rev. ed. 1962 pp. (Freedom from Hunger Campaign. Basic Studies, 2)
1964, 2nd printing, FAO $1.50
Report on the background and aims of the World Food Program, 1963- , arrangements under the auspices of FAO/UN to mobilize available surplus foodstuffs for distribution in areas of greatest need, particularly in the developing countries. Annexes include resolutions and report by an expert group. First edition issued in 1961; revised edition first issued in 1962.

558   DISPOSAL OF AGRICULTURAL SURPLUSES. Principles Recommended by FAO. 2nd ed. 23 pp.
1963, FAO

559   WORLD FOOD PROGRAM BASIC DOCUMENTS: General Regulations; Provisional Financial Procedures; Rules of Procedure of the United Nations/FAO Intergovernmental Committee. 2nd ed. rev. 31 pp.
1963, UN/FAO
First edition, January 1963; 2nd revised edition, October 1963.

560   FOOD AID AND OTHER FORMS OF UTILIZATION OF AGRICUL-
TURAL SURPLUSES: a Review of Programs, Principles and Consultations. 55 pp. (FAO Commodity Policy Studies, 15)
1964, FAO $1.00
Covers bilateral and multilateral programs for the utilization of agricultural surpluses since 1954, including international food aid; the
FAO Principles of Surplus Disposal and intergovernmental consultations on their application.

561   Report on the WORLD FOOD PROGRAM by the Executive Director.
100 pp. (P/WFP:R1)
1965, FAO $1.00
Covers origin and growth of the WFP through 1964; evaluation of its
activities; some conclusions of the study program: role of food aid
in development (WFP Studies 1-3) and operational problems and
scope of multilateral food aid (WFP Studies 4-5); a forward look and
recommendations. (For these studies, see entries 562-566.)

562   The DEMAND FOR FOOD, AND CONDITIONS GOVERNING FOOD AID
DURING DEVELOPMENT. By V.M. Dandekar. 69 pp. (World Food
Program Studies, 1; P/WFP:S1)
1965, UN/FAO $1.00
Surveys present and future demand in developing countries for food
aid (nutritional deficiencies, gaps between consumption and production of food, food import needs, gaps in balance of payments, size
and scope of food aid), and considerations governing the use of food
during development (circumstances under which food aid may be received, use to facilitate works programs, problem of sales proceeds,
etc.).

563   The IMPACT OF FOOD AID ON DONOR AND OTHER FOOD-EXPORT-
ING COUNTRIES. By G.R. Allen and R.G. Smethurst. 52 pp. (World
Food Program Studies, 2; P/WFP:S2)
1965, UN/FAO $1.00
Studies the effect of food aid in relation to prospective supply and
demand position in developed countries, possible agricultural policies confronting countries with actual or potential surpluses, and
third-party commercial exporters.

564   The LINKING OF FOOD AID WITH OTHER AID. By S. Chakravarty and
P.N. Rosenstein-Rodan. 39 pp. (World Food Program Studies, 3; P/
WFP:S3)
1965, UN/FAO $1.00
Papers seek to indicate the proper relationship between specific
commodity aid, in particular food aid, and the nonspecific aid given
in the form of freely expendable foreign exchange.

565   OPERATIONAL AND ADMINISTRATIVE PROBLEMS OF FOOD AID.
By D.A. FitzGerald. 63 pp. (World Food Program Studies, 4; P/WFP:
S4)
1965, UN/FAO $1.00
Surveys the organization and administration of bulk-supply food aid,
project-oriented food aid, and emergency food aid, and draws conclusions.

566   The ROLE OF MULTILATERAL FOOD AID PROGRAMS. By Jan Des-
sau. 38 pp. (World Food Program Studies, 5; P/WFP:S5)
1965, UN/FAO $1.00
Identifies the distinctive features and purposes of multilateral food

aid in order to provide the yardsticks against which the role of
multilateral food aid may be assessed, and to determine how and to
what extent multilateral food aid may complement bilateral programs.

567   WORLD FOOD PROGRAM; a Story of Multilateral Aid. 38 pp., illus.
       1967, UN/FAO
       Describes background of the World Food Program and what it does
       and how it works since it started operations under UN/FAO aus-
       pices on 1 January 1963.

568   The POWER OF FOOD; a Progress Report on the WORLD FOOD PRO-
       GRAMME. [32] pp., illus.
       1969, UN/FAO   Free on request to WFP, Rome
       A report in pictures and graphic diagrams and tables on WFP de-
       velopment projects throughout the world.

569   FOOD AID TO EDUCATION AND TRAINING. 44 pp., illus. (World
       Food Programme in Action)
       1970, UN/FAO   Free on request to WFP, Rome
       Describes the use of food aid to advance education by increasing
       school attendance and improving the diet of schoolchildren, a main
       activity of WFP carried out in close cooperation with UNESCO.

                              Bibliography

       See entry 1128.

FOOD STANDARDS

570   CODEX ALIMENTARIUS COMMISSION. Procedural Manual. 2nd ed.
       81 pp., fold. chart. (Joint FAO/WHO Food Standards Programme; Cx
       8/7 - 2nd ed.)
       196 , FAO/WHO
       Contains Statute and rules of procedure of the Codex Alimentarius
       Commission, General Principles of the Codex Alimentarius, pro-
       cedure for the elaboration of the Codex standards, procedure for
       the elaboration of milk and milk products standards, format of
       Codex commodity standards, guidelines for Codex committees and
       brief details on subsidiary bodies of the Commission.

571   Joint FAO/WHO CODEX ALIMENTARIUS COMMISSION. Report of the
       Sixth Session, Geneva, Switzerland, 4-14 March 1969. 125 pp. (Joint
       FAO/WHO Food Standards Programme; ALINORM 69/67)
       1969, FAO/WHO
       An FAO main document (reference no. 0629-69-MR).

                            Codex Standards

Canned fruit and vegetable products. See entry 575.

Canned Pacific salmon. See entry 593.

Fish and fishery products hygiene. See entry 576.

Food hygiene. See entry 574.

Frozen foods. See entry 594.

Labeling of prepackaged foods. See entry 592.

Milk and milk products. See entry 585.

Pesticide residues. See entry 1014.

FOOD HYGIENE

GENERAL

572   FOOD HYGIENE. Fourth Report of the Expert Committee on Environ-
       mental Sanitation. 28 pp. (WHO Technical Report Series, 104)
       1956, WHO $0.30
       Lays down principles applicable to the control of goods commonly
       known to have carried disease. Defines general objectives in terms
       of the degree of development of different areas, and discusses tech-
       nical and administrative procedures for improving conditions.

573   EUROPEAN TECHNICAL CONFERENCE ON FOOD-BORNE INFEC-
       TIONS AND INTOXICATIONS, Geneva, 16-21 February 1959. Report.
       18 pp. (WHO Technical Report Series, 184)
       1959, WHO   Out of print
       Reviews present knowledge of food-borne diseases in Europe and
       recommends measures for their notification, investigation, and pre-
       vention.

574   Recommended International Code of Practice: General Principles of
       FOOD HYGIENE. 14 pp.
       1969, FAO/WHO $1.00
       Code, advisory in nature, adopted by the Joint FAO/WHO Codex
       Alimentarius Commission. An FAO information document (refer-
       ence no 07176-69-MI).

CANNED FRUIT AND VEGETABLE PRODUCTS

575   Recommended International Code of Hygienic Practice for CANNED
       FRUIT AND VEGETABLE PRODUCTS. 14 pp. (CAC/RCP 2-1969)
       1969, FAO/WHO $1.00
       Code, advisory in nature, adopted by the Joint FAO/WHO Codex
       Alimentarius Commission. An FAO information document (refer-
       ence no. 07174-69-MI).

FISH AND FISHERY PRODUCTS HYGIENE

576   FISH INSPECTION AND QUALITY CONTROL. Ed. by Rudolf Kreuzer.
        xvi, 290 pp., fig.
        1971, Fishing News (Books) Limited (London) for FAO $30.00
        © FAO
        Papers (67) and discussions of the FAO Technical Conference on
        Fish Inspection and Quality Control, Halifax, N.S., 15-25 July 1969.
        Papers cover need for fish inspection (3), inspection programs (10),
        inspection of fish and fishery products (19), industrial and commer-
        cial aspects of quality control (9), methods of quality assessment
        (11), hygiene and safety aspects of quality control (10), training in
        inspection and quality control (3), and international cooperation in
        the promotion of quality control (2): on the work of the Codex
        Alimentarius Commission and its codes of practice. Two papers are
        in French.

MEAT HYGIENE

577   PLAN PARA UN MATADERO MUNICIPAL. Por Pedro Acha Jamet. 67
        pp., illus. (Publicaciones Científicas, 32)
        1957, PAHO   In Spanish only   Out of print
        Based on a plan for a municipal slaughterhouse to be erected in
        Chimbote, Peru. Contains many technical and practical details
        which should be taken into account in the construction and mainte-
        nance of a modern and efficient municipal slaughterhouse.

578   MEAT HYGIENE. By V.E. Albertsen and others. 527 pp., illus. (FAO
        Agricultural Studies, 34)
        1967, 2nd printing, FAO $9.00
        Advances and problems in the safe processing of meat from pro-
        ducer to consumer, including all public health aspects of processes
        involved in preparing, inspecting, and marketing meat and its
        products. An invaluable work for authorities in public health and in
        veterinary science. First issued in 1957.
        ————. 527 pp., illus. (WHO Monograph Series, 33)
        1957, WHO $10.00 cl.

579   ASPECTOS SANITARIOS A CONSIDERAR EN LA CONSTRUCCIÓN Y
        OPERACIÓN DE MATADEROS. Por Pedro Acha Jamet. 68 pp.
        (Publicaciones Científicas, 45)
        1960, PAHO   In Spanish only   Out of print
        Reviews briefly current literature on meat hygiene. Deals with the
        problems involved in the design, construction, and operation of a
        slaughterhouse with a view to achieving appropriate standards of hy-
        giene. Complements the author's Plan para un Matadero Municipal.
        (entry 577).

580   MEAT HANDLING IN UNDERDEVELOPED COUNTRIES: SLAUGHTER
        AND PRESERVATION. By I. Mann. 200 pp., illus. (FAO Agricultural
        Development Papers, 70)
        1963, 2nd printing, FAO   Out of print
        Provides a guide to the slaughtering of meat and poultry, including
        the design of abbatoirs, equipment and methods, environmental
        sanitation, and rendering of fats and oils. Intended for use in de-
        veloping countries, including those in tropical and subtropical area.
        Bibliography. First issued in 1960.

581   JOINT FAO/WHO EXPERT COMMITTEE ON MEAT HYGIENE. Second
       Report. 87 pp. (FAO Agricultural Studies, 58)
       1962, FAO $1.00
       ————. 87 pp. (WHO Technical Report Series, 241)
       1962, WHO $1.00
       Report of meeting in 1961 provides a detailed discussion of the
       measures to be taken to prevent outbreaks of meat-borne disease.
       It covers the entire field from the handling of the animal on the
       farm and its transportation to the abbatoir until the final meat prod-
       uct is consumed. Draws attention to certain recent problems, such
       as the use of irradiation and antibiotics for meat preservation, and
       the contamination of meat with radionuclides and with drugs and
       pesticide residues. A special section deals with problems of poultry
       meat hygiene. The first report, on the meeting of 1954 (1955) is out
       of print.

MILK HYGIENE

582   The PRINCIPLES OF MILK LEGISLATION AND CONTROL. By W.A.
       Lethem. 67 pp. (FAO Agricultural Development Papers, 59)
       1956, FAO   Out of print

583   JOINT FAO/WHO EXPERT COMMITTEE ON MILK HYGIENE. Third
       Report, Geneva, 22-28 April 1969. 82 pp. (FAO Agricultural Studies,
       83)
       1970, FAO $1.25
       ————. 82 pp. (WHO Technical Report Series, 453)
       1970, WHO $1.25
       Reviews reports of the first and second Joint FAO/WHO Joint Ex-
       pert Committees on Milk Hygiene (published in 1957 and 1960 and
       out of print), and considers newer problems which have arisen since
       their publication. The main objectives of the third Joint Expert
       Committee were to provide information and advice to those who are
       responsible for developing milk hygiene programs in countries
       where they are not yet in existence or are in an initial stage of de-
       velopment, including scientific and administrative problems, and to
       indicate where further research is needed.

584   MILK HYGIENE. Hygiene in Milk Production, Processing and Distribu-
       tion. By M. Abdussalam and others. 782 pp., illus. (WHO Monograph
       Series, 48)
       1962, WHO   Out of print
       This monograph (by 37 authors from all parts of the world) brings
       together the most recent information up to the time of publication on
       the safe handling of milk from production to consumer. Covers all
       aspects of milk hygiene: transmission and prevention of the major
       milk-borne diseases; handling and processing of milk on the farm
       and in the dairy, during collection, transport, and distribution, and
       under varying climatic and economic conditions; use of milks other
       than cow's milk; and administrative, legislative, and training as-
       pects.

585   Code of Principles concerning MILK AND MILK PRODUCTS and As-
       sociated Standards. (6th ed.) 146 pp. and corrigendum slip. (Joint
       FAO/WHO Food Standards Programme; Cx 8/6)
       1968, FAO/WHO
       Code of Principles; standards for milk products; standard refer-
       ence methods of sampling and analysis for milk products; status of
       acceptances; extracts from reports and resolutions. A main docu-
       ment of FAO (reference no. 02809-68-MR).

FOOD PROCESSING

GENERAL

---

586   A SHORT GUIDE TO FISH PRESERVATION, with Special Reference to
West African Conditions. By G.C. Rawson, with a chapter by Florence
A. Sai. 68 pp., illus.
1966, FAO $1.00
Preservation of fish on a domestic scale under West African conditions by salting, drying, and smoking; appendices on icing, fish protein concentrate production, canning, and a chapter on the nutritional
importance of fish.

587   The USE OF CENTRI-THERM, EXPANDING-FLOW AND FORCED-
CIRCULATION PLATE EVAPORATORS IN THE FOOD AND BIO-
CHEMICAL INDUSTRIES. By Bengt Hallström. 51 pp., illus. (UNIDO.
Food Industry Studies, 1; ID/SER.I/1)
1969, UN   Sales No. E.69.II.B.14 $0.75
Recent developments in processing techniques which may be applied
in developing countries in the food-processing, biochemical, and
pharmaceutical industries.

588   WATER-SAVING TECHNIQUES IN FOOD-PROCESSING PLANTS. By
Lavoslav Richter. 69 pp. (UNIDO. Food Industry Studies, 3; ID/SER.
I/3)
1969, UN   Sales No. E.69.II.B.16 $1.00

589   FOOD-PROCESSING INDUSTRY. Based on the Proceedings of the International Symposium on Industrial Development (Athens, November-
December 1967). 77 pp. (UNIDO Monographs on Industrial Development, 9; ID/40/9)
1969, UN   Sales No. E.69.II.B.39, vol. 9 $0.50
Discusses the nature of the food-processing industry, world food
trends, major factors to consider in establishing the food industry,
evaluating food processes, the food factory, food-processing investment factors; includes the recommendations of the Symposium, and
describes action by UNIDO to promote food-processing industries in
developing countries.

590   INTEGRATED FOOD PROCESSING IN YUGOSLAVIA. Report of
Seminar and Digest of Technical Papers, Novi Sad, Yugoslavia, 4-28
November 1968. 120 pp., fig. (ID/48)
1970, UN   Sales No. E.70.II.B.14 $2.00
Contains 13 chapters based on lectures by experts from agroindustrial combines varying in size from 20,000 to 140,000 hectares.
Some combines are integrated with ten to fifteen factories and have
their own distribution network. Other have a staff of from 10,000 to
25,000 and, in some cases, they cooperate with 10,000 to 20,000
private farmers.

591   PACKAGING AND PACKAGING MATERIALS WITH SPECIAL REFER-
ENCE TO THE PACKAGING OF FOOD. By Anton Petrišić. 56 pp.,
illus. (UNIDO. Food Industry Studies, 5; ID/SER.I/5)
1969, UN   Sales No. E.69.II.B.31 $0.75
Covers measures to protect food from microorganisms, insects, and
rodents; paper and paperboard packaging; metal and metal contain-
ers; other packaging materials; packaging techniques.

592   Recommended International General Standard for the LABELLING OF
PREPACKAGED FOODS. 12 pp. (Joint FAO/WHO Food Standards
Programme; CAC/RS 1-1969)
1969, FAO/WHO
Standard, advisory in nature, adopted by the Joint FAO/WHO Codex
Alimentarius Commission.

593   Recommended International Standard for CANNED PACIFIC SALMON.
12 pp. (Joint FAO/WHO Food Standards Programme; CAC/RS 3-1969)
1969, FAO/WHO $1.00
Standard, advisory in nature, adopted by the Joint FAO/WHO Codex
Alimentarius Commission.

FREEZING

594   Western European Markets for FROZEN FOODS. 230 pp.
1969, ITC UNCTAD/GATT $5.00
Survey of the Western European market for frozen foods including
quick-frozen foods carried out by the UNCTAD-GATT International
Trade Centre, Geneva, with a view to identifying trading opportunities that exist or may develop in the near future for processors and
exporters located in developing countries. Includes chapters on
process of manufacture and distribution of frozen foods and on
legislation thereon, and in an annex excerpts from the proposed
Standards for Quick-frozen Foods under consideration by the FAO/
WHO Codex Alimentarius Commission.

IRRADIATION

595   RADIATION CONTROL OF SALMONELLAE IN FOOD AND FEED
PRODUCTS. Report of a Panel on Irradiation Control of Harmful Organisms Transmitted by Food and Feed Products, with particular reference to Salmonellae, held in Vienna, 12-14 December 1962. 148 pp.
(Technical Reports Series, 22; STI/DOC/10/22)
1963, IAEA $3.00
Papers (12), a summary and conclusions, and recommendations of
the panel.

596   The TECHNICAL BASIS FOR LEGISLATION ON IRRADIATED FOOD.
Report of a Joint FAO/IAEA/WHO Expert Committee, Rome, 21-28
April 1964. [56] pp. (FAO Atomic Energy Series, 6)
1965, FAO $1.00
————. [56] pp. (WHO Technical Report Series, 316)
1966, WHO $1.00
General principles governing production and use of irradiated
food; recommendations for establishment of legislation and control; recommended technical procedures and tests required to
permit an evaluation of the safety for consumption of irradiated
food; microbiological aspects. (See also entry 605.)

597   APPLICATION OF FOOD IRRADIATION IN DEVELOPING COUNTRIES.
Report of a Panel on the Application of Food Irradiation in Developing
Countries, held in Vienna, 3-6 August 1964. 183 pp. (Technical Re-
ports Series, 54; STI/DOC/10/54)
1966, IAEA $4.00
Report and papers (16) on problems in developing countries con-

cerned with food preservation; insect control in stored products; destruction of pathogenic microorganisms and other infective agents, including parasites; extension of the period of use and distribution of fresh products; inhibition of sprouting of certain vegetables; and inhibition of maturing of certain fruits. Considers how ionizing radiation might be applied to some of these problems, and suggests areas for further investigation and research.

598   FOOD IRRADIATION. Proceedings of the International Symposium on Food Irradiation jointly organized by IAEA and FAO and held in Karlsruhe, 6-10 June 1966. 956 pp., illus. (Proceedings Series; STI/PUB/127)
1966, IAEA   Composite: E/F/R/S $20.00
Papers (68) on radiation sources and dosimetry; wholesomeness of irradiated food; chemical and physical effects of ionizing radiation; microbiology, virology, and quarantine problems; status of various irradiated commodities; programs and facilities for food irradiation; legislation and clearance of irradiated food.

599   MICROBIOLOGICAL PROBLEMS IN FOOD PRESERVATION BY IRRADIATION. Report of a Panel on Microbiological Problems in Food Preservation by Irradiation organized by the Joint FAO/IAEA Division of Atomic Energy in Food and Agriculture and held in Vienna, 27 June-1 July 1966. 148 pp., illus. (Panel Proceedings Series; STI/PUB/168)
1967, IAEA $3.00
Papers (14) on such topics as inactivation of microorganisms in seafood, studies on Clostridium botulinum, foot-and-mouth disease virus and salmonellae, and the effects of various additives and combination treatments; and summaries, conclusions, and recommendations of the panel.

600   RADIOISOTOPES AND RADIATION IN DAIRY SCIENCE AND TECHNOLOGY. Proceedings of a Seminar on Radioisotopes and Radiation in Dairy Science and Technology jointly organized by IAEA and FAO and held in Vienna, 12-15 July 1966. 258 pp., illus. (Proceedings Series; STI/PUB/135)
1966, IAEA   Composite: E/F $6.00
Papers (17) on such topics as application of radioisotopes and radiation to evaluation and control of dairy food processes, metabolic patterns in dairy organisms, removal of radionuclides from milk, analytical quality control and milk contamination survey, feeding practices to reduce the concentration of ^{137}Cs fallout in cow's milk, and trace element determination in milk by neutron activation analysis.

601   PRESERVATION OF FRUIT AND VEGETABLES BY RADIATION. Proceedings of a Panel on Preservation of Fruit and Vegetables by Radiation, Especially in the Tropics, organized by the Joint FAO/IAEA Division of Atomic Energy in Food and Agriculture and held in Vienna, 1-5 August 1966. 152 pp., illus. (Panel Proceedings Series; STI/PUB/149)
1968, IAEA $3.00
Papers (12) on topics such as radiation-induced delay in ripening of Alphonso mangoes, influence of ionizing radiation on flavanoid pigments of some berry fruits, effects of ionizing radiation on the storage properties of fruits, tissue texture and intermediary metabolism of irradiated fresh fruits and vegetables, and the role of calcium in softening and refirming irradiated plant tissues; includes conclusions and recommendations of the panel.

602   ELIMINATION OF HARMFUL ORGANISMS FROM FOOD AND FEED BY
       IRRADIATION. Report of a Panel on Elimination of Harmful Organisms
       from Food and Feed by Irradiation organized by the Joint FAO/IAEA
       Division of Atomic Energy in Food and Agriculture and held in Zeist,
       The Netherlands, 12-16 June 1967. 118 pp., illus. (Panel Proceedings
       Series; STI/PUB/200)
       1968, IAEA $3.50
          Papers (12) on elimination of Salmonella, inactivation of viruses,
          devitalization of parasites and elimination of clostridia by irradia-
          tion in different food and feed products; includes summaries, con-
          clusions, and recommendations of the panel.

603   ENZYMOLOGICAL ASPECTS OF FOOD IRRADIATION. Proceedings
       of a Panel on Enzymological Aspects of the Application of Ionizing Ra-
       diation to Food Preservation organized by the Joint FAO/IAEA Divi-
       sion of Atomic Energy in Food and Agriculture and held in Vienna, 8-
       12 April 1968. 110 pp., illus. (Panel Proceedings Series; STI/PUB/
       216)
       1969, IAEA $4.00
          Contains conclusions and recommendations of the panel and ten pa-
          pers on radiation inactivation of pure enzymes of importance in
          food technology, radiation inactivation of enzymes in situ (in animal
          and vegetable tissues used as foodstuffs), and methods to control en-
          zyme activity in irradiated food (combined treatments).

604   NUCLEAR LAW FOR A DEVELOPING WORLD. Lectures given at the
       Training Course on the Legal Aspects of Peaceful Uses of Atomic En-
       ergy, held by IAEA in Vienna, 16-26 April 1968, 329 pp. (Legal Series,
       5; STI/PUB/215)
       1969, IAEA $9.00
          Contains 33 lectures, including one by H.E. Goresline on "Prospects
          of Food Irradiation," pp. 249-258

605   MICROBIOLOGICAL SPECIFICATIONS AND TESTING METHODS FOR
       IRRADIATED FOOD. Report of a Panel of Experts organized by FAO
       and IAEA in collaboration with the International Association of Micro-
       biological Societies. 121 pp. (Technical Reports Series, 104; STI/
       DOC/10/104)
       1970, IAEA/FAO/IAMS $4.00
          The panel, after comparing microbiological methods in use in most
          countries, reviewing the literature, and pooling their individual ex-
          periences and knowledge of unpublished research, have consolidated
          this information into a limited number of acceptable, reproducible
          methods for the evaluation of food for specific microorganisms.
          The information is presented for voluntary use in evaluating radia-
          tion-treated foods or for use in promulgating legislation on irradi-
          ated food or feeds. Supplements The Technical Basis for Legisla-
          tion on Irradiated Food, 1965 (entry 596).

606   Training Manual on FOOD IRRADIATION TECHNOLOGY AND TECH-
       NIQUES. A joint undertaking by FAO and IAEA. 134 pp. (Technical
       Reports Series, 114; STI/DOC/114)
       1970, IAEA $4.00
          In two parts: (I): The basic part, covers general information and
          discussions on the applications; (II) is a section on laboratory exer-
          cises to demonstrate the principles of radiation processing and the
          effects of radiation on certain types of food.

607   WHOLESOMENESS OF IRRADIATED FOOD with special reference to Wheat, Potatoes and Onions. Report of a Joint FAO/IAEA/WHO Expert Committee. 44 pp. (WHO Technical Report Series, 451)
1970, WHO $1.00
Reviews investigations required for assessment of the safety of irradiated food, procedures for evaluation of the experimental data, and makes recommendations for future research. Annex 1: mutagenic and cytotoxic considerations; annex 2: monographs on irradited wheat, irradiated potatoes, and irradiated onions.

608   RADIATION AND RADIOISOTOPES FOR INDUSTRIAL MICROORGANISMS. Proceedings of a Symposium on Use of Radiation and Radioisotopes for Genetic Improvement of Industrial Microorganisms, held by IAEA in Vienna, 29 March-1 April 1971. 338 pp., fig., tables. (Proceedings Series; STI/PUB/287)
1971, IAEA $9.00
Papers (24) and discussions review the results of research on microbial mutagenesis and recombination, physiology, biochemistry, and allied topics in relation to microbial fermentation, and include a brief review (7 papers) of the current status and future outlook of applied microbiology in certain developing countries.

SPECIFIC COMMODITIES

609   Processing of CASSAVA AND CASSAVA PRODUCTS in Rural Industries. By L.W.J. Holleman and A. Aten. 115 pp., illus. (FAO Agricultural Development Papers, 54)
1966, 2nd printing, FAO $1.50
Outlines the essentials of processing tapioca flour, including technical information and descriptions of machines and other accessory equipment that may be used with advantage by individual farmers and farmer cooperatives in rural areas. Gives, in addition, some details regarding the manufacture of certain food products based on tapioca flour as raw material which can be processed both in farmers' homes and through rural industries. Includes also some essential methods of analyzing cassava and tapioca, as well as important applications of tapioca flour. First issued in 1956.

610   Advances in CHEESE TECHNOLOGY. By Frank V. Kosikowski and Germain Mocquot. 236 pp., illus. (FAO Agricultural Studies, 38)
1958, FAO $2.00
Considers results of research on problems on inhibitory or stimulatory patterns of milk toward bacteria, phage in lactic acid bacteria, and chromatographic studies of cheese ripening; also important contributions on antibiotic residues in milk for cheese, curd hydrolysis, cheese rheology, etc.

611   Industrial Processing of CITRUS FRUIT. By Zeki Berk. 45 pp., illus. (UNIDO. Food Industry Studies, 2; ID/SER.I/2)
1969, UN   Sales No. E.69.II.B.9 $0.75
Study designed for the use of the citrus fruit-processing industry of developing countries; describes processes and techniques which can be adopted with advantage.

612   Processing of Raw COCOA for the Market. By T.A. Rohan. 207 pp.,
illus. (FAO Agricultural Studies, 60)
1963, FAO $3.50
Reviews technological methods employed in all phases of the preparation of cocoa for the market. Begins with harvesting and follows
the succeeding processes, paying particular attention to those which
improve the quality of cocoa. Deals with processing by small holders as well as on a commercial scale, and is intended to assist
cocoa-producing countries to develop their industry on a sound
basis. Bibliography.

613   The COCONUT INDUSTRY OF ASIA. 126 pp. (ECAFE. Regional Plan
Harmonization and Integration Studies, 1; E/CN.11/887)
1969, UN   Sales No. E.69.II.F.4 $2.50
Contains a chapter on the coconut-processing industry of the ECAFE
region and a chapter on quality standardization of coconut products.
Most of the study deals with the production of coconut and domestic
marketing and international trade of coconut and coconut products.
It served as a background paper for the meeting in Bangkok in October 1968 of representatives of the coconut-producing countries of the
ECAFE region from which was developed the Asian Coconut Community.

614   COCONUT OIL Processing. By J.G. Thieme. 251 pp., illus. (FAO Agricultural Development Papers, 89)
1968, FAO $3.00
Deals with processing methods, preparatory equipment, oil extraction equipment, equipment for treating oil and cake, examples of
rural oil mills, refining of coconut oil, properties and usage of coconut oil, properties and uses of coconut cake. References.

615   DATES: Handling, Processing and Packing. By V.H.W. Dowson and A.
Aten. 392 pp., illus. (FAO Agricultural Development Papers, 72)
1962, FAO $4.00
Describes in full the operations to which dates are subjected at
various stages on the way from grower to wholesaler. Incorporates
the results of extensive research, experience, and observations on
the subject, and includes, in addition to more than 200 illustrations,
graphs, and tables, a large vocabulary in Arabic of names for every
part of the palm and the date, for the different varieties, and for the
date in each stage of its development. Extensive bibliography.

616   The Technology of FISH UTILIZATION: Contributions from Research.
Ed. by Rudolf Kreuzer. xxii, 280 pp., illus. ([Technology of Fish
Utilization, 1])
1965, FAO/Fishing News (Books) Ltd. (London) $21.00
Papers (60) at the Symposium on the Significance of Fundamental
Research in the Utilization of Fish, Husum, Federal Republic of
Germany, May 1964. In six sessions: rigor mortis; problems related to the preservation of fresh fish; control of deteriorative
changes in frozen fish; measuring the degree of freshness of fish;
production and storage of fish protein concentrate; dehydration and
canning of fish.

617   FREEZING AND IRRADIATION OF FISH. Ed. by Rudolf Kreuzer. xix,
528 pp., illus. (The Technology of Fish Utilization, 2)
1969, FAO/Fishing News (Books) Ltd. (London) $34.50
Composite: E/F/S
Contains papers (80) of the Technical Conference on the Freezing

and Irradiation of Fish, Madrid, September 1967. In six parts:
freezing fish at sea; freezing and processing frozen fish; economics
of producing and marketing frozen fish products; the quality of frozen fish products and its assessment; storing, packing, and distribution; preservation of fishery products by irradiation.

618   PRESERVATION OF FISH BY RADIATION. Proceedings of a Panel on
the Irradiation Preservation of Foods of Marine Origin organized by the
Joint FAO/IAEA Division of Atomic Energy in Food and Agriculture and
held in Vienna, Austria, 15-19 December 1969. 163 pp., fig. (Panel
Proceedings Series; STI/PUB/196)
1970, IAEA $5.00
Papers (10) and summaries, conclusions, and recommendations of
the panel.

619   RADURIZATION OF SCAMPI, SHRIMP AND COD. By G. Hannesson and
B. Dagbjartsson. Report of a Project organized and supervised by the
Joint FAO/IAEA Division of Atomic Energy in Food and Agriculture.
93 pp., tables. (Technical Reports Series, 124; STI/DOC/10/124)
1971, IAEA $3.00
Report on a project to demonstrate whether radiation preservation
is practical for seafood exports under the conditions prevailing in
the fishing industry and trade in Iceland. The investigation concentrated on cod, scampi (Norwegian lobster tails), and deep-sea
shrimp.

620   DISINFESTATION OF FRUIT BY IRRADIATION. Proceedings of a
Panel on the Use of Irradiation to Solve Quarantine Problems in the In-
ternational Fruit Trade, organized by the Joint FAO/IAEA Division of
Atomic Energy in Food and Agriculture, and held in Honolulu, Hawaii,
United States of America, 7-11 December 1970. 177 pp., fig., tables.
(Panel Proceedings Series; STI/PUB/299)
1971, IAEA $5.00
Papers (15) and panel summaries, conclusion, considerations, and
recommendations on the industrial application of irradiation in the
disinfestation of tropical fruits.

621   MILK PLANT LAYOUT. By H.S. Hall, Yngve Rosén and Helge Blombergsson. 156 pp., illus. (FAO Agricultural Studies, 59)
1968, 2nd printing, FAO $2.00
Aims to assist government officials and others in developing countries concerned with the design of milk plants by outlining the principles of such designs which have proved successful in countries
with a flourishing milk industry. First issued in 1963.

622   MILK STERILIZATION. By H. Burton, J. Piene, G. Thieulin and
others. 265 pp., illus. (FAO Agricultural Studies, 65)
1965, FAO $3.50
Provides technicians, health experts, and officials with full technical information on milk sterilization, including processes used, control methods, economic aspects, and distribution. A sequel to the
out-of-print Milk Pasteurization: Planning, Plant Operation and Control, by H.D. Kay and others, 1953, FAO (FAO Agricultural Studies,
23); WHO (WHO Monograph Series, 14).

623   The Development and Manufacture of Sterilized MILK CONCENTRATE. By M.E. Seehafer. 52 pp., illus. (FAO Agricultural Studies, 72)
1967, FAO $1.00
     Deals with sterilization and canning of milk concentrate. A research report by a marketing representative of the United States Steel Corporation on a technological process still being developed.

624   Modern Sterilization Methods for MILK Processing. By Sune Holm. 43 pp., illus. (UNIDO. Food Industries Studies, 4; ID/SER.I/4)
1969, UN   Sales No. E.69.II.B.29 $0.75
     Chapters cover general viewpoints on modern heating systems; different sterilizing units; description of equipment; intermediate aseptic tank; sterilized milk control; variety of products; and aseptic filling.

625   OLIVE OIL Processing in Rural Mills. By G. Frezzotti, M. Manni and A. Aten. 103 pp., illus. (FAO Agricultural Development Papers, 58)
1956, FAO   Out of print
     Describes methods and processes, tools, implements, and machinery necessary to process virgin olive oil, so that the material presented is useful for the rural factory manager, mill owner, and heads of the many cooperative organizations who possess oil mills. Also contains a chapter on how to analyze olives, oil, and residues.

626   Equipment for the Processing of RICE. By A. Aten, A.D. Faunce and Luther R. Ray. 55 pp., illus. (FAO Agricultural Development Papers, 27)
1953, FAO   Out of print
     Describes fundamental operations in rice milling. Indicates methods and procedures in broad outline only.

627   New Equipment for the Processing of TEA. By Luther R. Ray and Evert Beekman. 12 pp., illus. (FAO Agricultural Development Papers, 12)
1951, FAO   Out of print
     Includes descriptions and illustrations of machines and associated equipment used in the processing of tea. Indicates methods and procedures in broad outline only.

FOOD ADDITIVES

GENERAL

---

628   The PUBLIC HEALTH ASPECTS OF THE USE OF ANTIBIOTICS IN FOOD AND FEEDSTUFFS. Report of an Expert Committee. 30 pp. (WHO Technical Report Series, 260)
1963, WHO $0.30
     Considers residue levels of antibiotics in foods and investigates possible creation of resistant bacterial strains and sensitivity states in food handlers and consumers. It accepts that the use of certain antibiotics of medical value cannot yet be dispensed with, but stresses the desirability of using, where possible, those that are not of medical value. Recommends further research.

629   Second JOINT FAO/WHO CONFERENCE ON FOOD ADDITIVES. Re-
      port. 12 pp. (FAO Nutrition Meetings Report Series, 34)
      1963, FAO $0.50
      ————. 12 pp. (WHO Technical Report Series, 264)
      1963, WHO $0.50
         Reviews work of the Joint FAO/WHO Expert Committee on Food
         Additives and discusses its future program. The report of the
         first conference (1956, FAO Nutrition Meetings Report Series, 11;
         WHO Technical Report Series, 107. 14 pp.) is out of print. It had
         recommended setting up a Joint Committee.

630   PROCEDURES FOR INVESTIGATING INTENTIONAL AND UNINTEN-
      TIONAL FOOD ADDITIVES. Report of a WHO Scientific Group. 25 pp.
      (WHO Technical Report Series, 348)
      1967, WHO $0.60
         Reviews criteria used in establishing acceptable daily intakes of food
         additives and suggests further studies on toxicological procedures
         used for evaluation of food additives in order to establish their safety
         to the consumer.

JOINT FAO/WHO EXPERT COMMITTEE ON FOOD ADDITIVES: REPORTS
_____

631   Report of the JOINT FAO/WHO EXPERT COMMITTEE ON FOOD AD-
      DITIVES, Rome, 3-10 December 1956. General Principles Governing
      the Use of Food Additives. 22 pp. (FAO Nutrition Meetings Report
      Series, 15)
      1957, FAO   Out of print
      ————. WHO. (WHO Technical Report Series, 129) $0.30

632   Procedures for the Testing of Intentional Food Additives to Establish
      Their Safety for Use. Second Report of the JOINT FAO/WHO EXPERT
      COMMITTEE ON FOOD ADDITIVES, Geneva, 17-24 June 1957. 19 pp.
      (FAO, Nutrition Meetings Report Series; 17)
      1958, FAO $0.30
      ————. WHO (WHO Technical Report Series, 144) $0.30

633   SPECIFICATIONS FOR IDENTITY AND PURITY OF FOOD ADDITIVES.
      2 vol.
      1962-63, FAO $1.75 each vol.
         Vol. 1: Antimicrobial Preservatives and Antioxidants. 116 pp.
         Vol. 2: Food Colors. 136 pp.
      Vol. 1 contains the specifications resulting from the deliberations of
      the third (1958) meeting of the committee, and vol. 2, the specifications
      resulting from the fourth (1959) meeting, respectively, both as subse-
      quently revised.

634   Evaluation of the Carcinogenic Hazards of Food Additives. Fifth Report
      of the JOINT FAO/WHO EXPERT COMMITTEE ON FOOD ADDITIVES,
      Geneva, 12-19 December 1960. 33 pp. (FAO Nutrition Meetings Report
      Series, 29)
      1961, FAO   Out of print
      ————. WHO. (WHO Technical Report Series, 220) $0.60

635   Evaluation of the Toxicity of a Number of Antimicrobials and Antioxi-
      dants. Sixth Report of the JOINT FAO/WHO EXPERT COMMITTEE ON
      FOOD ADDITIVES, Geneva, 5-12 June 1961. 104 pp. (FAO Nutrition
      Meetings Report Series, 31)
      1962, FAO $1.25
      ————. WHO. (WHO Technical Report Series, 228)   Out of print

636   Specifications for the Identity and Purity of Food Additives and Their
        Toxicological Evaluation: Emulsifiers, Stabilizers, Bleaching and Ma-
        turing Agents. Seventh Report of the JOINT FAO/WHO EXPERT COM-
        MITTEE ON FOOD ADDITIVES, Rome, 18-25 February 1963. 189 pp.
        (FAO Nutrition Meetings Report Series, 35)
        1964, FAO $2.00
        ———. WHO. (WHO Technical Report Series, 281) $2.00

637   Specifications for the Identity and Purity of Food Additives and Their
        Toxicological Evaluation: Food Colours and Some Antimicrobials and
        Antioxidants. Eighth Report of the JOINT FAO/WHO EXPERT COM-
        MITTEE ON FOOD ADDITIVES, Geneva, 8-17 December 1964. 25 pp.
        (FAO Nutrition Meetings Report Series, 38)
        1965, FAO $0.60
        ———. WHO. (WHO Technical Report Series, 309) $0.60

638   SPECIFICATIONS FOR IDENTITY AND PURITY AND TOXICOLOGICAL
        EVALUATION OF SOME ANTIMICROBIALS AND ANTIOXIDANTS, Ge-
        neva, 8-17 December 1964. 89 pp. (FAO Nutrition Meetings Report
        Series, 38A; WHO/Food Add./24.65)
        1965, FAO   Available on request*
        Results from the deliberations of the eighth (1964) meeting.

639   SPECIFICATIONS FOR IDENTITY AND PURITY AND TOXICOLOGICAL
        EVALUATION OF SOME FOOD COLOURS. 212 pp. (FAO Nutrition
        Meetings Report Series, 38B; WHO/Food Add./66.25)
        1966, FAO   Free on request*
        Results from the deliberations of the eighth (1964) and tenth (1966)
        meetings.

640   Specifications for the Identity and Purity of Food Additives and Their
        Toxicological Evaluation: Some Antimicrobials, Antioxidants, Emulsi-
        fiers, Stabilizers, Flour-treatment Agents, Acids, and Bases. Ninth
        Report of the JOINT FAO/WHO EXPERT COMMITTEE ON FOOD AD-
        DITIVES, Rome, 13-20 December 1965. 24 pp. (FAO Nutrition Meet-
        ings Report Series, 40)
        1966, FAO $0.60
        ———. WHO. (WHO Technical Report Series, 339) $0.60

641   TOXICOLOGICAL EVALUATION OF SOME ANTIMICROBIALS, ANTI-
        OXIDANTS, EMULSIFIERS, STABILIZERS, FLOUR-TREATMENT
        AGENTS, ACIDS AND BASES. 169 pp. (FAO Nutrition Meetings Report
        Series, 40A,B,C; WHO/Food Add./67.29)
        1967, FAO   Available on request*
        Results from the deliberations of the ninth (1965) and tenth (1966)
        meetings.

642   Specifications for the Identity and Purity of Food Additives and Their
        Toxicological Evaluation: Some Emulsifiers and Stabilizers and Cer-
        tain Other Substances. Tenth Report of the JOINT FAO/WHO EXPERT
        COMMITTEE ON FOOD ADDITIVES, Geneva, 11-18 October 1966. 47
        pp. (FAO Nutrition Meetings Report Series, 43)
        1967, FAO $1.00
        ———. WHO. (WHO Technical Report Series, 373) $1.00

---

* From Food Additives, WHO (Geneva), or from Food Science and
Technology Branch, FAO (Rome).

643 Specifications for the Identity and Purity of Food Additives and Their
Toxicological Evaluation: Some Flavouring Substances and Non-nutritive Sweetening Agents. Eleventh Report of the JOINT FAO/WHO EX-
PERT COMMITTEE ON FOOD ADDITIVES, Geneva, 21-28 August 1967.
18 pp. (FAO Nutrition Meetings Report Series, 44)
1968, FAO $0.60
———. WHO. (WHO Technical Report Series, 383) $0.60

644 TOXICOLOGICAL EVALUATION OF SOME FLAVOURING SUBSTANCES
AND NON-NUTRITIVE SWEETENING AGENTS. 110 pp. (FAO Nutrition Meetings Report Series, 44A; WHO/Food Add./68.33)
1967, FAO Available on request*
Results from deliberations of the eleventh (1967) meeting.

645 SPECIFICATIONS AND CRITERIA FOR IDENTITY AND PURITY OF
SOME FLAVOURING SUBSTANCES AND NON-NUTRITIVE SWEETEN-
ING AGENTS. 70 pp. (FAO Nutrition Meetings Report Series, 44B;
WHO/Food Add./69.31)
1969, FAO Available on request*
Results from deliberations of the eleventh (1967) meeting.

646 Specifications for the Identity and Purity of Food Additives and Their
Toxicological Evaluation: Some Antibiotics. Twelfth Report of the
JOINT FAO/WHO EXPERT COMMITTEE ON FOOD ADDITIVES, Geneva, 1-8 July 1968. 49 pp. (FAO Nutrition Meetings Report Series,
45)
1969, FAO $1.00
———. 49 pp. (WHO Technical Report Series, 430) $1.00

647 SPECIFICATIONS FOR IDENTITY AND PURITY OF SOME ANTIBI-
OTICS. 107 pp. (FAO Nutrition Meetings Report Series, 45A; WHO/
Food Add./69.34)
FAO, 1969 Available on request*
Results from deliberations of the twelfth (1968) meeting.

648 Specifications for the Identity and Purity of Food Additives and Their
Toxicological Evaluation: Some Food Colours, Emulsifiers, Stabilizers,
Anti-caking Agents and Certain Other Substances. Thirteenth Report of
the JOINT FAO/WHO EXPERT COMMITTEE ON FOOD ADDITIVES,
Rome, 27 May-4 June 1969. 36 pp., and corrigendum slip. (FAO Nutrition Meetings Report Series, 46)
1970, FAO $1.00
———. 36 pp., and corrigendum slip. (WHO Technical Report
Series, 445) $1.00

649 TOXICOLOGICAL EVALUATION OF SOME FOOD COLOURS, EMULSI-
FIERS, STABILIZERS, ANTI-CAKING AGENTS AND CERTAIN OTHER
SUBSTANCES. 161 pp. (FAO Nutrition Meetings Report Series, 46A;
WHO/Food Add./70.36)
1969, FAO Available on request*
Results from deliberations of the thirteenth (1969) meeting.

650 SPECIFICATIONS FOR THE IDENTITY AND PURITY OF SOME FOOD
COLOURS, EMULSIFIERS AND STABILIZERS, ANTICAKING AGENTS

* From Food Additives, WHO (Geneva), or from Food Science and
Technology Branch, FAO (Rome).

AND CERTAIN OTHER SUBSTANCES. 138 pp. (FAO Nutrition Meetings Report Series, 46B; WHO/Food Add./70.37)
1970, FAO/WHO   Available on request*
Results from deliberations of the thirteenth (1969) meeting.

651   Evaluation of Food Additives: Specifications for the Identity and Purity of Food Additives and Their Toxicological Evaluation: Some Extraction Solvents and Certain Other Substances; and a Review of the Technical Efficacy of Some Antimicrobiol Agents. Fourteenth Report of the JOINT FAO/WHO EXPERT COMMITTEE ON FOOD ADDITIVES, Geneva, 24 June-2 July 1970. 36 pp. (FAO Nutrition Meetings Report Series, 48)
1971, FAO $1.00
———. 36 pp. (WHO Technical Report Series, 462) $1.00

652   TOXICOLOGICAL EVALUATION OF SOME EXTRACTION SOLVENTS AND CERTAIN OTHER SUBSTANCES. 131 pp. (FAO Nutrition Meetings Report Series, 48A; WHO/Food Add./70.39)
1971, FAO/WHO   Available on request*
Results from deliberations of fourteenth (1970) meeting.

653   SPECIFICATIONS FOR THE IDENTITY AND PURITY OF SOME EXTRACTION SOLVENTS AND CERTAIN OTHER SUBSTANCES. 124 pp. (FAO Nutrition Meetings Report Series, 48B; WHO/Food Add./70.40)
1971, FAO /WHO   Available on request*
Results from deliberations of fourteenth (1970) meeting.

654   A REVIEW OF THE TECHNOLOGICAL EFFICIENCY OF SOME ANTIMICROBIAL AGENTS. 61 pp. (FAO Nutrition Meetings Report Series, 48C; WHO/Food Add./70.41)
1971, FAO/WHO   Available on request*
Results from deliberations of fourteenth (1970) meeting.
The reports listed above contain general considerations, including the principles adopted for the evaluation, and a summary of the results of evaluation of a number of food additives. The documents which have resulted from the deliberations of the meetings contain additional information, such as biological data and toxicological evaluation, considered at the meetings.

NATIONAL CONTROL LEGISLATION

655   FOOD ADDITIVE CONTROL IN AUSTRALIA. by W.R. Jewell. 30 pp. (FAO Food Additive Control Series, 4)
1961, FAO   Out of print

656   FOOD ADDITIVE CONTROL IN CANADA. By L.I. Pugsley. 36 pp. (FAO Food Additive Control Series, 1)
1962, 2nd printing, FAO $0.50
First issued in 1959.

657   FOOD ADDITIVE CONTROL IN DENMARK. by E. Uhl and S.C. Hansen. 35 pp. (FAO Food Additive Control Series, 5)
1961, FAO   Out of print

* From Food Additives, WHO (Geneva), or from Food Science and Technology Branch, FAO (Rome).

658   FOOD ADDITIVE CONTROL IN THE FEDERAL REPUBLIC OF GER-
      MANY. By Volker Hamann. 41 pp. (FAO Food Additive Control Se-
      ries, 7)
      1963, FAO $1.00

659   FOOD ADDITIVE CONTROL IN FRANCE. By R. Truhaut and R.
      Souverain. 69 pp. (FAO Food Additive Control Series, 6)
      1965, 2nd printing, FAO $1.00
         First issued in 1963.

660   FOOD ADDITIVE CONTROL IN THE NETHERLANDS. By W. Meijer.
      41 pp. (FAO Food Additive Control Series, 3)
      1963, 2nd printing, FAO $0.50
         First issued in 1961.

661   FOOD ADDITIVE CONTROL IN THE U.S.S.R. By A.I. Stenberg, Y.I.
      Shillinger and M.G. Shevchenko. 45 pp. (FAO Food Additive Control
      Series, 8)
      1969, FAO $1.00

662   FOOD ADDITIVE CONTROL IN THE UNITED KINGDOM. By C.L.
      Hinton. 52 pp. (FAO Food Additive Control Series, 2)
      1962, 2nd printing, FAO   Out of print
         First issued in 1960.

      These national surveys comprise the background, procedures for
      adoption, and operation of control laws and regulations; current reg-
      ulations; a list of legislation and sources; and bibliography.

      Current Food Additives Legislation. See entry 1194.

NUTRITION

GENERAL

663   WORKERS' NUTRITION AND SOCIAL POLICY. 249 pp. (Studies and
      Reports, Series B (Economic Conditions), 23)
      1936, ILO   Out of print
         Study of nutrition of workers and their families, relationship between
         food consumption and health in general and productive efficiency in
         particular; assembles data on workers' diets; indicates methods to
         improve nutrition.

664   NUTRITION IN INDUSTRY. 177 pp., illus. (Studies and Reports, New
      Series, 4)
      1946, ILO   Out of print
         Study of steps taken in Canada, the United Kingdom, and the U.S.A.
         to safeguard the nutrition of workers in wartime.

665   NUTRITION AND SOCIETY. 54 pp. (World Food Problems, 1)
      1956, FAO   Out of print
         Text of 1955 lecture by André Mayer (1875-1956), brief biography of
         Mayer, and account of FAO's work in the field of nutrition.

666   NUTRITION AND WORKING EFFICIENCY. 47 pp., illus. (Freedom
from Hunger Campaign. Basic Studies, 5)
1966, 3rd printing, FAO $0.50
      Chapters cover dietary requirements for various activities; relation
between diet and working capacities; factors affecting food consumption; measures to improve workers' nutrition, including industrial
feeding programs. First issued in 1962.

667   NUTRITION IN PREGNANCY AND LACTATION. Report of WHO Expert
Committee. 54 pp. (WHO Technical Report Series, 302)
1965, WHO $1.00
      Examines physiological aspects of pregnancy and lactation in rela-
tion to changed nutritional needs during these states. Stresses need
for thorough and coordinated investigation.

668   RADIOISOTOPE TECHNIQUES IN THE STUDY OF PROTEIN METABO-
LISM. Findings of a Panel on Radioisotope Techniques in the Study of
Protein Metabolism held in Vienna, 1-5 June 1964. 258 pp., illus.
(Technical Reports Series, 45; STI/DOC/10/45)
1965, IAEA $5.50
      Papers (24) on methods of preparation of labeled proteins and pro-
tein-like substances; radioisotope techniques for the study of protein
metabolism and of gastrointestinal protein absorption and loss;
clinical applications of radioisotope techniques. Extensive references.

669   WHO ACTIVITIES IN NUTRITION, 1948-1964. 38 pp., illus.
1965, WHO $0.60
      Includes a general survey of WHO nutrition program, and sections
devoted to protein malnutrition; nutrition anaemias, endemic goiter;
avitaminosis A and xerophthalmia; food additives; education and
training in nutrition; and a proposed five-year program for promotion of better nutrition for populations in developing countries. Reprinted from WHO Chronicle, vol. 19, no. 10-12, October-December
1965, pp. 387-396, 429-443, 467-476.

670   Report of the JOINT FAO/WHO TECHNICAL MEETING ON METHODS
OF PLANNING AND EVALUATION IN APPLIED NUTRITION PRO-
GRAMS, Rome, Italy, 11-16 January 1965. 77 pp. (FAO Nutrition
Meetings Report Series, 39)
1966, FAO $0.80
      ————. 77 pp. (WHO Technical Report Series, 340) $0.80
      Gives background, characteristics, and present position of applied
nutrition programs; elaborates some basic concepts and working
definitions for future planning and evaluation, and outlines stages
of an applied nutrition program.

671   JOINT FAO/WHO EXPERT COMMITTEE ON NUTRITION. Seventh Report. 84 pp. (FAO Nutrition Meetings Report Series, 42)
1967, FAO   Out of print
      ————. 84 pp. (WHO Technical Report Series, 377)
1967, WHO $1.25
      Report of meeting in 1966 reviews knowledge in relation to nutri-
tion requirements and discusses a variety of health problems
arising from undernutrition. Critically examines the means
available for preventing and combating specific nutritional deficiencies. Considers also food additives, food standards, myco-
toxins in food, and the feeding of industrial workers. The first
to sixth reports, covering the sessions of 1949-1961, were issued during 1950-1962.

672 FOOD AND NUTRITION PROCEDURES IN TIME OF DISASTER. By G.
     B. Masefield. 96 pp. (FAO Nutritional Studies, 21)
     1968, 2nd printing, FAO $2.00
     Emphasizes the practical aspects of food management in times of
     disaster, including appraisal of existing stores, rationing, methods
     of distribution, price control, and first aid for nutritional relief.
     Considers primarily food supply in the event of earthquakes, hur-
     ricanes, volcanic eruptions, floods, crop failures due to drought, or
     interruption of food supplies in wartime. Deals with short-term,
     medium-term, and long-term emergencies, including the strategy of
     relief in each instance. First issued in 1967.

673 "Qualitative and Quantitative Aspects of Nutrition." By A.A. Pokrov-
     sky. Impact of Science on Society, vol. 20, no. 3, 1970, pp. 219-234.
     1970, UNESCO $1.00
     Examines in detail the problem of quality in the human diet, the de-
     velopment of the balanced-nutrition concept, the detrimental effects
     upon quality of insecticides and injurious chemical additives and
     excess of minerals derived from synthetic fertilizers; increase in
     the world protein supply; and quality problems in developing coun-
     tries.

674 METABOLIC ADAPTATION AND NUTRITION. Proceedings of the
     Special Session held during the Ninth Meeting of the PAHO Advisory
     Committee on Medical Research, 16 June 1970. 145 pp., fig. (Scien-
     tific Publications, 222)
     1971, PAHO $2.50
     Opening statements, papers (11), discussions, and summaries ex-
     plore the regulation of cell function in relation to nutrition.

Africa

675 NUTRITION ET ALIMENTATION TROPICALES. 3 vol. (Réunions de la
     FAO sur la Nutrition, Rapport no. 20)
     1957, FAO   In French only   Out of print
     Report of the lectures on nutrition for tropical Africa, organized by
     France, FAO, and WHO.

676 Report of the FAO/WHO Seminar on PROBLEMS OF FOOD AND NU-
     TRITION IN AFRICA SOUTH OF THE SAHARA, Lwiro, Bukavu (Congo),
     18-29 May 1959. 92 pp. (FAO Nutrition Meetings Report Series, 25)
     1961, FAO   Out of print

677 HUMAN NUTRITION IN TROPICAL AFRICA; a Textbook for Health
     Workers with special reference to Community Health Problems in East
     Africa. By Michael C. Latham. 268 pp., illus.
     1968, 2nd printing, FAO $2.50
     Sections cover public health aspects of nutrition (including food
     habits, assessment of nutritional status, and maternal and infant nu-
     trition); basic nutrition; disorders of malnutrition; diets and home
     preservation of food. Published under the auspices of FAO,
     UNICEF, and WHO. First issued in 1965.

Asia and the Far East

678 NUTRITION PROBLEMS OF RICE-EATING COUNTRIES IN ASIA. Re-
     port of the Nutrition Committee, Baguio, Philippines, February 1948.
     24 pp. (FAO Nutrition Meetings Report Series, 2)
     1948, FAO   Out of print

679   Report of the NUTRITION COMMITTEE FOR SOUTH AND EAST ASIA,
        Fourth Session, Tokyo, 25 September-2 October 1956. 50 pp. (FAO
        Nutrition Meetings Report Series, 14)
        1957, FAO $0.50
            Reports of the second and third meetings, 1950, 1953 (no. 3 and 6 in
            this series), are out of print.

Europe

680   NUTRITION WORK IN GREECE. By Andromache G. Tsongas. 67 pp.,
        illus. (FAO Nutritional Studies, 7)
        1951, FAO   Out of print
            Describes FAO technical assistance work in Greece during 1947-
            1950 aiding the government to initiate and develop a national nutri-
            tion program.

681   Report of the FAO NUTRITION MEETING FOR EUROPE, Rome, 23-28
        June 1958. 28 pp. (FAO Nutrition Meetings Report Series, 21)
        1958, FAO $0.50
            Report and recommendations concerning the study of food consump-
            tion with special reference to fat consumption, including the estab-
            lishment of food composition tables for fatty acids and cooperation
            with WHO in further studies; and education and training in nutrition.

Latin America

682   NUTRITION CONFERENCE, Montevideo, July 1948.... 316 pp. (FAO
        Nutrition Meetings Report Series, 1)
        1950, FAO   Bilingual: E/S   Out of print

683   Report of the Second Conference on NUTRITION PROBLEMS IN LATIN
        AMERICA, Rio de Janeiro, June 1950. 34 pp. (FAO Nutrition Meetings
        Report Series, 4)
        1950, FAO   Out of print

684   Report of the Third Conference on NUTRITION PROBLEMS IN LATIN
        AMERICA, Caracas, 19-28 October 1953. 60 pp. (FAO Nutrition
        Meetings Report Series, 8)
        1954, FAO   Out of print

685   Report of the Fourth Conference on NUTRITION PROBLEMS IN LATIN
        AMERICA, in Guatemala City, from 23 September to 1 October 1957.
        81 pp. (FAO Nutrition Meetings Report Series, 18)
        1959, FAO $1.00
            Reviews the development and utilization of food resources, espe-
            cially protein-rich foods; education in nutrition; national nutrition
            policy; evidence of the prevalence of deficiency diseases and malnu-
            trition in Latin America.

686   PUBLICACIONES CIENTÍFICAS DEL INSTITUTO DE NUTRICIÓN DE
        CENTRO AMÉRICA Y PANAMÁ. 5a Recopilación. 320 pp. (Publi-
        caciones Científicas, 136)
        1966, PAHO   In Spanish only $1.00
            Fifth compilation of INCAP papers on nutrition, including clinical
            studies, relationship of nutrition and various diseases, local food-
            stuffs, and diet, etc. The first to third compilations were published
            as Suplementos No. 1, 2, 3, respectively, to the Boletín de la Oficina
            Sanitaria Panamericana, and the fourth compilation was No. 59 of
            Publicaciones Científicas.

687   Report of the LATIN AMERICAN SEMINAR ON THE PLANNING AND
       EVALUATION OF APPLIED NUTRITION PROGRAMS, Popayan,
       Colombia, 10-17 November 1966. 72 pp. (Scientific Publications, 160)
       1967, PAHO $0.50
         Examines present procedures for planning and evaluation of applied
         nutrition programs in Latin America.

688   MATERNAL NUTRITION AND FAMILY PLANNING IN THE AMERICAS.
       Report of a PAHO Technical Group Meeting, Washington, D.C., 20-24
       October 1969. 47 pp. (Scientific Publications, 204)
       1970, PAHO $1.00
         General background; situation in the Americas; nutrition of pregnant
         and lactating women; ultimate goals and practical aims; local health
         activities in relation to maternal nutrition and family planning; con-
         clusions. Appendix: "Report of Current Studies of Maternal Nutri-
         tion Status in the U.S.A." By Edwin M. Gold.

688   GUIDELINES TO YOUNG CHILD FEEDING IN THE CONTEMPORARY
bis   CARIBBEAN. Report of a Meeting of the Caribbean Food and Nutrition
       Institute, Mona, Jamaica, 15-19 June 1970. 16 pp. (Scientific Publica-
       tions, 217)
       1970, PAHO $0.50
         Concerns maternal nutrition, preparation for breast feeding, nutri-
         tion of the newborn (establishment and maintenance of breast
         feeding), artificial feeding (including vitamin and mineral supple-
         ments), and weaning and transition to family diet.

Near and Middle East

689   Report of the NUTRITION COMMITTEE FOR THE MIDDLE EAST,
       First Session, Cairo, 18-26 November 1958. 55 pp. (FAO Nutrition
       Meetings Report Series, 24)
       1959, FAO $0.50
         Reviews regional nutrition problems, education and training in nu-
         trition, development of national nutrition services, and formulation
         of policy. Seminar was sponsored jointly by FAO and WHO.

SURVEYS

690   DIETARY SURVEYS: Their Technique and Interpretation. By Thelma
       Norris. 108 pp. (FAO Nutritional Studies, 4)
       1949, FAO   Out of print
         Designed to assist workers undertaking dietary studies.

691   Manual on HOUSEHOLD FOOD CONSUMPTION SURVEYS. By Emma
       Reh. 96 pp. (FAO Nutritional Studies, 18)
       1967, 2nd printing, FAO $1.00
         Practical handbook for personnel undertaking field surveys. Shows
         how such programs are made in actual practice and what they in-
         volve. The work is based largely on experience in Latin America,
         but it may also be applied to rural societies in many other develop-
         ing countries. First issued in 1962.

692   EXPERT COMMITTEE ON MEDICAL ASSESSMENT OF NUTRITIONAL
       STATUS. Report. 67 pp. (WHO Technical Report Series, 258)
       1963, WHO $1.00
         Deals mainly with clinical and biological methods of assessing nu-

tritional status but discusses also other approaches (social, economic, statistical, and anthropometric). Gives special attention to
the role of surveys and mentions the need to correlate nutrition
with communicable and degenerative diseases.

693   The ASSESSMENT OF THE NUTRITIONAL STATUS OF THE COM-
MUNITY. By D.B. Jelliffe. 271 pp., illus. (WHO Monograph Series,
53)
1966, WHO $6.00 cl.
Presents methods for nutritional assessment of entire communities
and determination of the magnitude and geographical distribution of
malnutrition. Methods comprise direct assessment of human groups
by clinical examination, anthropometric measurement, biochemical
and biophysical tests, indirect assessment by health statistics and
ecological factors. Describes procedures used in assessment of
ecological factors and procedures used in nutrition surveys. Examines special problems encountered in various groups of people.

FOOD COMPOSITION REQUIREMENTS

694   CALCIUM REQUIREMENTS. Report of an FAO/WHO Expert Group.
54 pp. (FAO Nutrition Meetings Report Series, 30)
1968, 2nd printing, FAO $0.50
————. 1962, WHO. (WHO Technical Report Series, 230) $0.60
Surveys intake and dietary sources of calcium in a number of
countries of different dietary patterns, and reviews evidence
showing the effect on health of varying intakes of calcium. Deals
with methods of estimating calcium requirements, and sets forth
a range of suggested practical allowances. Originally issued in
1962.

695   CALORIE REQUIREMENTS. Report of the Second Committee on
Calorie Requirements. 68 pp. (FAO Nutritional Studies, 15)
1968, 4th printing, FAO $0.75
Includes such subjects as relationship between activity, body size,
climate, etc., and calorie requirements; the contribution of alcohol
to calorie intake; requirements of children and the effect of aging on
adult requirements; and body fat content and its significance. Revised and extended version of the 1949 Calorie Committee's report
(FAO Nutritional Studies, 5). First issued in 1957.

696   HUMAN PROTEIN REQUIREMENTS AND THEIR FULFILMENT IN
PRACTICE. Proceedings of a Conference in Princeton, United States
(1955), sponsored jointly by FAO, WHO, Josiah Macy Jr. Foundation,
New York. Ed. by J.C. Waterlow and Joan M.L. Stephen. xi, 193 pp.,
illus. (FAO Nutrition Meetings Report Series, 12)
1960, 2nd printing, FAO   Out of print
Discussions on contributions from experimental work on man and
animals, human requirements at different ages, practical measures
to increase protein intakes. First issued in 1957.

697   PROTEIN REQUIREMENTS. Report of a Joint FAO/WHO Expert
Group. 71 pp. (FAO Nutrition Meetings Reports, 37)
1965, FAO $1.25
————. WHO. (WHO Technical Report Series, 301)   Out of print
Defines protein requirements of children and adults, using the

factorial method, in which all components of the requirements at each age are estimated separately for each physiological state. Discusses fully other factors affecting protein requirements, amino acid patterns, and protein values of foods, and suggests practical allowances to meet the various contingencies likely to arise in normal life.

698 REQUIREMENTS OF VITAMIN A, THIAMINE, RIBOFLAVINE AND NIACIN. Report of a Joint FAO/WHO Expert Group. 86 pp. (FAO Nutrition Meetings Report Series, 41)
1967, FAO $1.00
――――. WHO. (WHO Technical Report Series, 362) $1.00
Recommends intake levels that will ensure the health of the large majority of groups of individuals or populations, rather than suggesting figures for individuals. Gives attention also to differing requirements of adults and infants, factors that affect requirements (such as climate, physical activity, and pathological states), and factors that affect the vitamin content of foods (e.g., processing, preservation, and cooking).

699 REQUIREMENTS OF ASCORBIC ACID, VITAMIN D, VITAMIN B_{12}, FOLATE, AND IRON, Report of a Joint FAO/WHO Expert Group. 75 pp., fig. (FAO Nutrition Meetings Report Series, 47)
1970, FAO $1.25
――――. WHO. (WHO Technical Report Series, 452) $1.25
Reviews nutritional requirements for certain essential nutrients, deficiencies of which still present important public health problems in different parts of the world, recommends intakes necessary for maintenance of health, and makes suggestions for further research.

PROTEIN SOURCES

GENERAL

700 Feeding the Expanding World Population: INTERNATIONAL ACTION TO AVERT THE IMPENDING PROTEIN CRISIS. Report to the Economic and Social Council of the Advisory Committee on the Application of Science and Technology to Development. 106 pp. and corrigendum sheet (E/4343/Rev.1 and Corr.1)
1968, UN   Sales No. E.68.XIII.2 $1.50
Recommends a program of action for closing the gap between world protein needs and protein supplies and preventing even more widespread protein deficiency in future generations.

701 "Food Supply and New Protein Sources." By D.M. DeMaeyer. WHO Chronicle, vol. 22, no. 6, June 1968, pp. 225-325, illus.
1968, WHO $0.60
Examines conventional and unconventional sources of protein, including among the latter oleaginous seeds, soybeans, cottonseed, peanuts, fish protein concentrate, algae, yeast, and unicellular organisms; amino acid enrichment of food; and the activities of WHO and other international organizations in nutrition.

702   LIVES IN PERIL: PROTEIN AND THE CHILD. 52 pp., illus. (World
Food Problems, 12)
1970, FAO on behalf of the FAO/WHO/UNICEF $1.00
Protein Advisory Group
Surveys the problem of protein malnutrition as it affects children,
the most vulnerable group in the population. Describes the nutritional needs of the pregnant woman, the infant, and the preschool
child and the consequences to them of malnutrition. Describes the
efforts to overcome dietary deficiencies in developing areas through
feeding programs, especially through the use of protein-rich food
mixtures, through education in the basic nutrition needs of children,
and through efforts to make better use of local food resources. Outlines the international role of the Protein Advisory Group (PAG).

703   Strategy Statement on ACTION TO AVERT THE PROTEIN CRISIS IN
THE DEVELOPING COUNTRIES. Report of the Panel of Experts on the
Protein Problem confronting Developing Countries, United Nations
Headquarters, 3-7 May 1971. vii, 27 pp. (E/5018/Rev.1-ST/ECA/144)
1971, UN   Sales No. E.71.II.A.17 $1.00
The report indicates the substantive, institutional, and financial
steps which must be undertaken to avert the protein crisis in the de-
veloping countries.

CEREAL GRAINS

704   MAIZE AND MAIZE DIETS; a Nutritional Survey. 94 pp., illus. (FAO
Nutritional Studies, 9)
1968, 3rd printing, FAO $1.00
Covers the nutrient content of maize; the effect of household preparation, processing, and storage upon its nutritive value; nutrition of
maize-eating populations and its improvement; maize and pellagra;
improvement of the nutritional value of maize. Extensive references. First issued in 1953.

705   RICE AND RICE DIETS; a Nutritional Survey. Rev. ed. 78 pp. (FAO
Nutritional Studies, 1)
1965, 2nd printing, FAO $0.75
Deals primarily with the nutritive value of rice and rice diets and
with ways and means of improving nutrition in rice-eating areas.
References. The revised edition was first issued in 1954; the first
edition was issued in 1948 and reprinted in 1951 and 1952.

706   RICE ENRICHMENT IN THE PHILIPPINES. By W.C. Aalsmeer and
others. 109 pp. (FAO Nutritional Studies, 12)
1954, FAO   Out of print
Report of a 1952 survey of results of the experimental introduction
into the Philippines of artificially enriched rice.

707   RICE—Grain of Life. 93 pp., illus. (World Food Problems, 6)
1968, 2nd printing, FAO $1.50
Describes the main features of the world rice economy, as well as
the situation and outlook in 1966 (International Rice Year), including
processing and storage and economic and institutional factors. Reprint of a special chapter in The State of Food and Agriculture 1966
(entry 1174). First issued in 1966.

708   WHEAT IN HUMAN NUTRITION. By W.R. Aykroyd and Joyce Doughty.
       163 pp., illus. (FAO Nutritional Studies, 23)
       1970, FAO $3.50
       Covers the history of wheat, its nutritive value, and a survey of
       wheat production; forms in which wheat is eaten; consumption; mil-
       ling and nutrition; enrichment; wheat, health, and disease; trends in
       production and consumption; wheat as a food for young children; en-
       couraging wheat consumption; extraction levels and enrichment.
       Bibliography.

709   AMINO-ACID CONTENT OF FOODS AND BIOLOGICAL DATA ON
       PROTEINS.... By the Food Policy and Food Science Service, Nutrition
       Division, FAO. x, 285 pp. (FAO Nutritional Studies, 24)
       1970, FAO   Trilingual: E/F/S $12.00
       Table of data on the amino acid content of foods and the biological
       measures of protein quality, resulting from a complete review of
       all available data by Dr. M. Cresta and Dr. W.A. Odentaal. Table is
       in two parts: (1) amino acids: the level of amino acids in food and
       the chemical scores being presented in two sections; and (2) biologi-
       cal data, the data relating to biological value, digestibility, net pro-
       tein utilization, and protein efficiency ratio being presented in three
       sections. Extensive bibliography.

710   IMPROVING PLANT PROTEIN BY NUCLEAR TECHNIQUES. Pro-
       ceedings of a Symposium on Plant Protein Resources: Their Improve-
       ment through the Application of Nuclear Techniques, jointly organized
       by the IAEA and FAO and held in Vienna, 8-12 June 1970. 458 pp.
       (Proceedings Series; STI/PUB/258)
       1970, IAEA   Composite: E/F/R/S $12.00
       Papers (44) on developing and utilizing plant protein resources, the
       role of induced mutations in breeding for improved protein in cereal
       and noncereal crops, and the use of nuclear techniques.

FRUITS AND VEGETABLES

711   LEGUMES IN HUMAN NUTRITION. By W.R. Aykroyd and Joyce
       Doughty. 138 pp. (FAO Nutritional Studies, 19)
       1966, 2nd printing, FAO $3.00
       Gives an account of the grain legumes and their contribution to
       human diets and nutrition, based on the available technical litera-
       ture. Deals also with measures to increase the production and con-
       sumption of this group of foods. Extensive references. First issued
       in 1964.

712   FRUITS AND VEGETABLES IN WEST AFRICA. By H.D. Tindall, with
       a chapter by Florence A. Sai. 259 pp., illus.
       1965, FAO $2.50
       Manual on the cultivation of fruits and vegetables, school gardening,
       soil conservation, and extension techniques, with a chapter on the
       nutritional value of West African fruits and vegetables. A joint pub-
       lication of FAO/WHO/UNICEF.

713   GRAIN LEGUMES IN AFRICA. By W.R. Stanton, with chapters by Joyce
       Doughty and R. Orraca-Tetteh, and W. Steele. 183 pp., illus.
       1966, FAO $1.50
       Manual on the classification and cultivation of grain legumes with
       chapters on improved varieties and their nutritional value. A joint
       publication of FAO/WHO/UNICEF.

MILK AND MILK PRODUCTS

714   MILK AND MILK PRODUCTS IN HUMAN NUTRITION. By S.K. Kon.
        76 pp. and corrigendum sheet. (FAO Nutritional Studies, 17)
        1959, FAO   Out of print
        Handbook on processing and use of milk in different parts of the
        world, including an outline of measures being taken to supply milk to
        those in need of it.

715   THE ECONOMICS OF FILLED MILK; a Case Study. 33 pp. (Commod-
        ity Bulletin Series, 35)
        1962, FAO $0.50
        Study of the role of filled milk, a product made from skim milk with
        vegetable fats or oils substituted for animal fats, in the Philippines,
        where it now meets over half of the market demand for milk.

FISH

716   FISH: THE GREAT POTENTIAL FOOD SUPPLY. By D.B. Finn. 47
        pp., illus. (World Food Problems, 3)
        1960, FAO   Out of print
        Reviews activities intended to increase the world catch of fish and
        the problems which hinder and hamper fish production.

717   FISH IN NUTRITION. Ed. by Eirik Heen and Rudolf Kreuzer. 464 pp.,
        illus.
        1962, FAO/Fishing News (Books) Ltd. (London) $18.00
        Technical papers (70) of the International Conference on Fish in
        Nutrition, Washington, D.C., 19-27 September 1961, organized by
        FAO. They cover the role of fish in world nutrition; chemical com-
        ponents of fish and their changes under treatment; contribution of
        fish and fish products to national diets; fish and fishery products in
        animal nutrition; demand for fish as human food and possibilities
        for increased consumption.

718   ENCOURAGING THE USE OF PROTEIN-RICH FOODS. By John
        Fridthjof. 103 pp., illus.
        1967, 3rd printing, FAO $1.00
        Serves as guide for persons participating in campaigns to introduce
        protein-rich foodstuffs as a valuable means of fighting malnutrition
        and improving nutritional standards, especially in the less developed
        regions. Based in large part on the author's experience in cam-
        paigns to promote fish consumption in Brazil, Chile, Mexico, Mo-
        rocco, and Yugoslavia. First issued in 1962.

719   PRODUCTION OF FISH-PROTEIN CONCENTRATE. Report and Pro-
        ceedings of the Joint UNIDO/FAO Expert Group Meeting, Rabat, Mo-
        rocco, 8-12 December 1969. 2 vols. (ID/60-ID/WG.48/17/Rev.1)
        1971, UN
        Vol. 1: Part I: Report of the Meeting. 38 pp.
        Sales No. E.71.II.B.7 $0.75
        Includes general recommendations on FPC, recommendations for
        the SONAFAP plant in Agadir, Morocco, a summary of discus-
        sions, and statements describing the situation in Morocco and ef-
        forts made elsewhere to produce FPC.
        Vol. 2: Proceedings. In preparation.

NUTRITION DEFICIENCY

GENERAL

720   PREVENTION AND TREATMENT OF SEVERE MALNUTRITION IN
        TIMES OF DISASTER. Report approved by the Joint FAO/WHO Com-
        mittee on Nutrition.... 56 pp. (WHO Technical Report Series, 45)
        1951, WHO   Out of print
        Deals with food management; physiological, clinical, and therapeutic
        aspects of severe malnutrition and starvation; and organizational
        aspects of general relief activities.

721   PROTEIN MALNUTRITION. Proceedings of a Conference in Jamaica
        (1953) sponsored jointly by FAO, WHO and the Josiah Macy Jr. Foun-
        dation, New York. Ed. by J.C. Waterlow. xvi, 227 pp., 24 plates.
        ([FAO Nutrition Meetings Report Series, 10])
        1955, FAO, printed by Cambridge U.P.   Out of print
        Discussions deal with biological aspects, pathology, clinical aspects
        and treatment, and epidemiology and prevention of protein malnutri-
        tion.

722   PROTEIN MALNUTRITION IN BRAZIL. By J.C. Waterlow and A. Ver-
        gara. 40 pp. (FAO Nutritional Studies, 14)
        1956, FAO   Out of print
        A 1953 study of protein malnutrition in five states of Brazil.

723   RADIOISOTOPES IN TROPICAL MEDICINE. Proceedings of the Sym-
        posium on the Use of Radioisotopes in the Study of Endemic and Tropi-
        cal Diseases jointly organized by IAEA and WHO and held in Bangkok,
        12-16 December 1960. 379 pp., illus. (Proceedings Series; STI/PUB/
        31)
        1962, IAEA $7.00 cl.
        Papers (20) on nutrition, protein metabolism and deficiencies, tropi-
        cal sprue, hematological problems, iron metabolism, blood loss
        caused by parasites, hemolytic anemia, endemic goiter, water and
        electrolytic balance, entomological problems, insect biochemistry,
        parasitology, helminth life cycles, protozoa.

724   MALNUTRITION AND DISEASE. 47 pp., illus. (Freedom from Hunger
        Campaign. Basic Studies, 12)
        1963, WHO   Out of print
        Describes advances in the knowledge of malnutrition; discusses its
        relation to child health, to infections, and to infestations; and re-
        views deficiency diseases.

725   NUTRITION AND INFECTION. Report of a WHO Expert Committee.
        30 pp. (WHO Technical Report Series, 314)
        1965, WHO $0.60
        Reviews the evidence on relationship between malnutrition and in-
        fectious deseases, concluding that malnutrition is synergistic with
        some infections but sometimes antagonistic to others. Recommends
        the direction further studies should take.

726   ''Protein-Calorie Malnutrition and Psychobiological Development in Children.'' By Joaquín Cravioto. Boletín de la Oficina Sanitaria Panamericana, English edition, Selections from 1966, pp. 34-52.
1967, PAHO
Studies show that children surviving acute protein malnutrition show retardment of growth, physiological disorders, and retardment in some aspects of their biochemical maturation. First published in Spanish in Boletín de la Oficina Sanitaria Panamericana, vol. 61, no. 4, octubre de 1966, pp. 285-306.

727   BIOMEDICAL CHALLENGES PRESENTED BY THE AMERICAN INDIAN. 191 pp. (Scientific Publications, 165)
1968, PAHO $2.50
Papers and discussions of the special session of the seventh meeting of the PAHO Advisory Committee on Medical Research. Contains chapters on origins and dispersal of Indians in North and South America; biological subdivisions of Indians on the basis of physical anthropology and of genetic traits; the American Indian in the International Biological Program; survey of the unacculturated Indians of Central and South America; medical problems of newly contacted Indian groups, including special studies on gallbladder disease, hyperglycemia, malaria, food and nutrition, iodine deficiency without goiter, and endemic goiter. Detailed tabulations of phenotype and gene frequencies for 11 different genetic systems studied in the American Indian are presented in an appendix.

728   ''The Influence of Nutrition and Environment on Brain Function.'' IBRO Bulletin, vol. 7, no. 3, August 1968, pp. 69-74.
1968, UNESCO   Out of print
Summary, prepared by C.A. Canosa, of a round table of the Symposium on Brain Research and Human Behaviour organized by UNESCO and the International Brain Research Organization (IBRO), Paris, 11-15 March 1968. There is excellent evidence that in infants and preschool children, irreversible damage to physical growth is produced by protein-calorie malnutrition (PCM); preliminary evidence in man suggests that malnutrition during the early phases of central nervous system development may impair mental performance.

729   INFANT NUTRITION IN THE SUBTROPICS AND TROPICS. By D.B. Jelliffe. 2nd ed. 336 pp., (WHO Monograph Series, 29)
1968, WHO $9.00
Gives a comprehensive picture of feeding practices in different regions of the world, of infant feeding problems, and of the diseases that result. Discusses in detail approaches to improving infant nutrition in the developing countries and to overcoming the effects of malnutrition. First edition issued in 1955.

730   ''A Maternal and Child Nutrition Program in Costa Rica.'' By Carlos Díaz Amador. Boletín de la Oficina Sanitaria Panamericana, English edition, Selections from 1968, pp. 30-34.
1969, PAHO
Describes the results of a program to bring about the recovery of severely undernourished children and to help keep them well fed by providing health, nutritional, and social education to the family. First published in Spanish in Boletín de la Oficina Sanitaria Panamericana, vol. 64, no. 6, junio de 1968, pp. 471-476.

731 PERINATAL FACTORS AFFECTING HUMAN DEVELOPMENT. Proceedings of the Special Session held during the Eighth Meeting of the PAHO Advisory Committee on Medical Research, 10 June 1969. 253 pp., illus. (Scientific Publications, 185)
1969, PAHO $3.50
Papers (30) on factors affecting intrauterine development of the fetus (including maternal nutritional deficiencies), acute effects of labor on the fetus, and effects of intrapartum fetal asphyxia and acidosis and methods of treatment.

732 INTERACTIONS OF NUTRITION AND INFECTION. By Nevin S. Scrimshaw, Carl E. Taylor and John E. Gordon. 329 pp. (WHO Monograph Series, 57)
1968, WHO $9.00 cl.
This study is based on an extensive review of the literature (listing over 1,500 references), on personal field studies of the authors, and on observations and ideas from many parts of the world.

733 CONQUEST OF DEFICIENCY DISEASES: Achievements and Prospects. By W.R. Aykroyd. 98 pp., illus. (Freedom from Hunger Campaign. Basic Studies, 24)
1970, FAO/WHO $3.00
Surveys general advances in knowledge concerning deficiency diseases and their total impact upon human health, and assesses the possibility of eliminating them or of greatly reducing their prevalence. Chapters on beriberi, pellagra, rickets and osteomalacia, scurvy, protein-calorie malnutrition, vitamin-A deficiency, endemic goiter, nutritional anemias, and other deficiency diseases.

734 "Assessment of Some Biochemical Parameters related to Protein-Calorie Nutrition in Children." By W.K. Simmons and M. Bohdal. Bulletin of the World Health Organization, vol. 42, no. 6, 1970, pp. 897-906.
1970, WHO $3.25
Results of a cross-sectional nutrition survey of children aged 1-15 years in five different rural areas of Kenya compared with those from a dietary and anthropometric survey, in order to determine the most useful biochemical tests to establish differences between communities associated with varying adequacy of protein-calorie intake.

NUTRITIONAL ANEMIAS

735 IRON DEFICIENCY ANAEMIA. Report of a Study Group. 15 pp. (WHO Technical Report Series, 182)
1959, WHO   Out of print
Discusses present status of the problem from the point of view of etiology, evaluation, and prevention; recommendations.

736 IRON METABOLISM AND ANEMIA. Proceedings of a Symposium held during the Eighth Meeting of the PAHO Advisory Committee on Medical Research, 9 June 1969. 89 pp., illus. (Scientific Publications, 184)
1969, PAHO $1.50
Papers (10) on iron metabolism, its biochemistry, absorption of iron, epidemiological and therapeutic aspects of iron-deficiency anemia, including the effect of protein depletion, question of iron requirements, how anemia affects pregnancy and infancy, and the relationship of anemia and hookworm disease.

737   "Anaemias, 1926-66." World Health Statistics Report, vol. 22, no. 7, 1969, pp. 409-427.
    1969, WHO   Bilingual: E/F $2.75

ENDEMIC GOITER

738   "Control of Endemic Goitre." Bulletin of the World Health Organization, vol. 9, no. 2, 1953, pp. 171-309.
    1953, WHO   Composite: E/F $3.25
    Articles (11) on physiology, iodization, and prevention and prophylaxis; and final report of a Study Group on Endemic Goitre, London, 8-12 December 1952.

739   ENDEMIC GOITRE. By F.W. Clements and others. 471 pp., illus. (WHO Monograph Series, 44)
    1960, WHO   Out of print
    This monograph (with contributions by 18 authors from ten countries) provides a comprehensive review of the subject. Includes chapters on the history, prevalence, and geographical distribution of endemic goiter; health significance, physiology, pathological anatomy, and etiology of the disease; experimental studies on goiter; technique of endemic goiter surveys; therapy and prophylaxis of endemic goiter; iodization of salt; principles and practice of endemic goiter control; and legislation on iodine prophylaxis.

740   ENDEMIC GOITRE. Legislation on Iodine Prophylaxis. 14 pp.
    1960, WHO $0.30
    A comparative study of laws providing for the iodization of salt in various countries. Discusses also labeling and packing requirements, as well as provisions for controlling standards of iodized salt existing in some countries. Reprint from International Digest of Health Legislation, vol. 11, no. 3, 1960, pp. 387-398.

741   ENDEMIC GOITER. Report of the Meeting of the PAHO Scientific Group on Research on Endemic Goiter held in Puebla, Mexico, 27 to 29 June 1968. Comp. and ed. by John B. Stanbury. xv, 447 pp. (Scientific Publications, 193)
    1969, PAHO $3.50
    Papers (34) on endemic goiter and cretinism—general aspects; endemic goiter in the Congo, New Guinea, Argentina, Paraguay, Brazil, Chile, Colombia, Ecuador, Mexico, and Peru.

KWASHIORKOR

742   KWASHIORKOR IN AFRICA. By J.F. Brock and M. Autret. 78 pp., illus. (WHO Monograph Series, 8)
    1952, WHO   Out of print
    Definition and description of individual clinical characters; general clinical picture; nutritional cirrhosis and primary carcinoma of the liver; frequency of the syndrome; treatment; kwashiorkor and diet; etiology; prevention: increasing the supply of proteins, supplementary feeding of infants and young children, education, immediate action. Originally published in English in Bulletin of the World Health Organization, vol. 5, no. 1, pp. 1-71, 1952—still available at $3.25.
    ————. FAO. (FAO Nutritional Studies, 8)   Out of print

743   SÍNDROME POLICARENCIAL INFANTIL (KWASHIORKOR) AND ITS
       PREVENTION IN CENTRAL AMERICA. By Marcel Autret and Moise
       Behar. 81 pp., illus. (FAO Nutritional Studies, 13)
       1954, FAO   Out of print

FEEDING PROGRAMS

INDUSTRIAL FEEDING

744   FOOD AND CATERING ON BOARD SHIP. 37 pp. (International Labour
       Conference, 28th (maritime) session, 1946. Report IV)
       1946, ILO $0.20

745   WELFARE FACILITIES FOR WORKERS. 39 pp. (International Labour
       Conference, 39th session, 1956. Report V (1))
       1955, ILO $0.40
       Extracts from the report of the Committee on Welfare Facilities,
       1955, concerning welfare facilities for workers, including feeding
       facilities; and proposed recommendation.

746   WELFARE FACILITIES FOR WORKERS. 73 pp. (International Labour
       Conference, 39th session, 1956. Report V (2))
       1956, ILO $0.50
       Summarizes replies from governments concerning the proposed rec-
       ommendation concerning welfare facilities for workers, including
       feeding facilities.

747   Report of the Joint Symposium on INDUSTRIAL FEEDING AND CAN-
       TEEN MANAGEMENT IN EUROPE, Rome, Italy, 2-7 September 1963.
       Published under the auspices of FAO, ILO, WHO. 44 pp. (FAO Nutri-
       tional Meetings Report Series, 36)
       1965, FAO $0.50
       Reviews the current situation of industrial feeding in Europe and
       suggests some steps to be taken if the specialized agencies are to
       assist member countries in the solution of their more urgent prob-
       lems.

748   ...Summary of Reports on Selected Recommendations (Article 19 of the
       Constitution). HEALTH, WELFARE AND HOUSING OF WORKERS. 181
       pp. (International Labour Conference, 54th session, 1970. Report III
       (Part 2))
       1970, ILO $2.50
       Summarizes reports by governments of member states of ILO on
       recommendations on the health, welfare, and housing of workers, in-
       cluding Welfare Facilities Recommendation, 1956 (No. 102), which
       relates in part to feeding facilities.

SCHOOL FEEDING

749   SCHOOL FEEDING: Its Contribution to Child Nutrition. By Marjorie L.
       Scott. 129 pp., illus. (FAO Nutritional Studies, 10)
       1953, FAO   Out of print
       Deals with the historical development of school feeding; meal pat-

terns and special food supplements; effects on health; administration and organization; educational and social aspects; relationship to food supply and distribution; relief feeding; and the preschool child.

750   Report of the FAO/UNICEF Regional SCHOOL FEEDING SEMINAR FOR ASIA AND THE FAR EAST, Tokyo, 10-19 November 1958. 52 pp. (FAO Nutrition Meetings Report Series, 22)
1959, FAO   Out of print

751   Report of the Regional Seminar on SCHOOL FEEDING IN SOUTH AMERICA, Bogotá (Colombia), 27 October-8 November 1958. Sponsored by FAO and UNICEF in co-operation with the Government of Colombia. 60 pp. (FAO Nutrition Meetings Report Series, 23)
1959, FAO   Out of print

752   FOOD AID AND EDUCATION. Prepared by The World Food Program on behalf of the Secretary-General of the United Nations and the Director-General of FAO. 22 pp. (World Food Program Studies, 6; P/WFP:S6)
1965, UN/FAO $1.00
Analyzes costs and benefits of school feeding programs in primary schools and of food aid to boarding facilities for secondary and higher-level institutions, adult literacy and vocational training programs, and to persons participating in community development projects.

NUTRITION EDUCATION AND TRAINING

753   Report of an International Seminar on EDUCATION IN HEALTH AND NUTRITION, Baguio, Philippines, 13 October-3 November 1955. By F.W. Clements. 92 pp. (FAO Nutrition Meetings Report Series, 13)
1958, 2nd printing, FAO $1.00
First issued in 1956. Seminar jointly sponsored by FAO and WHO.

754   Report of the Technical Meeting on HOME ECONOMICS FOR SOUTH AND EAST ASIA, Tokyo, Japan, 5-12 October 1956. 32 pp. (FAO Nutrition Meetings Report Series, 16)
1957, FAO $0.40

755   Report of the Symposium on EDUCATION AND TRAINING IN NUTRITION IN EUROPE, Bad Homburg, Germany, 2-11 December 1959. 56 pp. (FAO Nutrition Meetings Report Series, 26)
1960, FAO $0.50
Symposium sponsored by FAO and WHO.

756   Report of the European Seminar on EVALUATION OF HOME ECONOMICS EXTENSION PROGRAMS, Vienna, 9-21 May 1960. 50 pp. (FAO Nutrition Meetings Report Series, 27)
1960, FAO $0.50

757   EDUCATION AND TRAINING IN NUTRITION. 56 pp., illus. (Freedom from Hunger Campaign. Basic Studies, 6)
1967, 3rd printing, FAO $0.50
Notes the value of programs of education in nutrition in modifying

folk beliefs about food and in changing and improving food habits; discusses the organization and evaluation of such programs and the training of personnel. First issued in 1962.

758   NUTRITION IN RELATION TO AGRICULTURAL PRODUCTION. A Manual for Extension Supervisors in Agriculture, Home Economics, Rural Health, Community Development and for Workers in Other Fields Related to Nutrition in Nigeria and the Countries of West Africa. By I.S. Dema. 123 pp., illus.
        1965, FAO   Out of print
        Covers basic and applied nutrition; medical evidence of malnutrition in Nigeria; environmental and social factors related to malnutrition; food objectives in agricultural planning; teaching and extension of better nutrition; place of nutrition in integrated community development. Bibliography. Publication financed by UNICEF under a program jointly sponsored by FAO/WHO/UNICEF.

759   OUR FOODS: a Handbook for Educationists in Africa. By H. and M. Dupin. 110 pp., illus.
        1965, FAO $2.00
        Provides elementary information on food requirements, on the value of local foodstuffs, and on efforts that should be made to develop the food resources of the countries of West Africa. Translation (by G. Goodman) and adaptation of Nos aliments, by H. and M. Dupin, Paris, Editions sociales françaises, 1962. Publication financed by UNICEF under a project jointly sponsored by FAO/WHO/UNICEF.

760   CHILD CARE: A HANDBOOK FOR VILLAGE WORKERS AND LEADERS. By Mary Elizabeth Keister. 58 pp., illus.
        1968, 2nd printing, FAO $2.00
        Outlines in simple language the general principles underlying the care and upbringing of children.

761   PLANNING, BUILDING AND EQUIPPING HOME ECONOMICS CENTERS. By Elsa Haglund and Lars Erik Magnusson. 101 pp., drawings.
        1967, FAO $2.00
        Manual prepared by a home economist and an architect on planning and equipment of home economics education centers.

762   LEARNING BETTER NUTRITION: a Second Study of Approaches and Techniques. By J.A.S. Ritchie. 264 pp., illus. (FAO Nutritional Studies, 20)
        1969, 2nd printing, FAO $4.00
        Reexamines the techniques used in programs of education and training in nutrition within the context of economic and social change. Deals also with the need for such education, the background information on which it should be based, its planning, support, and evaluation, and the educational channels and methods which seem to show greatest promise. Contains an extensive bibliographical list and appendices. First issued in 1967. Replaces her Teaching Better Nutrition, 1950 (FAO Nutritional Studies, 6)

763   "Conference on the Training of Public Health Nutritionists-Dieticians. Final Report." Boletín de la Oficina Sanitaria Panamericana, English edition, Selections from 1967, pp. 22-32.
        1968, PAHO
        The report of a conference held in Caracas, 24-30 July 1966, under the auspices of PAHO and the Government of Venezuela, describes the role of the nutritionist-dietician and the training required to pre-

pare this type of professional worker. Makes recommendations on expanding and improving training programs. First published in Spanish in Boletín de la Oficina Sanitaria Panamericana, vol. 62, no. 4, abril de 1967, pp. 303-313.

764 VISUAL AIDS IN NUTRITION: a Guide to Their Preparation and Use. By Alan C. Holmes. 154 pp., illus.
1969, 2nd printing, FAO $3.00
Practical guide to the selection and preparation of audiovisual aids for use in nutrition programs. First issued in 1968.

765 "The Planning of Nutrition Education in School Programs." By Concha Barnoya de Asturias. Boletín de la Oficina Sanitaria Panamericana, English edition, Selections from 1968, pp. 57-64.
1969, PAHO
Explains the absence of nutrition programs in the schools of Central America and Panama as a vicious circle in which nutrition is not taught for want of teachers trained in the subject and the teachers are not trained because no program exists. It is up to educational planners to break this circle by preparing nutrition curricula relevant to the needs and resources of each country. First published in Spanish in Boletín de la Oficina Sanitaria Panamericana, vol. 65, no. 3, septiembre de 1968, pp. 187-196.

DEMOGRAPHY

GENERAL

765 MAIN TRENDS OF RESEARCH IN THE SOCIAL AND HUMAN SCI-
bis ENCES. Part One: Social Sciences. xlvii, 819 pp.
1970, Mouton/UNESCO (The Hague) $30.00 cl.
© UNESCO
In three sections: (1) main trends of research in the different social sciences, including a chapter on demography by Jean Bourgeois-Pichat; (2) interdisciplinary aspects of research; and (3) science policy and development or research in the social sciences.

UNITED NATIONS PROGRAMS

766 POPULATION COMMISSION. Report of the Fifteenth Session (3-14 November 1969). 67 pp. (Economic and Social Council. Official Records, 48th session, 1970, Supplement No. 3 (E/4768-E/CN.9/235))
1970, UN   No sales number $2.00
Fifteenth report to the Economic and Social Council. Reviews United Nations action in the fields of population since the previous session of the Commission; recommends future actions, including the five-year and the two-year programs of work. The reports of the first fourteen sessions cover the period 1947-1967.

767 STATISTICAL COMMISSION. Report of the Sixteenth Session (5-15 October 1970). 52 pp. (Economic and Social Council. Official Records, 50th session, 1971, Supplement No. 2 (E/4938-E/CN.3/417)).
1971, UN   No sales number $1.50
Sixteenth report to the Economic and Social Council. Covers Com-

mission recommendations on United Nations work programs and priorities in the field of statistics, including demographic and housing statistics. The reports of the first fifteen sessions cover the period 1947-1968.

EDUCATION AND TRAINING

768   DEMOGRAPHY. Ed. by D.V. Glass. A survey prepared under auspices of the International Union for the Scientific Study of Population. 200 pp. (The University Teaching of Social Sciences)
1957, UNESCO $2.00
Part 1: General survey; part 2: Area and national surveys covering 29 countries.

769   Report of the INTERREGIONAL WORKSHOP ON PROGRAMMES OF TRAINING IN THE FIELD OF POPULATION, Elsinore, Denmark, 19-30 June 1967. 38 pp. (ST/TAO/SER.C/98-E/CN.9/207-E/CN.9/CONF.4/1)
1967, UN

POPULATION STUDIES

770   WORLD POPULATION TRENDS, 1920-1947. 16 pp. (Population Studies, 3; ST/SOA/SER.A/3)
1949   Sales No. 49.XIII.3   Out of print
Estimates of population, birth and death rates, life expectancy and age structure of the population, for the world and its principal regions.

771   Proceedings of the WORLD POPULATION CONFERENCE, 1954, Rome, 31 August-10 September 1954. SUMMARY REPORT. 207 pp. (E/CONF.13/412)
1955, UN   Sales No. 55.XIII.8   Out of print
In three parts: (1) organization of the conference; (2) formal details of each of the 32 meetings; (3) summaries of the discussion at each of the meetings, prepared by a specially appointed rapporteur.

772   Proceedings of the WORLD POPULATION CONFERENCE, 1954. Rome, 31 August-10 September 1954. PAPERS.... 6 vol.
1955, UN   Sales No. 55.XIII.8, vol. 1-6   Out of print
Papers in original language; summaries in E and F
a. Vol. 1: Meetings 2, 4, 6, 8. 1,040 pp. (E/CONF.13/413)
b. Vol. 2: Meetings 10, 12, 17, 19, 11. 1,016 pp. (E/CONF.13/414)
c. Vol. 3: Meetings 13, 14, 16, 18, 29. 906 pp. (E/CONF.13/415)
d. Vol. 4: Meetings 3, 5, 9, 27. 1,073 pp. (E/CONF.13/146)
e. Vol. 5: Meetings 20, 22, 24, 26. 1,115 pp. (E/CONF.13/417)
f. Vol. 6: Meetings 7, 15, 21, 23, 25, 28. 1,047 pp. (E/CONF.13/418)
Subjects covered include: Mortality and fertility trends (vol. 1); international and internal migration, population distribution, legislation and administrative programs and services relevant to population (vol. 2); population trends and projections, aging, recruitment and training for demographic research and teaching (vol. 3); demographic statistics: evaluation, analytical techniques, concepts, and definitions (vol. 4); demographic aspects of economic development: population in relation to

nonbiological resources and to agriculture, capital formation, investment and employment, development planning (vol. 5); new census inquiries, problems and methods in demographic studies of preliterate peoples, design and control of demographic field studies, relation of population changes to the distribution of genetic factors, relations, between intelligence and fertility, social aspects of population changes (vol. 6).

773   "Population Studies." [119] pp. International Social Science Journal, vol. 17, no. 2, 1965, pp. 213-331.
1965, UNESCO $2.00
Articles on problems and methods in demography, population, and policy: Internal migration in Latin America; demography of the Australian Aborigines; estimating population size and growth from inadequate data; social strategies of family formation—some comparative data for Scandinavia, the British Isles, and North America. Economics of health in conditions of low population growth (Sierra Leone); migrations in Europe; the demographic variables in the assessment of educational needs. Select bibliography of works published between 1959 and 1964.

774   WORLD POPULATION: CHALLENGE TO DEVELOPMENT. Summary of the Highlights of the World Population Conference, Belgrade, Yugoslavia, 30 August to 10 September 1965.... 48 pp. (E/CONF.41/1)
1966, UN   Sales No. 66.XIII.4 $0.75

775   Proceedings of the WORLD POPULATION CONFERENCE, Belgrade, 30 August-10 September 1965. 4 vol. (E/CONF.41/2 to 5)
1966-67, UN   Sales No. 66.XIII.5,6,7,8   Out of print
a. Vol. 1: Summary Report. 349 pp. (E/CONF.41/2)
b. Vol. 2: Selected Papers and Summaries—Fertility, Family Planning, Mortality. 509 pp. (E/CONF.41/3)
c. Vol. 3: Selected Papers and Summaries—Projections, Measurement of Population. 435 pp. (E/CONF.41/4)
d. Vol. 4: Selected Papers and Summaries—Migration, Urbanization, Economic Development. 557 pp. (E/CONF.41/5)

Africa

776   "Recent Demographic Levels and Trends in Africa." Economic Bulletin for Africa, vol. 5 (E/CN.14/325), January 1965, pp. 30-79.
1965, UN   Sales No. 65.II.K.6   Out of print
Incorporates some recent data and outlines demographic levels and trends in Africa since 1945.

777   The POPULATION OF RUANDA-URUNDI. 37 pp. (Population Studies, 15; ST/SOA/SER.A/15)
1953, UN   Sales No. 53.XIII.4   Out of print

778   The POPULATION OF TANGANYIKA. 151 pp. (Population Studies, 2; ST/SOA/SER.A/2)
1949, UN   Sales No. 49.XIII.2   Out of print
Movement and structure of the African population, population distribution and migration, economic basis for demographic development, demographic prospects and problems of the future, and the European and Asian population.

779   ADDITIONAL INFORMATION ON THE POPULATION OF TANGANYIKA.
       32 pp. (Population Studies, 14; ST/SOA/SER.A/14)
       1952, UN   Sales No. 52.XIII.3   Out of print
       Supplement based on the results of the 1948 census.

Asia and the Far East

780   The MYSORE POPULATION STUDY. Report of a Field Survey carried
       out in Selected Areas of Mysore State, India. A Co-operative Project of
       the United Nations and the Government of India. xxvii, 443 pp., illus.
       (Population Studies, 34; ST/SOA/SER.A/34)
       1961, UN   Sales No. 61.XIII.3   Out of print
       Field study as experiment in the use of a sampling survey of house-
       holds to measure the trends and characteristics of the population and
       to investigate their interrelations with the processes of economic
       and social change in an area undergoing economic development.
       Covers methods used in obtaining data; findings; assessment of ac-
       curacy and reliability of the data. (See also entry 867.)

Latin America

781   "The Demographic Situation in Latin America." Economic Bulletin for
       Latin America, vol. 6, no. 2, October 1961, pp. 13-52.
       1961, UN   Sales No. 61.II.G.5   Out of print

782   METHODS OF USING CENSUS STATISTICS FOR THE CALCULATION
       OF LIFE TABLES AND OTHER DEMOGRAPHIC MEASURES, WITH
       APPLICATION TO THE POPULATION OF BRAZIL.
       60 pp. (Population Studies, 7; ST/SOA/SER.A/7)
       1950, UN   Sales No. 50.XIII.3   Out of print
       Describes techniques for computing life tables and estimating the
       birth rate. Includes also methods for computing marriage rates, re-
       production rates, and other indices.

Oceania

783   The POPULATION OF WESTERN SAMOA. 61 pp. (Population Studies,
       1; ST/SOA/SER.A/1)
       1948, UN   Sales No. 48.XIII.1   Out of print
       History of population growth, racial groups, factors of population
       change, population density and distribution, relation of population
       growth to economic development, and future prospects.

POPULATION ESTIMATES

784   METHODS OF ESTIMATING TOTAL POPULATION FOR CURRENT
       DATES. 45 pp. and corrigendum sheet. (Manuals on Methods of Esti-
       mating Population, 1, Population Studies, 10; ST/SOA/SER.A/10 and
       Corr.1)
       1952, UN   Sales No. 52.XIII.5   Out of print
       Methods of constructing population estimates for a current date and
       of evaluating their quality.

785 METHODS OF APPRAISAL OF QUALITY OF BASIC DATA FOR POPU-
LATION ESTIMATES. 67 pp., illus. (Manuals on Methods of Estimating Population, 2, Population Studies, 23; ST/SOA/SER.A/23)
1955, UN   Sales No. 56.XIII.2   Out of print
Methods for examining the accuracy of censuses, vital statistics,
statistics on age, and migration records.

786 GUANABARA DEMOGRAPHIC PILOT SURVEY. A Joint Project of the
United Nations and the Government of Brazil. Report of a Field Survey
carried out in the State of Guanabara, Brazil. Prepared by the United
Nations Regional Centre for Demographic Training and Research in
Latin America (CELADE), Santiago, Chile. 77 pp., illus. (Population
Studies, 35; ST/SOA/SER.A/35)
1964, UN   Sales No. 64.XIII.3   Out of print
Pilot survey especially designed to test out a method of obtaining
demographic information— specifically, fertility and mortality
rates, as well as information on nuptiality and migration.

787 SAMPLE SURVEYS OF CURRENT INTEREST (Eleventh Report). 374
pp. (Statistical Papers, Series C, No. 12; ST/STAT/SER.C/12)
1967, UN   Sales No. 67.XVII.11   Out of print
Reports of sample surveys carried out in 47 countries; bibliography.
Includes surveys of demographic subjects, manpower and housing.
The first ten reports were issued in 1949-1964.

788 The CONCEPT OF A STABLE POPULATION: APPLICATION TO THE
STUDY OF POPULATIONS OF COUNTRIES WITH INCOMPLETE
DEMOGRAPHIC STATISTICS. 237 pp., illus. (Population Studies, 39;
ST/SOA/SER.A/39)
1968, UN   Sales No. E.65.XIII.3 $3.50
Methodological study of the concepts of stable population, semi-
stable population, and quasi-stable population as particular cases of
Malthusian populations, and the calculation of schemes of quasi-stable
populations.

789 Methodology and Evaluation of POPULATION REGISTERS and Similar
Systems. 66 pp. (Studies in Methods, 15; ST/STAT/SER.F/15)
1969, UN   Sales No. E.69.XVII.15 $1.50
Comparative review of information available on continuous population registers and similar systems operating in 65 countries as of
the end of 1967. It is designed to provide some indication of the possibilities of utilizing such registers for obtaining statistical information on individuals, households, families, and other units of analysis for potential use in demographic, biologic, medical, sociological, and genetic studies.

790 METHODS OF MEASURING INTERNAL MIGRATION. x, 72 pp., maps.
(Manuals on Methods of Estimating Population, 6; Population Studies,
47; ST/SOA/SER.A/47)
1970, UN   Sales No. E.70.XIII.3 $1.50
Deals with the analytic study of internal migration, and especially
with the use of population census data for such purposes. The draft
version of the manual was prepared by the IUSSP Committee on Internal Migration with the assistance of various experts.

791   METHODOLOGY OF DEMOGRAPHIC SAMPLE SURVEYS. Report of
the Interregional Workshop on Methodology of Demographic Sample
Surveys, Copenhagen, Denmark, 24 September-3 October 1969. 31 pp.,
tables. (Statistical Papers, Series M, No. 51; ST/STAT/SER. M/
51-ST/TAO/SER.C/119)
1971, UN   Sales No. E.71.XVII.11 $5.00
Report and recommendations of the workshop, and 15 technical papers.

POPULATION CENSUSES

METHODS: GENERAL

792   PRINCIPLES AND RECOMMENDATIONS FOR NATIONAL POPULA-
TION CENSUSES. 29 pp. (Statistical Papers, Series M, No. 27; ST/
STAT/SER.M/27)
1958, UN   Sales No. 58.XVII.5   Out of print
Guide to the determination of a national census design, containing
widely recognized principles of efficient census planning and administration, and specific recommendations concerning census topics, definitions, classifications, and tabulations.

793   HANDBOOK OF POPULATION CENSUS METHODS. 2nd ed. 3 vol.
(Studies in Methods, 5/Rev.1; ST/STAT/SER.F/5/Rev.1, vol. 1-3)
1958, UN   Sales No. 58.XVII.6, vol. 1-3   Out of print
a. Vol. 1: General Aspects of a Population Census. 164 pp.
b. Vol. 2: Economic Characteristics of the Population. 79 pp.
c. Vol. 3: Demographic and Social Characteristics of the Population.
78 pp.
Manual for use in countries planning and preparing population censuses and for regional training centers. Revises and supersedes a
provisional edition, 1949, and first edition, 1954 (ST/STAT/SER.F/
5); Population Census Methods, 1949 (Population Studies, 4; ST/SOA/
SER.A/4); Fertility Data in Population Censuses, 1949 (ST/SOA/SER.
A/6); Data on Urban and Rural Population in Recent Censuses, 1958
(ST/SOA/SER.A/8); Application of International Standards to Census
Data on the Economically Active Population, 1952 (ST/SOA/SER.A/9).
It is to be superseded by the Handbook of Population and Housing Cen-
sus Methods (entry 798).

794   "Evaluation of the Population Census Data of Malaya." Economic Bulletin for Asia and the Far East, vol. 13, no. 2, September 1962, pp. 23-
44.
1962, UN   Out of print
Evaluation of the coverage of the population census of 1957 of the
Federation of Malaya and Singapore.

795   NATIONAL PROGRAMMES OF ANALYSIS OF POPULATION CENSUS
DATA AS AN AID TO PLANNING AND POLICY-MAKING. 67 pp.
(Population Studies, 36; ST/SOA/SER.A/36)
1964, UN   Sales No. 64.XIII.4   Out of print
Suggestions for utilizing population census results for the purpose of
planning in the economic and social fields, tested in seminar discussions and reflecting comments by numerous experts.

796 POPULATION CENSUSES AND NATIONAL SAMPLE SURVEYS IN DE-
VELOPING COUNTRIES. By B. Gil and E.N. Omaboe. 15 pp.
1965, ILO $0.15
   Reprinted from International Labour Review, vol. 92, no. 3, September 1965, pp. 169-183.

797 THE USE OF ELECTRONIC COMPUTERS FOR THE PROCESSING OF
POPULATION STATISTICS. Report of the Joint Meeting of the Working
Groups on Population Censuses and Electronic Data Processing,
Washington, May 1968. [223] pp. including annexes. (Conference of
European Statisticians. Statistical Standards and Studies, 12; ST/
CES/12)
1968, UN   Sales No. E.69.II.E/Mim.2 $2.50
   Papers (18) on national plans for processing 1970 data in the U.S.A.,
   the Federal Republic of Germany, and the U.S.S.R.; processing sys-
   tems and techniques; communication between subject-matter statis-
   ticians and processing specialists, and other related subjects.

798 HANDBOOK OF POPULATION AND HOUSING CENSUS METHODS.
1969- , UN

798A Part I: Planning, Organization and Administration of Population and
Housing Censuses.
In preparation

798B Part II: Topics and Tabulations for Population Censuses.
In preparation

798C Part III: Topics and Tabulations for Housing Censuses. 111 pp.
(Studies in Methods, 16; ST/STAT/SER.F/16)
1969, UN   Sales No. E.70.XVII.6 $2.00

798D Part IV: Survey of Population and Housing Census Experience, 1955-
1964.
In preparation

798E Part V: Methods of Evaluating Population and Housing Census Results.
In preparation

798F Part VI: Sampling in connexion with Population and Housing Censuses.
In preparation

798G Part VII: Statistical Cartography (with particular attention to Census
Cartography).
In preparation
This Handbook supersedes the Handbook of Population Census Methods,
2nd ed., 1958 (entry 793), and is intended to give assistance to governments in the implementation of the Principles and Recommendations
for the 1970 Population Censuses (entry 799) and the Principles and
Recommendations for the 1970 Housing Censuses (entry 921). It elaborates on the various aspects of census work that are briefly presented in the two sets of recommendations and suggests approaches to
problems frequently encountered in the planning and execution of population and housing censuses.

METHODS: 1970 CENSUSES

---

799 PRINCIPLES AND RECOMMENDATIONS FOR THE 1970 POPULATION
CENSUSES. 163 pp., illus. (Statistical Papers, Series M, No. 44; ST/
STAT/SER.M/44)
1969, 2nd printing with changes, UN   Sales No. E.67.XVII.3 $2.50
Principles and recommendations approved by the United Nations
Statistical Commission in 1966 and intended to serve as a guide for
countries planning and carrying out population censuses to be taken
around 1970. Covers the definition, essential features, uses, planning, organization, and administration of population censuses; use of
sampling; unit, place and time of enumeration; topics to be investi-
gated; and tabulations to be prepared. First issued in 1967.

800 ASIAN RECOMMENDATIONS FOR THE 1970 POPULATION CEN-
SUSES. 49 pp. (E/CN.11/773)
1967, UN   Sales No. 67.II.F.3 $1.00
Based on recommendations of the ECAFE Working Group on Censuses of Population and Housing, which met in Bangkok 2-8 December 1964 and 7-12 March 1966.

801 EUROPEAN RECOMMENDATIONS FOR THE 1970 POPULATION
CENSUSES. Regional Variant of Parts V and VI of the World Recommendations for the 1970 Population Censuses. [106] pp., including
annexes. (Conference of European Statisticians. Statistical Standards
and Studies, 13; ST/CES/13)
1969, UN   Sales No. E.69.II.E/Mim.17 $1.20
Revised European recommendations, approved by the CES in 1968,
concerning definitions and classification of topics and the scope of
the tabulation program.

POPULATION PROJECTIONS

802 METHODS FOR POPULATION PROJECTIONS BY SEX AND AGE. 81
pp. and corrigendum sheet; illus. (Manuals on Methods of Estimating
Population, 3, Population Studies, 25; ST/SOA/SER.A/25)
1956, UN   Sales No. 56.XIII.3   Out of print
Manual on methods of calculating estimates of future population
taking into account the factors of fertility, mortality, and migration.

803 GENERAL PRINCIPLES FOR NATIONAL PROGRAMMES OF POPULA-
TION PROJECTIONS AS AIDS TO DEVELOPMENT PLANNING. 60 pp.
(Population Studies, 38; ST/SOA/SER.A/38)
1965, UN   Sales No. 65.XIII.2 $0.75
Sets forth general principles which will be useful to government
agencies and research institutions in connection with the formulation of population projections for the purposes of development planning.

804 WORLD POPULATION PROSPECTS AS ASSESSED IN 1963. 149 pp.,
maps (Population Studies, 41; ST/SOA/SER.A/41)
1966, UN   Sales No. 66.XIII.2 $2.00
Presents future population estimates for the world, major areas, and
regions for the period 1960-2000, and for each country in the period
1960-1980. Also contains retrospective population estimates for the
period 1920-1960.

805   A CONCISE SUMMARY OF THE WORLD POPULATION SITUATION IN
       1970. 35 pp. and fold. map, tables. (Population Studies, 48; ST/SOA/
       SER.A/48)
       1971, UN   Sales No. E.71.XIII.2 $1.00
       Survey of trends in eight chapters: the historical perspective; popu-
       lation prospects, 1970-2000; fertility; mortality; natural increase;
       the changing role of migration; urbanization; and population policies.

Africa

806   "The Demographic Situation in Western Africa." Economic Bulletin
       for Africa, vol. 6, no. 2 (E/CN.14/400), July 1966, pp. 89-102.
       1966, UN   Sales No. 66.II.K.10 $2.00
       Discusses the contemporary demographic structure of West Africa,
       and analyzes recent population trends and their projections.

Asia and the Far East

807   "Population Trends and Related Problems of Economic Development in
       the ECAFE Region." Economic Bulletin for Asia and the Far East, vol.
       10, no. 1, June 1959, pp. 1-45.
       1959, UN   Out of print

Latin America

808   HUMAN RESOURCES OF CENTRAL AMERICA, PANAMA AND MEX-
       ICO, 1950-1980, IN RELATION TO SOME ASPECTS OF ECONOMIC
       DEVELOPMENT. Prepared by Louis J. Ducoff. 155 pp., illus. (ST/
       TAO/K/LAT/1-E/CN.12/548)
       1960, UN   Sales No. 60.XIII.1   Out of print
       Study based on demographic problems of Central American economic
       integration and the relation between population growth and economic
       growth, extended to include Panama and Mexico. Considers popula-
       tion trends and composition, future population prospects including
       urban-rural distribution, projections and utilization of the labor
       force. Notes on methodology.

809   SOCIAL CHANGE AND SOCIAL DEVELOPMENT POLICY IN LATIN
       AMERICA. vi, 318 pp., tables. (E/CN.12/826/Rev.1)
       1970, UN   Sales No. E.70.II.G.3 $4.00
       Survey of social trends in Latin America and relationship with social
       development policy. Contains an introduction and two parts: (I) so-
       cial trends: a policy-oriented diagnosis (in 8 chapters), and (II) so-
       cial development policy (in 11 chapters). Chapter IV, in Part I,
       deals with demographic trends and prospects: population growth,
       components of population growth, age distribution, future prospects
       for growth components, and population projections for 1970-1985,
       chapter XVII, in Part II, concerns population policy. Chapters VI,
       VIII and IX, in Part I, deal, respectively, with urban social change
       and consumption patterns, employment, and youth; and chapters XIII,
       XIV and XV, in Part II, cover, respectively, housing policy, social
       security policy, and health policy: food and nutrition policy.

810   "Population Trends and Policy Alternatives in Latin America." Eco-
       nomic Bulletin for Latin America, vol. 16, no. 1, 1971, pp. 71-103.
       1971, UN   Sales No. E.71.II.G.5 $3.00
       Summarizes the present demographic situation of Latin America,
       with particular attention to the likelihood of continuity or important
       change in the past trends serving as bases for statistical projec-
       tions. Discusses the main social and economic factors exerting a
       significant influence on population change.

VITAL AND HEALTH STATISTICS

STATISTICS

       World Health Statistics Annual. See entry 1176.

       World Health Statistics Report. See entry 1197.

811   INTERNATIONAL WORK IN HEALTH STATISTICS, 1948-1958. By
       H.S. Gear, Y. Biraud and S. Swaroop. 56 pp.
       1961, WHO $0.60
       Describes the work of WHO between 1948 and 1958 in international
       health statistics. Contains a historical introduction, a review of the
       problem of collecting reliable and accurate international statistics,
       and a sketch of some of the specific statistical studies undertaken by
       WHO. Reprint of a series of articles that appeared in WHO Chron-
       icle, vol. 13, no. 2, 5, 6, 9/10-12, passim, and vol. 14, no. 2-5, pas-
       sim, February 1959 through May 1960.

812   FACTS ON HEALTH PROGRESS. 51 pp., illus. (Scientific Publica-
       tions, 166)
       1968, PAHO $0.50
       Evaluative report on action taken to fulfill the health goals of the
       Charter of Punta del Este for the decade 1961-1971. Its chapters
       deal with population, child health, communicable diseases, nutrition,
       environmental sanitation, health services, health personnel, life ex-
       pectancy, and reductions in mortality.

813   HEALTH CONDITIONS IN THE AMERICAS, 1965-1968. xxi, 175 pp.
       (Scientific Publications, 207)
       1970, PAHO $1.50
       Sixth report. General vital statistics, child mortality, common dis-
       eases, health services, hospital services, environmental sanitation,
       health manpower. The first five reports, covering the years 1950-
       1964, were issued in 1956-1966.

METHODS

---

814   PRINCIPLES FOR A VITAL STATISTICS SYSTEM. Recommendations
       for Improvement and Standardization of Vital Statistics. 28 pp. (Sta-
       tistical Papers, Series M, No. 19; ST/STAT/SER.M/19)
       1953, UN   Sales No. 53.XVII.8   Out of print
       Covers all phases of the system, with emphasis on definition and
       classification of the items of information to be collected and the de-
       velopment of tabulation program. Revised edition in preparation.

815   HANDBOOK OF VITAL STATISTICS METHODS. 258 pp. (Studies in
       Methods, 7; ST/STAT/SER.F/7)
       1955, UN   Sales No. 55.XVII.1   Out of print
       Surveys vital statistics systems of 65 countries, with special empha-
       sis on methodology, and discusses these in relation to international
       recommended standards for registration of vital events and compila-
       tion of vital statistics.

816   METHODS OF ESTIMATING BASIC DEMOGRAPHIC MEASURES FROM
       INCOMPLETE DATA. 126 pp., illus. (Manuals on Methods of Esti-
       mating Population, 4, Population Studies, 42; ST/SOA/SER.A/42)
       1967, UN   Sales No. 67.XIII.2 $2.00
       Covers methods of estimating mortality and fertility rates from data
       on population growth and distribution by age. Prepared by Profes-
       sors Ansley J. Coale and Paul Demeny, Office of Population Re-
       search, Princeton University, with the assistance of other experts.

817   Manual of the INTERNATIONAL STATISTICAL CLASSIFICATION OF
       DISEASES, INJURIES, AND CAUSES OF DEATH. Based on the Recom-
       mendations of the Eighth Revision Conference, 1965, and Adopted by the
       Nineteenth World Health Assembly. 2 vol.
       1967-68, WHO $10.00 per set cl.
       Vol. 1: [Tabular List of Inclusions] xxxiii, 478 pp.
       Vol. 2: Alphabetical Index. xiv, 616 pp.
       Gives nomenclature to be used in preparing and publishing morbidity
       and mortality statistics. Also gives rules to be followed in selecting
       the underlying cause of death for deaths due to multiple causes, to-
       gether with additional instructions, an international form of medical
       certificate of cause of death, and the text of WHO Regulations re-
       garding nomenclature with respect to diseases and causes of death.
       They are intended to ensure greater uniformity and comparability of
       morbidity and mortality statistics between countries.

818   Report of the SEMINAR ON CIVIL REGISTRATION AND VITAL
       STATISTICS FOR ASIA AND THE FAR EAST, Copenhagen, Denmark,
       1968. 96 pp. and corrigendum sheet. (Statistical Papers, Series M,
       No. 50; ST/STAT/SER.M/50-E/CN.11/908 and Corr.1)
       1970, UN   Sales No. E.70.XVII.15 $2.00
       Contains report of the seminar and eight selected papers.

MORTALITY

819   "Study of the Influence of the Decline of Mortality on Growth of Popula-
       tion." By V. Kannisto and M. Pascua. Epidemiological and Vital Sta-
       tistics Report, vol. 5, no. 4, 1952, pp. 191-222.
       1952, WHO   Bilingual: E/F $0.50
       Detailed analysis covering the period between 1900 and 1945.

820   FOETAL, INFANT AND EARLY CHILDHOOD MORTALITY. 2 vol.
       (Population Studies, 13; ST/SOA/SER.A/13, vol. 1-2)
       1954, UN
       Vol. 1: THE STATISTICS. 137 pp. Sales No. 54.IV.7   Out of print
       Vol. 2: BIOLOGICAL, SOCIAL AND ECONOMIC FACTORS. 44 pp.
       Sales No. 54.IV.8   Out of print
       Vol. 1: Sources of statistical data on early life mortality; techniques of
       measurement; data for various countries, 1915-1949; chapter on medi-
       cal causes of this mortality. Vol. 2: Evaluates nonmedical causes of

the voluminous wastage of early life; mortality differentials in terms of biological, social, and economic factors; probable economic and social costs to society.

821 AGE AND SEX PATTERNS OF MORTALITY: MODEL LIFE-TABLES FOR UNDER-DEVELOPED COUNTRIES. 33 pp., illus. (Population Studies, 22; ST/SOA/SER.A/22)
1955, UN   Sales No. 55.XIII.9   Out of print
Series of 40 model life-tables which provide a time- and labor-saving method of approximating the most probable mortality level, by sex and age groups, for any population for which the infant, or better still, the early childhood, mortality rate is known with a certain degree of accuracy.

822 Population Bulletin of the United Nations. No. 6: 1962, with special reference to THE SITUATION AND RECENT TRENDS IN MORTALITY IN THE WORLD. 210 pp., illus. (ST/SOA/SER.N/6)
1963, UN   Sales No. 62.XIII.2   Out of print
Includes a factor analysis of sex-age-specific death rates (pp. 149-201).

823 "Maternal Mortality, 1950-1960." Epidemiological and Vital Statistics Report, vol. 16, no. 11, 1963, pp. 630-669.
1963, WHO   Bilingual: E/F   Out of print

824 "Infant and Child Mortality in Selected Countries, 1951-1962." Epidemiological and Vital Statistics Report, vol. 17, no. 11, 1964, pp. 536-642.
1964, WHO   Bilingual: E/F $2.75

825 TRENDS IN THE STUDY OF MORBIDITY AND MORTALITY. By D. Curiel and others. 196 pp. (Public Health Papers, 27)
1965, WHO $2.75
Papers on ways in which morbidity and mortality statistics are obtained in various countries. The volume discloses the great variety in the methods of reporting such statistics and discusses the specific problems encountered in certain developed and developing countries. It draws attention to the desirability of making morbidity and mortality data internationally comparable.

826 "Infant Deaths (Five Latest Available Years)." Epidemiological and Vital Statistics Report, vol. 19, no. 3, 1966, pp. 100-137.
1966, WHO   Bilingual: E/F $1.75

827 "Foetal Mortality, 1945-1963." Epidemiological and Vital Statistics Report, vol. 19, no. 6, 1966, pp. 262-334.
1966, WHO   Bilingual: E/F $1.75

828 PATTERNS OF URBAN MORTALITY: Report of the Inter-American Investigation of Mortality. [By] Ruth Rice Pufler [and] G. Wynne Griffith. xiii, 353 pp., illus. (Scientific Publications, 151)
1967, PAHO   Out of print
Results of an investigation of 43,298 adult deaths in twelve cities for a two-year period (ten Latin American cities, San Francisco, California, and Bristol, England). Analyzes causes and rates of death. List of references.

829   Programmes of Analysis of MORTALITY TRENDS AND LEVELS. Report of a Joint United Nations/WHO Meeting. 36 pp. (WHO Technical Report Series, 440)
1970, WHO $1.00
    Identifies existing gaps in the knowledge of levels and trends in mortality and their determinants and consequences, and makes research recommendations. Previously circulated to the fifteenth session of the Population Commission (UN), as mimeographed document E/CN.9/221, 16 June 1969.

FERTILITY

830   RECENT TRENDS IN FERTILITY IN INDUSTRIALIZED COUNTRIES. 182 pp. and corrigendum; illus. (Population Studies, 27; ST/SOA/SER. A/27 and Corr.1)
1958, UN   Sales No. 57.XIII.2   Out of print
    Comparative analysis of birth statistics of 20 countries from 1900 through 1954, with emphasis on the "baby boom" of the late 1940s and early 1950s; evaluation of those changes in the birth rate which are explained by demographic factors.

831   CULTURE AND HUMAN FERTILITY. A Study of the Relation of Cultural Conditions to Fertility in Non-industrial and Transitional Societies. By Frank Lorimer; with special contributions by Meyer Fortes; K.A. Busia; A.L. Richards and others. Foreword by Frank W. Notestein. 514 pp. (Population and Culture)
1958, 2nd printing, UNESCO   Out of print
    Study in two parts: (1) General theory; (2) special studies on Ashanti, the Gold Coast, Buganda and Buhaya, and Brazil. Carried out under the auspices of the International Union for the Scientific Study of Population (Committee on Population Problems of Countries in Process of Industrialization).

832   Population Bulletin of the United Nations. No. 7: 1963, with special reference to CONDITIONS AND TRENDS OF FERTILITY IN THE WORLD. 151 pp., illus. (ST/SOA/SER.N/7)
1965, UN   Sales No. 64.XIII.2   Out of print
    Survey of conditions and trends of fertility in the world.

833   HUMAN FERTILITY AND NATIONAL DEVELOPMENT: a Challenge to Science and Technology. x, 140 pp., illus., fold. map. (ST/ECA/138)
1971, UN   Sales No. E71.II.A.12 $2.50
    Report in nine chapters: origin and nature of the problem; fertility patterns and problems in developing countries; biological and health aspects of fertility and fertility control; social and cultural factors influencing reproductive behavior; organizational, logistic, and related aspects of family planning programs; communication for motivation of family planning; demographic statistics and analyses; five-year program in the field of population for the United Nations system; development of population programs in the United Nations system and their financing. Prepared for the United Nations Advisory Committee on the Application of Science and Technology to Development, with contributions from the United Nations, UNICEF, ILO, FAO, UNESCO and WHO, coordinated by a staff member of the United Nations Fund for Population Activities.

HUMAN REPRODUCTION
AND POPULATION GENETICS

834   RESEARCH IN POPULATION GENETICS OF PRIMITIVE GROUPS; Report of a WHO Scientific Group. 26 pp. (WHO Technical Report Series, 279)
1964, WHO $0.30
Report on the threat to many primitive communities of immediate disintegration, with consequent far-reaching biological changes, and suggestions for approaching this problem through studies of the population genetics of long-standing, but now rapidly changing, human indigenous populations. (See also entry 838.)

835   BIOLOGY OF HUMAN REPRODUCTION. Report of a WHO Scientific Group. 30 pp. (WHO Technical Report Series, 280)
1964, WHO   Out of print
Reviews developments concerning comparative and neuroendocrine aspects of reproduction, biology of the gonads and gametes, gestation, biochemistry of the sex steroids, immunological and pharmacological aspects of reproduction; makes research recommendations.

836   Mechanism of Action of SEX HORMONES AND ANALOGOUS SUBSTANCES. Report of a WHO Scientific Group. 24 pp. (WHO Technical Report Series, 303)
1965, WHO $0.60
Reviews present knowledge of the fundamental mechanisms involved in normal reproduction and of the effects of endogenous and exogenous steroids on the reproductive processes, with detailed research recommendations.

837   NEUROENDOCRINOLOGY AND REPRODUCTION IN THE HUMAN. Report of a WHO Scientific Group. 19 pp. (WHO Technical Report Series, 304)
1965, WHO $0.60
Reviews the developments in this new discipline and makes research recommendations.

838   Research on HUMAN POPULATION GENETICS. Report of a WHO Scientific Group. 32 pp. (WHO Technical Report Series, 387)
1968, WHO $0.60
Reviews present state of research on the population genetics of the remaining primitive populations of the world since the guidelines suggested by the WHO Scientific Group of 1962 (entry 834), and makes further recommendations for such studies.

839   BIOLOGICAL COMPONENTS OF HUMAN REPRODUCTION: Studies of Their Variations in Population Groups. Report of a WHO Scientific Group. 41 pp. (WHO Technical Report Series, 435)
1969, WHO $1.00
Outlines scope and purposes of studies of variation in the biological components of reproductive phenomena; considers some of the methodological principles that apply to their study in population groups; and gives illustrations of methodological problems as they arise in studies of specific reproductive phenomena.

840 ENDOCRINE REGULATION OF HUMAN GESTATION. Report of a WHO Scientific Group. 32 pp. (WHO Technical Report Series, 471) 1971, WHO $1.00
Assesses current information on the hormonal factors that may be involved in implantation and in the maintenance and termination of gestation in man, and reviews recent developments in the diagnosis and mangement of pathologic forms of pregnancy. Recommendations for further study.

POPULATION POLICIES

GENERAL

841 Report on the UNITED NATIONS TRUST FUND FOR POPULATION ACTIVITIES AND THE ROLE OF THE UNITED NATIONS IN POPULATION ACTION PROGRAMMES. By Richard Symonds. 35 pp. (ST/SOA/SER.R/10) 1969, UN
Abbreviated version of a report prepared in September 1968. The Fund, established in July 1967, is presently entitled United Nations Fund for Population Activities.

FAMILY PLANNING

842 Variables and Questionnaire for COMPARATIVE FERTILITY SURVEYS. Prepared by the Committee on Comparative Studies of Fertility and Fertility Planning of the International Union for the Scientific Study of Population in collaboration with the United Nations Secretariat. 104 pp. (Population Studies, 45; ST/SOA/SER.A/45) 1970, UN Sales No. E.69.XIII.4 $2.00
Report of an attempt to develop a list of variables (presented as a core list, a short list, and an expanded list) and a questionnaire that will serve as a basis for internationally comparable fertility and family planning surveys. Includes the Committee's work on the variables issued earlier as a document, Variables for Comparative Fertility Studies (E/CN.9/212), 1967, along with the questionnaire and additional notes and comments with respect to the questionnaire.

843 HEALTH ASPECTS OF FAMILY PLANNING. Report of a WHO Scientific Group. 50 pp. (WHO Technical Report Series, 442) 1970, WHO $1.00
Surveys the impact of family planning on the health of the mother, father, child, and family; delineates family planning care in the context of different health service activities.

844 FAMILY PLANNING IN HEALTH SERVICES. Report of a WHO Expert Committee. 65 pp. (WHO Technical Report Series, 476) 1971, WHO $1.25
After a brief review of family planning programs, dicusses health service implications of various objectives (health, human rights, population control) and of related legislation; planning; implementation of programs; process of evaluation. Contains general conclusions of the expert committee, and its recommendations for future work.

Asia and the Far East

845 ADMINISTRATIVE ASPECTS OF FAMILY PLANNING PROGRAMMES.
Report of a Working Group. 64 pp. (ECAFE. Asian Population Studies
Series, 1; E/CN.11/742)
1966, UN   Sales No. 66.II.F.10 $1.00
Report of a meeting which provided an exchange of recent experience
in dealing with the administrative problems of family planning programs and the development of new ideas. Participants included officials engaged in population programs in ten countries of the
ECAFE region and experts from a number of international agencies.

846 COMMUNICATIONS IN FAMILY PLANNING. Report of the Working
Group on Communications Aspects of Family Planning Programmes,
and Selected Papers (held at Singapore, 5-15 September 1967). 1964 pp.
(ECAFE. Asian Population Studies Series, 3; E/CN.11/830)
1968, UN   Sales No. E.68.II.F.17   Out of print
Report and selected papers (9) on the use of communications media
in family planning programs to provide information and motivation in
such programs. Participants from 13 countries of the ECAFE region took part, along with experts from international organizations.

847 Report of the Expert Group on ASSESSMENT OF ACCEPTANCE AND
USE-EFFECTIVENESS OF FAMILY PLANNING METHODS (held at
Bangkok, Thailand, 11-21 June 1968). 69 pp. (ECAFE. Asian Population Studies Series, 4; E/CN.11/882)
1969, UN   Sales No. E.69.II.F.15 $1.00
Report of the Expert Group, with recommendations on national program evaluation, regional cooperation and regional activities, studies, and research.

848 An EVALUATION OF THE FAMILY PLANNING PROGRAMME OF THE
GOVERNMENT OF INDIA. Prepared for the Government of India by a
United Nations Advisory Mission appointed under the United Nations
Programme of Technical Cooperation. 1; iii, 109 pp. (ST/SOA/SER.
11-TAO/IND/50)
1969, UN
In 1969 the mission evaluated the organization, progress, and
problems of the Indian family planning program and made recom-
mendations. An annex includes the recommendations of the first
United Nations family planning mission to India (1965).

849 Report on an EVALUATION OF THE FAMILY PLANNING PRO-
GRAMME OF THE GOVERNMENT OF PAKISTAN. Prepared for the
Government of Pakistan by a Joint United Nations-WHO Advisory Mission. [169] pp. (ST/SOA/SER.R/9-TAO/PAK/28)
1969, UN
Evaluation in 1968 by a technical assistance mission of the national
family planning program in mid term—it began in the latter half of
1965 and forms an integral part of the Pakistan Third Five-year
Plan. Report in three parts: (1) The program and its demographic
and sociological background and Pakistan's early experience in
family planning; (2) particular aspects and problems of the family
planning program; (3) conclusions and recommendations.

850   [EVALUATION OF FAMILY PLANNING PROGRAMMES.] Report of
       the Regional Seminar on Evaluation of Family Planning Programmes
       (held at Bangkok, Thailand, 24 November-12 December 1969). 95 pp.
       (ECAFE. Asian Population Studies Series, 5; E/CN.11/936)
       1970, UN   Sales No. E.70.II.F.20 $2.00
       In two parts: (1) report of the Seminar; (2) selected papers (7)

FERTILITY CONTROL: METHODS AND DEVICES
---

851   CLINICAL ASPECTS OF ORAL GESTOGENS; Report of a WHO Scien-
       tific Group. 24 pp. (WHO Technical Report Series, 326)
       1966, WHO $0.60
       Covers all aspects of the action of oral gestogens on the system and
       on individual organs and considers the benefits and possible hazards
       involved in their use.

852   BASIC AND CLINICAL ASPECTS OF INTRA-UTERINE DEVICES; Re-
       port of a WHO Scientific Group. 25 pp. (WHO Technical Report Series,
       332)
       1966, WHO $0.60
       Evaluation of intrauterine devices as a method of conception control.

853   BIOLOGY OF FERTILITY CONTROL BY PERIODIC ABSTINENCE;
       Report of a WHO Scientific Group. 20 pp. (WHO Technical Report
       Series, 360)
       1967, WHO $0.60
       Report on method of fertility control by abstinence from sexual con-
       tact during the fertile portion of the menstrual cycle.

854   HORMONAL STEROIDS IN CONTRACEPTION; Report of a WHO Sci-
       entific Group. 28 pp. and corrigenda slip. (WHO Technical Report
       Series, 386)
       1968, WHO $0.60
       Reviews information since 1965 on use of combined and sequential
       ostrogenic/progestogenic products and new oral and injectable com-
       pounds for hormonal contraception, and suggests guidelines for the
       use and study of steroid contraceptives.

855   INTRA-UTERINE DEVICES: PHYSIOLOGICAL AND CLINICAL AS-
       PECTS; Report of a WHO Scientific Group. 32 pp. (WHO Technical
       Report Series, 397)
       1968, WHO $0.60
       Reviews developments in the use of intrauterine devices for contra-
       ception and most recent results of basic research on their mode of
       action and biological effects.

856   DEVELOPMENTS IN FERTILITY CONTROL; Report of a WHO Sci-
       entific Group. 36 pp. (WHO Technical Report Series, 424)
       1969, WHO $1.00
       Reviews certain developments in present approaches to fertility con-
       trol, and suggests guidelines for the clinical application of knowledge
       derived from animal studies. Selected bibliography.

857   METHODS OF FERTILITY REGULATION: Advances in Research and
       Clinical Experience; Report of a WHO Scientific Group. 48 pp. (WHO
       Technical Report Series, 473)
       1971, WHO $1.00
       Reviews recent developments in fertility control, with special em-
       phasis on advances in knowledge of the mechanisms of action of
       hormonal contraceptive steroids and intrauterine devices, clinical
       experience of their effectiveness and side effects, and research on
       the development of improved methods. Recommendations for re-
       search on additional problems.

AGRICULTURAL CENSUSES

METHODS: GENERAL

858   SAMPLING METHODS AND CENSUSES. By S.S. Zarkovich. 213 pp.
       1965, FAO $3.50
       Deals with main uses of sampling methods, and reviews as well the
       sample and the auxiliary sample census, use of sampling methods in
       tabulation, adjustment of sample results, sampling errors, and sub-
       sequent survey work. First draft was issued in 1957 under the title
       Sampling Methods and Censuses, vol. 1: Collecting Data and Tabula-
       tion.

859   QUALITY OF STATISTICAL DATA. By S.S. Zarkovich. 395 pp.
       1966, FAO $6.00
       Aims to spread awareness of the quality problem of statistical data
       and to promote interest in quality checks as a source of guidance on
       the adequate uses of data and on the ways and means of improving
       the methods used. Explains in some detail what happens if errors
       are introduced into the data collected and what problems arise as a
       result. In presenting techniques which might be involved in checking
       the quality of data, the author emphasizes the explanation of the logic
       of the procedures rather than the analysis of various techniques that
       were actually used under specific circumstances. Bibliography.
       First draft was issued in 1963 under the title Sampling Methods and
       Censuses, vol. 2: Quality of Statistical Data.

METHODS: 1970 CENSUSES

860   1970 WORLD CENSUS OF AGRICULTURE.

860A   ———REGIONAL PROGRAM FOR AFRICA. 105 pp.
       1967, FAO

860B   ———REGIONAL PROGRAM FOR ASIA AND THE FAR EAST. 100
       pp.
       1967, FAO

860C   ———REGIONAL PROGRAM FOR EUROPE. 108 pp.
       1967, FAO

860D ———————REGIONAL PROGRAM FOR THE NEAR EAST. 97 pp.
1967, FAO
Working papers on methodology, each consisting of two parts: (1)
the World Program, and (2) the regional supplement containing
modifications in the World Program adapting it to the conditions and
needs prevailing in the respective region. In both parts there are
sections concerned with: holdings, holders, tenure, and type of
holdings; employment in agriculture; farm population; and other
agricultural data.

REPORTS: 1950 AND 1960 CENSUSES

861 Report on the 1950 WORLD CENSUS OF AGRICULTURE.
1955-58, FAO
a. Vol. 1: Census Results by Countries.
1955 $2.00
b. Vol. 2: Census Methodology. 168 pp.
1958 $2.00

862 Report on the 1960 WORLD CENSUS OF AGRICULTURE.
1966-71, FAO
a. Vol. 1, Part A: Census Results by Countries. 234 pp. and corrigenda
slip.
1966 $3.00
b. Vol. 1, Part B: Census Results by Countries. 319 pp.
1967 $6.00
c. Vol. 1, Part C: Census Results by Countries. 249 pp.
1970 $11.00
d. Vol. 2: Programme, Concepts and Scope. 196 pp.
1969 $8.00
e. Vol. 3: Methodology. 414 pp., illus.
1969 $5.00
f. Vol. 4: Processing and Tabulation. 147 pp.
1968 $3.00
g. Vol. 5: Analysis and International Comparison of Census Results.
xvi, 239 pp.
1971 $6.00
Tables of results by countries and territories in the world censuses of
agriculture in 1950 (77 countries) and 1960 (50 countries) include:
holder, holding, and tenure; employment in agriculture—in addition to
other agricultural data.

POPULATION AND DEVELOPMENT PLANNING

863 The DETERMINANTS AND CONSEQUENCES OF POPULATION
TRENDS. 404 pp. (Population Studies, 17; ST/SOA/SER.A/17)
1953, UN   Sales No. 53.XIII.3   Out of print
Comprehensive study of relationship between population changes and
economic and social conditions. Considers population trends in relation to resources, labor supply, consumption, and per capita output
in both industrialized and developing countries.

864   POPULATION GROWTH AND THE STANDARD OF LIVING IN UNDER-
        DEVELOPED COUNTRIES. 9 pp. (Population Studies, 20; ST/SOA/
        SER.A/20)
        1954, UN   Sales No. 54.XIII.7   Out of print
        Reviews outstanding facts in <u>The Determinants and Consequences of
        Population Trends</u> that relate to problems of economic and social
        development of developing countries.

865   Report on INTERNATIONAL DEFINITION AND MEASUREMENT OF
        STANDARDS AND LEVELS OF LIVING. Report of a Committee of Ex-
        perts convened by the Secretary-General of the United Nations jointly
        with ILO and UNESCO. 95 pp. (E/CN.3/179-E/CN.5/299)
        1954, UN   Sales No. 54.IV.5   Out of print
        Discusses methodology, choice of components and indicators, and
        needed improvements in the measurement of levels of living.

866   The AGING OF POPULATIONS AND ITS ECONOMIC AND SOCIAL
        IMPLICATIONS. 168 pp., illus. (Population Studies, 26; ST/SOA/
        SER.A/26)
        1956, UN   Sales No. 56.XIII.6   Out of print
        Trends and differentials of aging of population in various regions;
        causes, economic and social implications of aging; future trends;
        historical tables of comparative census statistics on the age struc-
        ture of the population in 26 countries.

867   FAMILY LIVING STUDIES—A SYMPOSIUM. 280 pp. (Studies and Re-
        ports, New Series, 63)
        1961, ILO   Out of print
        Describes six family budget surveys in various European countries;
        a food consumption survey in the United Kingdom; a survey of under-
        employment in Puerto Rico; a population study in Mysore State in
        India (<u>entry 780</u>); a health survey in Japan; surveys of consumer fi-
        nances in the U.S.A.; and four family living surveys of a more com-
        prehensive type, of multiple-purpose surveys, in four countries.

868   INTERNATIONAL DEFINITION AND MEASUREMENT OF LEVELS OF
        LIVING: AN INTERIM GUIDE. 18 pp. (E/CN.3/270/Rev.1-E/CN.5/353)
        1961, UN   Sales No. 61.IV.7 $0.35
        Current thinking of the United Nations and the specialized agencies
        on the system of components and indicators and on the basic infor-
        mation appropriate for the international definition of levels of living.
        Issued as a joint undertaking of UN, ILO, FAO, UNESCO and WHO.

869   HANDBOOK OF HOUSEHOLD SURVEYS: A PRACTICAL GUIDE FOR
        INQUIRIES ON LEVELS OF LIVING. (Provisional ed.) 172 pp.
        (Studies in Methods, 10; ST/STAT/SER.F/10)
        1964, UN   Sales No. 64.XVII.13   Out of print
        Serves as a practical guide for persons concerned with general
        (multipurpose) sample household surveys aimed at measuring sev-
        eral different aspects of the levels-of-living conditions of the popu-
        lation. Consists of nine chapters, including the household as a unit
        of enumeration; demographic characteristics; health; food consump-
        tion and nutrition. Selected references. Prepared jointly by United
        Nations, ILO, FAO, UNESCO, WHO.

870 COMPENDIUM OF SOCIAL STATISTICS: 1967 (Data Available as of 1
November 1966).... 662 pp. (Statistical Papers, Series K, No. 3;
ST/STAT/SER.K/3)
1968, UN   Sales No. 67.XVII.9 $8.75
                  Bilingual: E/F
Second issue. Statistical tables include: population and vital statistics; health conditions; food consumption and nutrition; housing; education and cultural activities; labor force and conditions of employment; income and expenditure; consumer prices. A joint undertaking
of the United Nations, ILO, FAO, UNESCO and WHO. The first issue, Compendium of Social Statistics: 1963 (586 pp., sales no. 63.
XVII.3, $7.00), was issued in 1963.

871 1967 Report on the WORLD SOCIAL SITUATION. 208 pp. and corrigendum sheet. (E/CN.5/417/Rev.1-ST/SOA/81, and Corr.1)
1969, UN   Sales No. E.68.IV.9 $3.00
Fifth periodic report. Chapter I-IX review on a sectoral basis
changes of levels of living and of policies and measures adopted to
improve social conditions (world population trends; developments
in family planning, 1960-1966; health; food and nutrition; urban and
physical planning, housing, and building; education; employment,
manpower, and income; income security and social security; social
welfare, community development, rehabilitation of the handicapped,
and crime and delinquency). Chapters X-XVI are devoted to regional
surveys of trends and programs in Asia, Latin America, Africa, the
Middle East, European socialist countries and the U.S.S.R., Western
Europe, and North America. The report should be read in conjunc-
tion with the Compendium of Social Statistics: 1967 (entry 871). A
preliminary report was issued in 1952, and the periodic reports for
the years 1957, 1961, 1963, and 1965 were issued in 1957-1966.

872 INTERNATIONAL STANDARD INDUSTRIAL CLASSIFICATION OF ALL
ECONOMIC ACTIVITIES. [2nd rev. ed.] 48 pp. (Statistical Papers,
Series M, No. 4/Rev.2; ST/STAT/SER.M/4/Rev.2)
1968, UN   Sales No. E.68.XVII.8 $1.00
In four parts: (1) Underlying principles and application of the ISIC;
(2) list of major divisions, divisions, and major groups; (3) the detailed classification; (4) differences between the present and the
preceding versions of the ISIC. The second revision of the ISIC.
The ISIC was first published in 1949; the first revised edition was
published in 1958.

Africa

873 "Demographic Factors Related to Social and Economic Development
in Africa." Economic Bulletin for Africa, Vol. 2, no. 2 (E/CN.14/171),
June 1962, pp. 59-81.
1962, UN   Sales No. 62.II.K.3 $1.50

874 "The Demographic Situation in Western Africa." Economic Bulletin
for Africa, vol. 6, no. 2 (E/CN.14/400), July 1966, pp. 89-102.
1966, UN   Sales No. 66.II.K.10
                                                         $2.00

Asia and the Far East

875   Report of the ASIAN POPULATION CONFERENCE and Selected Papers (held at New Delhi, India, 10-20 December 1963). 207 pp. (E/CN.11/670)
1965, UN   Sales No. 65.II.F.11   Out of print
Major problems of planning for economic and social development arising from present and prospective trends in the growth, composition, and geographical distribution of population within the ECAFE area.

Latin America

876   "Geographic Distribution of the Population of Latin America and Regional Development Priorities." Economic Bulletin for Latin America, vol. 8, no. 1, March 1963, pp. 51-63.
1963, UN   Sales No. 63.II.G.8   Out of print

POPULATION AND THE LABOR FORCE

877   EMPLOYMENT OBJECTIVES IN ECONOMIC DEVELOPMENT; Report of a Meeting of Experts. 255 pp. (Studies and Reports, New Series, 62)
1961, ILO $3.00
Analyzes problems of employment creation in the context of economic development, including overpopulation, underemployment, and food shortage. Appendices include case studies of employment problems and policies in Brazil, Ghana, India, Italy, Japan, the Philippines, Poland, and Egypt, and an analysis of employment objectives in the development programs of Ceylon, Greece, and Pakistan.

878   DEMOGRAPHIC ASPECTS OF MANPOWER. Report 1: SEX AND AGE PATTERNS OF PARTICIPATION IN ECONOMIC ACTIVITIES. 81 pp., illus. (Population Studies, 33, ST/SOA/SER.A/33)
1962, UN   Sales No. 61.XIII.4   Out of print
Report is concerned with the principal demographic factors affecting the size and composition of the economically active population. Sex, age, marital status, and fertility, with data derived from recent censuses of population (and some sample surveys) for nearly 100 geographical areas.

879   "The World's Labour Force and Its Industrial Distribution 1950 and 1960." By Samuel Baum. International Labour Review, vol. 95, no. 1/2, January/February 1967, pp. 96-112.
1967, ILO $0.60
Investigates changes, both relative and absolute, which have taken place in the distribution of the labor force between the three main sectors of economic activity (agriculture, industry, services).

880   METHODS OF ANALYSING CENSUS DATA ON ECONOMIC ACTIVITIES OF THE POPULATION. 143 pp. and corrigendum; illus. (Population Studies, 43; ST/SOA/SER.A/43 and Corr.1)
1968, UN   Sales No. E.69.XIII.2 $2.50
Technical manual prepared by John D. Durand and Ann R. Miller, of the Population Studies Center of the University of Pennsylvania, with

an annex contributed by Abdel Fattah Nassef, also of the Center: "A Complete Table of Economically Active Life, United Arab Republic, Males, 1960." Serves as a guide to the use of population census statistics and related data for studies of the growth and composition of the national labor force.

881 "Changes in the Industrial Distribution of the World Labour Force, by Region, 1880-1960." By P. Bairoch and J.-M. Limbor. International Labour Review, vol. 98, no. 4, October 1968, pp. 311-335.
1968, ILO $0.60
Synthesizes existing data, and supplements them with estimates whenever necessary, in a set of tables accompanied by an analytical commentary on the world labor force and its distribution by economic activity and major geographical region.

882 "Sex-Age Patterns of Labour Force Participation by Urban and Rural Populations." By Ettore Denti. International Labour Review, vol. 98, no. 6, December 1968, pp. 525-550.
1968, ILO $0.60

882 bis METHODS OF PROJECTING THE ECONOMICALLY ACTIVE POPULATION. viii, 119 pp., tables. (Manuals on Methods of Estimating Population, 5; Population Studies, 46; ST/SOA/SER.A/46)
1971, UN   Sales No. E.70.XIII.2 $1.50
Describes methods of projecting the economically active population with regard to both labor supply and demand. A cooperative project of the United Nations and the International Labour Organisation. The ILO contribution was completed with the collaboration of Mr. Claude Vimont, Directeur de Recherches, Institut national d'études démographiques (Paris).

Africa

883 "The Population and Labour Force in Africa." International Labour Review, vol. 84, no. 6, December 1961, pp. 499-514.
1961, ILO   Out of print

884 POPULATION GROWTH AND MANPOWER IN THE SUDAN. A Joint Study by the United Nations and the Government of the Sudan.... 150 pp., illus. (Population Studies, 37; ST/SOA/SER.A/37)
1964, UN   Sales No. 64.XIII.5 $2.00
Pilot study of size and growth of the population; population projections, 1956-1971; population distribution and internal migration; manpower resources for economic development.

Asia and the Far East

885 "Population Growth and Problems of Employment in the ECAFE Region." Economic Bulletin for Asia and the Far East, vol. 12, no. 2, September 1961, pp. 1-28.
1961, UN   Out of print
Outlines conditions of work and employment in Asian countries, discusses expected growth of the labor force based on available projections of total population by sex and age. Study is based on four countries: Federation of Malaya, India, Japan, and Thailand.

886   The POPULATION AND LABOUR FORCE OF ASIA, 1950-80. 21 pp.
         1962, ILO $0.15
            Reprint from International Labour Review, vol. 86, no. 4, October
            1962, pp. 348-368.

887   POPULATION GROWTH AND MANPOWER IN THE PHILIPPINES. A
         Joint Study by the United Nations and the Government of the Philippines.
         National Economic Council of the Philippines, Manila. 66 pp., illus.
         (Population Studies, 32; ST/SOA/SER.A/32)
         1960, UN   Sales No. 61.XIII.2   Out of print
            Pilot study of growth and structure of the population; future popula-
            tion; urban and rural population; demography of manpower; future
            labor force; implications for economic development.

Europe

888   "Population and Labour Force in Eastern Europe and the U.S.S.R.:
         Structure and Recent Trends." By Gh. Lungu. International Labour
         Review, vol. 91, no. 2, February 1965, pp. 135-148.
         1965, ILO $0.60

889   "Industrialisation and Structural Changes in Employment in the Social-
         ist Countries." By Antoni Rajkiewicz. International Labour Review,
         vol. 94, no. 3, September 1966, pp. 286-302.
         1966, ILO $0.60

890   MANPOWER ASPECTS OF RECENT ECONOMIC DEVELOPMENTS IN
         EUROPE. 175 pp.
         1969, ILO $2.50
            Deals with economic growth and structural changes in employment,
            1950-1965, and the outlook for 1965-1980; manpower aspects of
            economic and social policy; problems of adaptability of the labor
            force; international migration for employment; education and train-
            ing; problems of special categories of workers: women, older work-
            ers, disabled persons. A statistical appendix contains a projection
            of population and labor force by sex and age, 1965-1980, for 25
            European countries.

Latin America

891   "Structural Changes in Employment within the Context of Latin
         America's Economic Development." By Zygmunt Slawinski. Economic
         Bulletin for Latin America, vol. 10, no. 2, October 1965, pp. 163-187.
         1966, UN   Sales No. 66.II.G.3 $1.50

892   "Underutilisation of Manpower and Demographic Trends in Latin
         America." By Gavin W. Jones. International Labour Review, vol. 98,
          no. 5, November 1968, pp. 451-469.
         1968, ILO $0.60

POPULATION AND EDUCATIONAL PLANNING

893   ECONOMIC AND SOCIAL ASPECTS OF EDUCATIONAL PLANNING.
         264 pp.
         1966, 2nd printing, UNESCO   Out of print
           This handbook contains contributions from economists, educators,

sociologists, demographers, and administrators engaged in the day-to-day tasks of educational planning. Of special value to scholars, specialists, and students concerned with assisting developing countries to find solutions in their problems of social and economic development. First issued in 1964.

894   ESTIMATING FUTURE SCHOOL ENROLMENT IN DEVELOPING COUNTRIES; a Manual of Methodology. By Bangnee Alfred Liu. 156 pp., illus. (Population Studies, 40; ST/SOA/SER.A/40)
1966, UN   Sales No. 66.XIII.3 $2.00
————. UNESCO. (Statistical Reports and Studies, 10) $2.00
Suggests methods useful to the educational statistician seeking to make estimates of future school enrolment for educational planning. Three chapters are devoted to case studies of estimates in Colombia, the Philippines, and the Sudan, and another chapter to actual projections from three selected developed countries: France, New Zealand, and the U.S.A.

895   DEMOGRAPHIC ASPECTS OF EDUCATIONAL PLANNING. By Ta Ngoc Châu. 84 pp. (Fundamentals of Educational Planning, 9)
1969, UNESCO/IIEP $2.00
Explains the way in which the structure of the population can effect educational development at all levels. Analyzes the significance, usefulness, and relative value of demographic data to educational planners.

Unesco Statistical Yearbook. See entry 1184.

Asia and the Far East

896   An ASIAN MODEL OF EDUCATIONAL DEVELOPMENT: PERSPECTIVES FOR 1965-1980. 126 pp. and List of tables and figures, 2 pp.
1966, UNESCO $2.00
Visualizes in quantitative terms the prospects of educational development in Asia through 1980 and draws attention to some of the more important implications for educational development that become evident when specific data are examined systematically as has been done in the model.

Latin America

897   SITUACIÓN DEMOGRÁFICA, ECONÓMICA, SOCIAL Y EDUCATIVA DE AMÉRICA LATINA. 144 pp.
1966, Librería Hachette S.A. (Buenos Aires)   420 pesos argentinos
Prepared jointly by UNESCO, UN Economic Commission for Latin America, UN Bureau of Social Affairs, and the Centro Latinoamericano de Demografía (CELADE), for the Conference on Education and Economic and Social Development in Latin America, Santiago, Chile, 5-19 March 1962. First published in Proyecto Principal de Educación Unesco América Latina, Boletín Trimestral, no. 13, enero/marzo 1962, pp. 5-136 (out of print).

898   METHODOLOGIES OF EDUCATIONAL PLANNING FOR DEVELOPING COUNTRIES. By J.D. Chesswas. 2 vol. and corrigenda slip.
1969, UNESCO/IIEP $7.00 for the set; vols. not sold separately
Vol. 1: Text. 106 pp., illus.
Vol. 2: Tables. 84 pp.
Manual intended to help present and future practitioners of educational

planning to master techniques involved in making a quantitative assessment of the situation and trends of educational services and in preparing plans for future development. Includes demographic aspects of educational projections.

URBANIZATION AND INDUSTRIALIZATION

899   Handbook for SOCIAL RESEARCH IN URBAN AREAS. Ed. by Philip M. Hauser. 214 pp. (Technology and Society Series)
1967, 2nd printing, UNESCO $5.00 cl.
In two parts: (1) Social research data and procedures, including "Basic Statistics and Research," by Giuseppe Parenti; and (2) Types of studies, including "Demographic Trends in Urban Areas," by Judah Matras, and "Migration and Acculturation," by Gino Germani. First issued in 1965.

900   INDUSTRIALIZATION AND SOCIETY. Ed. by Bert F. Hoselitz and Wilbert E. Moore. 437 pp.
1968, 3rd printing, UNESCO/Mouton (The Hague) $11.00 cl.
Collection of fourteen papers prepared for the North American Conference on the Social Implications of Industrialization and Social Change, Chicago, September 1960, a summary of substantive finding, and four appended papers on social research. Topics covered are: concepts; entrepreneurship and innovation; consumption, savings, and investment; government and public administration; urbanization, population, and the family (3 papers); education and communication. First issued in 1963.

901   PLANNING OF METROPOLITAN AREAS AND NEW TOWNS. Meeting of the United Nations Group of Experts on Metropolitan Planning and Development, Stockholm, 14-30 September 1961; United Nations Symposium on the Planning and Development of New Towns, Moscow 24 August-7 September 1964. x, 255 pp., illus., maps. (ST/SOA/65)
1967, UN   Sales No. 67.IV.5 $3.50
Papers (31) and the conclusions and recommendations of the two meetings on metropolitan planning, including three papers on urbanization: processes of world urbanization, urbanization in Latin America, and urbanization in South East Asia.

902   URBANIZATION: DEVELOPMENT POLICIES AND PLANNING. 130 pp., illus. (International Social Development Review, No. 1; ST/SOA/SER.X/1)
1968, UN   Sales No. E.68.IV.1 $2.00
Four articles: demographic aspects of urban growth and population distribution; social and economic problems and processes of urbanization in developing countries; policies and planning related to urban growth and population distribution; the problem of slums and shantytowns.

903   GROWTH OF THE WORLD'S URBAN AND RURAL POPULATION, 1920-2000. 124 pp. (Population Studies, 44; ST/SOA/SER.A/44)
1967, UN   Sales No. E.69.XIII.3 $2.00
Surveys recent urbanization trends; considers the problem of statistical definition of "urban," "metropolitan area," "agglomerations," "rural," "small town" concepts; makes tentative assessments of possible future trends.

904   URBAN AND REGIONAL RESEARCH. Proceedings of the Conference
       of Senior Officials of National Bodies concerned with Urban and Re-
       gional Research.... held in Stockholm, Sweden, 24 April-1 May 1968.
       3 vol. (ST/ECE/HOU/37, vol. 1-3)
       1969, UN   Sales No. E.69.II.E/Mim.16 $4.20 the set
       Papers and discussion on urban and regional research: scope,
       character, and trends; methods and new techniques; scope and meth-
       ods for promoting systematic and direct collaboration between
       national research bodies. Papers from Czechoslovakia, France,
       Hungary, Italy, Poland, Sweden, Ukrainian S.S.R., U.S.S.R., United
       Kingdom, U.S.A., and various international organizations.

905   URBANIZATION IN THE SECOND UNITED NATIONS DEVELOPMENT
       DECADE. 39 pp. (ST/ECA/132)
       1970, UN   Sales No. E.70.IV.15 $0.75
       Emphasizes the social and economic problems of urbanization in re-
       lation to national development planning. Slightly revised text of a
       report of the Secretary-General entitled Housing, Building and Plan-
       ning in the Second United Nations Decade (E/C.6/90), prepared in
       cooperation with Barbara Ward, and submitted to the sixth session of
       the Committee on Housing, Building and Planning, 1969.

906   ADMINISTRATIVE ASPECTS OF URBANIZATION. Based upon a Com-
       parative Study carried out with the co-operation of the Institute of
       Public Adminstration of New York, and on the United Nations Work-
       shop on Adminstrative Aspects of Urbanization, held at the Institute of
       Social Studies, at The Hague, Netherlands, 11-20 November 1968. 228
       pp. (ST/TAO/M/51)
       1970, UN   Sales No. E.71.II.H.1 $3.50
       Report on the administrative aspects of urbanization; annexes in-
       clude a glossary of terms, studies of 11 selected cities, and a
       selected bibliography.

Africa

907   SOCIAL IMPLICATIONS OF INDUSTRIALIZATION AND URBANIZA-
       TION IN AFRICA SOUTH OF THE SAHARA. Prepared under the
       auspices of UNESCO by the International African Institute, London.
       143 pp. (Tensions and Technology Series)
       1964, 2nd printing, UNESCO $11.00 cl.
       Analyzes social effects of industrialization in tropical Africa from
       various standpoints: geographical setting, sociodemographic struc-
       ture, labor conditions, aptitudes and training of Africans. Also con-
       tains the conclusions of a conference jointly organized at Abidjan in
       1954 by UNESCO and the Commission for Technical Co-operation in
       Africa South of the Sahara, and an account of surveys in various
       parts of Africa. First issued in 1956.

Asia and the Far East

908   URBANIZATION IN ASIA AND THE FAR EAST. Proceedings of the
       Joint UN/UNESCO Seminar.... on Urbanization in the ECAFE Region,
       Bangkok, 8-18 August 1956. Ed. by Philip M. Hauser. 286 pp.
       (Tensions and Technology Series)
       1957, Calcutta, UNESCO Research Centre on the Social Implications
       of Industrialization in Southern Asia   Out of print
       Rapporteur's summary report, conclusions of the seminar and

papers (9) on urbanization in the ECAFE region and its social aspects and in relation to economic development, internal migration, manpower, and urban and regional planning. The seminar was held in cooperation with the International Labour Organisation.

Latin America

909 URBANIZATION IN LATIN AMERICA. Proceedings of a Seminar...on Urbanization Problems in Latin America, Santiago (Chile), 6 to 18 July 1959. Ed. by Philip M. Hauser. 331 pp., tables. (Technology and Society Series)
1967, 2nd printing, UNESCO $6.00 cl.
Rapporteur's report, conclusions of the seminar and selected papers (11) on demographic aspects of urbanization, internal migration, and manpower in relation to economic development in Argentina, Brazil, Ecuador, and Peru. First issued in 1961.

910 "The Urbanization of Society in Latin America." Economic Bulletin for Latin America, vol. 13, no. 2, November 1968, pp. 76-93.
1968, UN   Sales No. E.68.II.G.11 $2.00
Preliminary attempt to form a more adequate conceptual framework for the study of urbanization.

Mediterranean Region

911 Report of the URBANIZATION SURVEY MISSION IN THE MEDITERRANEAN REGION, November to December 1959. [143] pp., including annexes, fold. map. (ST/TAO/SER.C/51-ST/SOA/SER.T/1)
1962, UN   Out of print
Findings, conclusions, and recommendations of the Mission; five annexes: a sketch of the historical development and general characteristics of urbanization in the Mediterranean region, and summaries of information collected concerning Greece, Egypt, Morocco, and Libya.

POPULATION AND HOUSING

HOUSING STATISTICS AND PROGRAMS

912 WORLD HOUSING CONDITIONS AND ESTIMATED HOUSING REQUIREMENTS. 58 pp. (ST/SOA/58)
1965, UN   Sales No. 65.IV.8 $0.75
Based on analyses of housing conditions in the major regions of the world, the study presents estimates of housing requirements in these regions for the period 1960-1975.

913 CONSTRUCTION STATISTICS. 169 pp. (Studies in Methods, 13; ST/STAT/SER.F/13)
1965, UN   Sales No. 66.XVII.4 $2.00
Studies the collection of data on construction, including construction of dwellings.

914   Methods of ESTIMATING HOUSING NEEDS. 99 pp. (Studies in Methods, 12; ST/STAT/SER.F/12)
1967, UN   Sales No. 67.XVII.15 $2.00
Provisional technical manual intended as a guide for estimating housing needs.

915   INTERNATIONAL RECOMMENDATIONS FOR CONSTRUCTION STATISTICS. 57 pp. (Statistical Papers, Series M, No. 47; ST/STAT/SER.M/47)
1968, UN   Sales No. E.68.XVII.11 $1.00
International recommendations presented in two versions, one applicable to countries with developed construction statistics, and the other applicable to countries which are beginning to develop or are in the process of developing their construction statistics.

Asia and the Far East

916   Report of the SEMINAR ON HOUSING STATISTICS AND PROGRAMMES FOR ASIA AND THE FAR EAST, Copenhagen, Denmark, 25 August-14 September 1963. 73 pp. (E/CN.11/677/Rev.1)
1965, UN   Sales No. 65.II.F.12 $1.00

Europe

917   A STATISTICAL SURVEY OF THE HOUSING SITUATION IN EUROPEAN COUNTRIES AROUND 1960.... 75 pp., illus. (ST/ECE/HOU/12)
1965, UN   Sales No. 65.II.E.7   Out of print
Trilingual: E/F/R
Statistical survey on the basis of recent population and housing censuses of information from 25 European countries, Cyprus, and the U.S.A.

918   The HOUSING SITUATION AND PERSPECTIVES FOR LONG-TERM HOUSING REQUIREMENTS IN EUROPEAN COUNTRIES. 116 pp., illus. (ST/ECE/HOU/32)
1968, UN   Sales No. E.68.II.E.6 $2.00
Examines, on an all-European basis, changes that have taken place in the housing situation since World War II, and analyzes estimated dwelling shortages and perspective for long-term housing requirements. Also analyzes and appraises different norms and methods employed by countries when making their estimates.

Annual Bulletin of Housing and Building Statistics for Europe. See entry 1177.

Latin America

919   Report of the LATIN AMERICAN SEMINAR ON HOUSING STATISTICS AND PROGRAMMES, Copenhagen, Denmark, 2-25 September 1962. 89 pp. (E/CN.12/647/Rev.1)
1963, UN   Sales No. 63.II.G.14   Out of print

HOUSING CENSUSES

920   GENERAL PRINCIPLES FOR A HOUSING CENSUS, 14 pp. (Statistical Papers, Series M, No. 28; ST/STAT/SER.M/28)
1958, UN   Sales No. 58.XVII.8   Out of print
Guide to countries planning to take housing censuses or to collect housing information in connection with national population censuses.

921   PRINCIPLES AND RECOMMENDATIONS FOR THE 1970 HOUSING
       CENSUSES. 140 pp., illus. (Statistical Papers, Series M, No. 45; ST/
       STAT/SER.M/45)
       1967, UN   Sales No. 67.XVII.4 $2.50
       Principles and recommendations approved by the United Nations
       Statistical Commission in 1966 and intended to serve as a guide for
       countries planning and carrying out housing censuses to be taken
       around 1970.

922   ASIAN RECOMMENDATIONS FOR THE 1970 HOUSING CENSUSES. 43
       pp. (E/CN.11.772)
       1967, UN   Sales No. 67.II.F.9 $1.00
       Variant of the Principles and Recommendations for the 1970 Housing
       Censuses based on the ECAFE Working Group on Censuses of Popu-
       lation and Housing (Bangkok, 1964 and 1966).

923   EUROPEAN RECOMMENDATIONS FOR THE 1970 HOUSING CEN-
       SUSES. Regional Variant of Parts IV, V and VI of the World Recom-
       mendations for the 1970 Housing Censuses. 51 pp. (Conference of
       European Statisticians. Statistical Standards and Studies, 15; ST/CES/
       15-ST/ECE/HOU/39)
       1969, UN   Sales No. E.69.II.E/Mim.25 $0.50
       Variant of the Principles and Recommendations for the 1970 Housing
       Censuses based on the recommendations of the joint Working Group
       on Housing Censuses (Conference of European Statisticians/ECE
       Committee on Housing, Building and Planning), approved by the CES
       in 1968.

INTERNATIONAL MIGRATION

924   PROBLEMS OF MIGRATION STATISTICS. 65 pp. (Population Studies,
       5; ST/SOA/SER.A/5)
       1950, UN   Sales No. 50.XIII.1   Out of print
       Description of methods of collecting data in various countries and of
       tabulations provided in published migration statistics by 69 coun-
       tries. Analyzes major difficulties involved in collecting data and in
       attaining comparable statistics for different countries.

925   SEX AND AGE OF INTERNATIONAL MIGRANTS: STATISTICS FOR
       1918-1947. 281 pp. (Population Studies, 11; ST/SOA/SER.A/11)
       1953, UN   Sales No. 53.IV.15   Out of print
       Data on 74 countries and territories. For later statistics, consult
       the Demographic Yearbook (entry 1170).

926   INTERNATIONAL MIGRATION STATISTICS. 25 pp. (Statistical
       Papers, Series M, No. 20; ST/STAT/SER.M/20)
       1953, UN   Sales No. 53.XVII.10   Out of print
       Analyzes major problems of statistical organization and operation.
       Recommendations for improvement of migration statistics; applica-
       tions of sampling methods.

927   FLIGHT AND RESETTLEMENT. By H.B.M. Murphy and others. 229
       pp., illus. (Population and Culture)
       1955, UNESCO   Out of print
       Fifteen chapters on psychological problems of refugees and dis-
       placed persons in camp, in flight, and in resettlement.

928 ECONOMIC CHARACTERISTICS OF INTERNATIONAL MIGRANTS:
STATISTICS FOR SELECTED COUNTRIES, 1918-1954. 314 pp.
(Population Studies, 12; ST/SOA/SER.A/12)
1958, UN   Sales No. 58.XIII.3   Out of print
 Migration statistics, classified by occupation, industry, etc., for 36
 countries.

929 INTERNATIONAL MIGRATION, 1945-1957. 414 pp. (Studies and Reports, New Series, 54)
1959, ILO $4.00
 Describes political migration in Germany and other parts of Europe,
 in Israel and the Arab States, in India and Pakistan, and in the Far
 East, and also the major postwar currents of economic migration
 within and between the continents.

930 The CULTURAL INTEGRATION OF IMMIGRANTS. A Survey based on
the Papers and Proceedings of the Unesco Conference held in Havana,
April 1956. By W.D. Borrie and others. 297 pp. (Population and Culture)
1959, UNESCO $3.00
 In two parts: (1) Survey of concepts and practices based principally
 upon the conference papers and sessional reports, with illustrative
 statistical material to 1954; (2) four case studies of migration in
 Europe, assimilation of immigrants in Brazil, in Israel, and immi-
 gration and group settlement with special reference to Europeans in
 Australia, Canada, and the U.S.A.

931 The POSITIVE CONTRIBUTION BY IMMIGRANTS. A Symposium pre-
pared [under the auspices of] UNESCO, the International Sociological
Association and the International Economic Association. By Oscar
Handlin and others. 199 pp. (Population and Culture)
1960, 2nd printing, UNESCO $2.25
 A series of national studies (U.S.A., United Kingdom, Australia,
 Brazil, Argentina) which discusses the economic aspects separately.
 First issued in 1955.

ENVIRONMENTAL HEALTH

GENERAL

932 EXPERT COMMITTEE ON ENVIRONMENTAL SANITATION. Third
Report. 25 pp. (WHO Technical Report Series, 77)
1954, WHO $0.25
 Contains a detailed consideration of sanitation in rural areas and
 communities and gives information on the improvement of sanitary
 conditions in developing countries. (For the fourth report, see entry
 572; for the fifth, entry 948.)

933 EXPERT COMMITTEE ON THE PUBLIC HEALTH ASPECTS OF HOUS-
ING. First Report. 60 pp. (WHO Technical Report Series, 225)
1961, WHO $0.60
 Expounds the fundamentals of a healthful residential environment, in
 terms of the design and arrangements of dwellings, household ser-

vices, physiological requirements, protection and maintenance, availability of community facilities, and town and country planning. Gives attention also to housing in rural areas and for groups (the elderly and the handicapped), the role of public health agencies, community action, training of personnel, and research needs.

934   "Environmental Health," Bulletin of the World Health Organization, vol. 26, no. 4, 1962.
1962   Out of print
Papers on water pollution in the U.S.S.R. and other Eastern European countries; oxidation ditches for treatment of sewage from small communities; sanitary engineering and nuclear energy development; measuring sulfur dioxide, dustfall, and suspended matter in city air; development of community water supplies in Ghana; recent information on water pollution in Europe; sewage treatment methods in Germany; experience with stabilization ponds in the U.S.A.; research into drinking water purification and disinfection.

935   ENVIRONMENTAL CHANGE AND RESULTING IMPACTS ON HEALTH. Report of a WHO Expert Committee. 23 pp. (WHO Technical Report Series, 292)
1964, WHO $0.60
Reviews health problems connected with ionizing radiation, air pollution, liquid and solid wastes, food hygiene, and other manifestations of environmental change in modern life; and makes recommendations.

936   "Some World Problems of Environmental Health." By N.F. Izmerov. WHO Chronicle, vol. 24, no. 4, April 1970, pp. 145-152.
1970, WHO $0.60
Notes our debt to individual contributors for work in environmental sanitation since the rise of modern industrialism and surveys the work of WHO in the field of environmental health. Based on a lecture delivered at the London School of Hygiene and Tropical Medicine on 2 December 1969.

937   ENVIRONMENTAL HEALTH ASPECTS OF METROPOLITAN PLANNING AND DEVELOPMENT; Report of a WHO Expert Committee. 66 pp. (WHO Technical Report Series, 297)
1965, WHO $1.25
Presents the views of the Expert Committee on problems of metropolitan planning and development, including the specific problems of water supply, drainage, sanitary waste disposal, air and land pollution, and noise and vibration.

938   RESEARCH INTO ENVIRONMENTAL POLLUTION; Report of Five WHO Scientific Groups. 83 pp. (WHO Technical Report Series, 406)
1968, WHO $1.25
Survey of existing knowledge of biological aspects of environmental pollution and guidance as to the further research needs. The principal aspects considered by the five groups during 1963-1965 were microchemical pollution of water systems, biological estimation of water pollution levels, identification and measurement of air pollution, long-term effects on health of new pollutants, and research into environmental pollution.

939   Methods for Establishing TARGETS AND STANDARDS FOR HOUSING
      AND ENVIRONMENTAL DEVELOPMENT. 59 pp. (ST/SOA/76)
      1968, UN   Sales No. E.68.IV.5 $1.00
          Report of a consultant on methods of formulating the goals, targets
          and standards for housing and environmental development, of use
          primarily to technicians and administrators of housing programs in
          developing countries.

940   NATIONAL ENVIRONMENTAL HEALTH PROGRAMMES: Their Plan-
      ning, Organization, and Administration. Report of a WHO Expert Com-
      mittee. 56 pp. (WHO Technical Report Series, 439)
      1970, WHO $1.00
          Outlines present trends in environmental health and considers desir-
          able methods of planning, organizing, and administering national
          programs, the role of international collaboration on common prob-
          lems, and technical needs and problems requiring future action by
          WHO.

941   NUCLEAR TECHNIQUES IN ENVIRONMENTAL POLLUTION. Proceed-
      ings of a Symposium on the Use of Nuclear Techniques in the Measure-
      ment andControl of Environmental Pollution, held by IAEA in Salzburg,
      26-30 October 1970. 807 pp. and corrigendum sheet, fig., tables. (Pro-
      ceedings Series; STI/PUB/268)
      1970, IAEA   Composite: E/F/R $22.00
          Papers and discussions on nuclear techniques and pollution studies
          (7 papers); measurement of air pollutants and their dispersions (14);
          identification and analysis of water pollutants (6); environmental
          contamination of mercury and other metallic compounds (7); move-
          ment of pollutants in water and soil (6); tracer techniques in coastal
          pollution studies (7); nuclear techniques and ecosystems (4); and
          sludge and waste treatment (3).

942   "A Message to Our $3\frac{1}{2}$ Billion Neighbours on Planet Earth: S.O.S. En-
      vironment." Unesco Courier, vol. 24, no. 7, July 1971, pp. 4-32.
      1971, UNESCO $0.50
          Cover title of issue, with several articles on aspects of environ-
          mental pollution and measures for its control or prevention; illus-
          trations.

943   "Environmental Health." World Health, August/September 1971.
      48 pp., illus.
      1971, WHO $0.50
          Issue devoted to environmental health, with articles on pollution of
          the environment, and clean water supply.

EDUCATION AND TRAINING
_____

944   The EDUCATION OF ENGINEERS IN ENVIRONMENTAL HEALTH; Re-
      port of a WHO Expert Committee. 26 pp. (WHO Technical Report
      Series, 376)
      1967, WHO $0.60
          Survey of the changing scope of environmental health and needs in
          education and training of engineers, architects, and city planners in
          this field.

945   The EDUCATION AND TRAINING OF ENGINEERS FOR ENVIRON-
       MENTAL HEALTH. By John Cassel and others. 152 pp.
       1970, WHO $6.00 cl.
            Contributions cover the work of the environmental health engineer
            (2 papers); trends in modern engineering education (5); new devel-
            opments in environmental health engineering (4); examples of edu-
            cational programs in environmental health engineering in different
            regions of the world. Indicates those trends likely to influence the
            planning of future curricula. An outcome of the Expert Committee
            report of 1967 (entry 944).

                               Latin America

946   "Comments on the Teaching of Sanitary Engineering in Latin America."
       By Ildeu Duarte Filho. Boletín de la Oficina Sanitaria Panamericana,
       English edition, Selections from 1967, pp. 17-21.
       1968, PAHO
            First published in Boletín de la Oficina Sanitaria Panamericana, vol.
            62, no. 6, junio de 1967, pp. 531-537.

AIR POLLUTION

GENERAL
_____

947   AIR POLLUTION. 442 pp., illus. (WHO Monograph Series, 46)
       1961, WHO $11.00 cl.
            Fourteen papers by experts from seven countries on various aspects
            of air pollution and its prevention and control.

948   AIR POLLUTION. Fifth Report of the Expert Committee on Environ-
       mental Sanitation. 26 pp. (WHO Technical Report Series, 157)
       1958, WHO $0.30
            Report, including recommendations concerning recognition, evalua-
            tion, prevention, and control of man-made air pollution, including the
            administration aspects of control.

949   ATMOSPHERIC POLLUTANTS, Report of a WHO Expert Committee.
       18 pp. (WHO Technical Report Series, 271)
       1964, WHO $0.30
            Review of progress in air pollution control, excluding pollution by
            radioactive materials.

950   The HEALTH EFFECTS OF AIR POLLUTION. Report on a Symposium,
       Prague, 6-10 November 1967. 73 pp., illus. (EURO 1143).
       1968, WHO Regional Office for Europe
            Report and conclusions on physical aspects of air pollution, its ef-
            fects upon man, and criteria, guides, and standards of air quality.

951   SOLID SMOKELESS FUELS. 36 pp. (ST/ECE/COAL/22)
       1967, UN   Sales No.E.68.II.E/Mim.5 $0.75
            Study by the ECE Group of Experts on Coking of solid smokeless
            fuels based on information from fourteen European countries and the

U.S.A. with respect to the production of these fuels, and from eight
European countries on air pollution standards with a bearing upon
the characteristics of smokeless fuels.

952   AIR POLLUTION BY COKING PLANTS. 65 pp. (ST/ECE/COAL/26)
Mimeographed.
1968, UN   Sales No. E.68.II.E/Mim.1 $0.75
Study by the ECE Group of Experts on Coking of air pollution caused
by coke-oven operations proper. Covers methods and results of
measuring pollutants; danger thresholds—maximum allowable concentrations; measures taken to reduce pollution; outlay required to
reduce pollution. Draws upon information from nine European countries and the U.S.A.

953   AIR POLLUTION BY THE CHEMICAL DEPARTMENT OF COKING
PLANTS. (Consolidated Report) By J.E. Barker and A.R. Jack.
37 pp., fold. tables. (ST/ECE/COAL/36)
1968, UN   Free on request to ECE
Contains from 11 European countries and the U.S.A. on toxic materials released into the air and methods used to detect their presence
and to measure concentrations of such materials in the air in the vi-
cinity of coking plants.

954   AIR POLLUTANTS, METEOROLOGY AND PLANT INJURY. By E.I.
Mukammal and others. 73 pp. (Technical Notes, 96)
1968, WMO   WMO-No.234.TP.127 $3.50
Reviews present knowledge of the subject; with extensive bibliogra-
phy. Report of the Working Group on Plant Injury and Reduction of
Yield by Non-radioactive Air Pollutants of the WMO Commission for
Agricultural Meteorology.

955   MEASUREMENT OF AIR POLLUTANTS: GUIDE TO THE SELECTION
OF METHODS. By Morris Katz. 123 pp.
1969, WHO $5.00
Discusses some fundamental principles underlying air sampling and
analyzes and outlines some methods now used in different countries
for identifying and estimating the most common pollutants. Extensive references.

956   URBAN AIR POLLUTION WITH PARTICULAR REFERENCE TO MO-
TOR VEHICLES; Report of a WHO Expert Committee. 53 pp. (WHO
Technical Report Series, 410)
1969, WHO $1.00
Report on the contribution of the motor vehicle to air pollution, including methods of sampling and analysis; effects of vehicle exhausts
on health and on the environment, and methods of control.

957   "Meteorology and Urban Air Pollution." By Robert A. McCormick.
WMO Bulletin, vol. 18, no. 3, July 1969, pp. 155-165.
1969, WMO $1.25

958   NUCLEAR ENERGY AND THE ENVIRONMENT. Addendum to the
Agency's Report to the Economic and Social Council of the United Nations for 1969-70. 36 pp., illus. (INFCIRC/139/Add.1)
1970, IAEA   Free on request to IAEA
Brief account of the potential effects of nuclear power on the environment and human health and of the safeguards devised to protect
the public, the use of nuclear tracer techniques to help control pollution, and the reduction of pollution through the replacement by nuclear power of conventional forms of power generation.

959 METEOROLOGICAL ASPECTS OF AIR POLLUTION. 69 pp., illus., maps. (Technical Notes, 106)
1970, WMO WMO-No.251.TP.139 $4.50
Three papers—"Meteorological Aspects of Air Pollution in Urban and Industrial Areas" by Robert A. McCormick; "The Importance of Investigating Global Background Pollution" by Erik Eriksson; "Le Réseau belge de mesure de la pollution atmosphérique per les oxydes de soufre et la fumée" by J. Grandjean and J. Bouquiaux—presented at the twenty-first session of the WMO Executive Committee, 1966.

960 METEOROLOGICAL FACTORS IN AIR POLLUTION. By A.G. Forsdyke. x, 32 pp., fig., tables. (Technical Notes, 114)
1970, WMO WMO-No.274.TP.153 $3.50
Introductory text designed specifically for the nonspecialist and ordinary meteorologist seeking information on the meteorological aspects of air pollution.

961 PROBLEMS OF AIR AND WATER POLLUTION ARISING IN THE IRON AND STEEL INDUSTRY. viii, 86 pp., tables. (E/ECE/STEEL/32)
1970, UN Sales No. E.70.II.E.6 $1.50
Study of the causes of air and water pollution arising in the iron and steel industry and a description of the measures taken to combat the effects of pollution, including both the actual cleaning facilities in the industry and improvement of the technology of production and of equipment, and cost aspects thereof. An annex briefly reviews the legal regulations in 16 European countries, Japan, and the U.S.A.

962 WMO Operations Manual for SAMPLING AND ANALYSIS TECHNIQUES FOR CHEMICAL CONSTITUENTS IN AIR AND PRECIPITATION. Loose-leaf, with expandable cover.
1971- , WMO WMO-No.299 $4.00
———Preface, General Introduction, and Part I: The Mimimum Programme at Regional Air Pollution Stations. 1971. [41] p., including appendixes.
———Part II: The Minimum Programme at Baseline Air Pollution Stations.
In preparation
Provides detailed guidance for the establishment and operation of a WMO background air pollution station.

962 DESULPHURIZATION OF FUELS AND COMBUSTION GASES. Pro-
bis ceedings of the First Seminar organized by the Working Party on Air Pollution Problems of the United Nations Economic Commission for Europe and held in Geneva, 16-20 November 1970. 4 vols. (ST/ECE/AIR POLL/1 and Add. 1-3)
1971, UN Sales No. E.71.II.E/Mim.21
———Report on the Proceedings of the Seminar. [126] pp. including annexes. (ST/ECE/AIR POLL/1) $2.70
———Addendum I: Presentation and Discussion of Expert Papers on Hydrosulphurization of Residual Oils and Coal. (ST/ECE/AIR POLL/1/Add.1) In preparation
———Addendum II: Presentation and Discussion of Expert Papers on Economic Methods of Fuel Gasification and Fluidized Bed Combustion. [233] pp., fig., tables. (ST/ECE/AIR POLL/1/Add.2) $3.80
———Addendum III: Presentation and Discussion of Expert Papers on Removal of Sulphur Oxides from Combustion Gases by Dry and Wet Methods. [239] pp., fig., tables. (ST/ECE/AIR POLL/1/Add.3)
                 $4.20

Europe

963 Conference on PUBLIC HEALTH ASPECTS OF AIR POLLUTION IN
EUROPE, Milan, 6-14 November 1957. Report. 318 pp., illus. (EURO/
AP/1)
1958, WHO Regional Office for Europe   Out of print
   Summary report; papers on the situation in Europe and the U.S.A,
   and on biological, physical, chemical, engineering, administrative,
   educational, and other aspects of air pollution.

964 AIR POLLUTION IN EUROPE. Report on the Eighth European Seminar
for Sanitary Engineers...Brussels, 2-9 October 1962. 56 pp. (EURO
9.8)
1963, WHO Regional Office for Europe   Out of print
   Surveys current air pollution problems and central programs in
   Europe.

965 EPIDEMIOLOGY OF AIR POLLUTION; Report on a Symposium. By
P.J. Lawther, A.E. Martin [and] E.T. Wilkins. 32 pp. (Public Health
Papers, 15)
1962, WHO $0.30
   Brief review of epidemiological techniques and their potentialities in
   relation to air pollution problems in Europe.

CONTROL LEGISLATION

966 AIR POLLUTION: A SURVEY OF EXISTING LEGISLATION. 45 pp.
1963, WHO $0.60
   General survey of legislation and information on national legislation
   of 13 countries, with bibliographical and legislative references. Reprint from International Digest of Health Legislation, vol. 14, no. 2,
   1963, pp. 187-229.

Food and Agricultural Legislation. See entry 1195.

International Digest of Health Legislation. See entry 1196.

IONIZING RADIATION

EFFECTS

967 EFFECT OF RADIATION ON HUMAN HEREDITY. Report of a Study
Group convened by WHO together with papers presented by various
members of the Group. 168 pp.
1959, 2nd printing, WHO $4.00 cl.
   Report of the Study Group and papers (12) on the genetic effects of
   ionizing radiation on man. First issued in 1957.

968 EFFECT OF RADIATION ON HUMAN HEREDITY: INVESTIGATIONS
OF AREAS OF HIGH NATURAL RADIATION. First Report of the
Expert Committee on Radiation. 47 pp. (WHO Technical Report
Series, 166)
1959, WHO $0.30
   Considers the general principles of planning investigations of high-

radiation areas anywhere in the world, and illustrates their application to the specific situation in Kerala State and adjoining areas of Madras State in India where deposits of radioactive monazite are found. The second, fourth, and fifth reports of the Expert Committee concern exposure to ionizing radiation in radiation work. (For the third report, see entry 971.)

969   IONIZING RADIATION AND HEALTH. By B. Lindell and R. Lowry Dobson. 81 pp. (Public Health Papers, 6)
1961, WHO $1.00
Study of the effects of radiation upon health.

970   The USE OF VITAL AND HEALTH STATISTICS FOR GENETIC AND RADIATION STUDIES. Proceedings of the Seminar sponsored by the United Nations and WHO, held in Geneva, 5-9 September 1960. 259 pp. (A/AC.82/Seminar)
1962, UN   Sales No. 61.XVII.8 $7.50
Papers (22) and consensus of the seminar on how best to use and adapt national civil registration practices and vital and health statistics for genetic and other studies related to the radiation problem.

971   RADIATION HAZARDS IN PERSPECTIVE. Third Report of the Expert Committee on Radiation. 37 pp. (WHO Technical Report Series, 248)
1962, WHO $0.60
Compares the genetic effects of radiation, chemicals, and temperature and makes recommendations for protection programs.

972   Report of the UNITED NATIONS SCIENTIFIC COMMITTEE ON THE EFFECTS OF ATOMIC RADIATION. 165 pp. and corrigendum sheet; illus. (General Assembly. Official Records, 24th session, 1969, Supplement No. 13 (A/7613 and Corr.1))
1969, UN   No sales number $4.00
Fifth report to the General Assembly. Surveys the work of the Committee during its seventeenth and eighteenth sessions, 1967,1968. Includes technical annexes in which the Committee discusses in detail the scientific information on which it rests its conclusions on radioactive contamination of the environment by nuclear tests; effects of ionizing radiation on the nervous system; and radiation-induced chromosome aberrations in human cells. Includes list of reports received by the Committee. The first four reports, covering the first sixteen sessions, 1956-1966, were issued in 1958-1966.

CONTAINMENT AND SITING OF NUCLEAR POWER AND RESEARCH INSTALLATIONS

973   REACTOR SAFETY AND HAZARDS EVALUATION TECHNIQUES. Proceedings of the Symposium on Reactor Safety and Hazard Evaluation Techniques, sponsored by IAEA and held in Vienna, 14-18 May 1962. 2 vols. (544, 544 pp.) (Proceedings Series; STI/PUB/57)
1962, IAEA   Composite: E/F/R/S $10.00 per vol. cl.
Papers and discussions: (vol. 1) on review of reactor accidents and incidents (5 papers); some examples of national practice and principles (5); safety through good design and construction (8), through good siting and containment (9), and through control devices and instrumentation (10) — (vol. 2) on safety through good administration (8 papers); hazards evaluation (23); and safety assessment (5).

974 SITING OF REACTORS AND NUCLEAR RESEARCH CENTRES. Proceedings of the Symposium on Criteria for Guidance in the Selection of Sites for the Construction of Reactor and Nuclear Research Centres, held by IAEA at Bombay, 11-15 March 1963. 511 pp., fig., tables. (Proceedings Series; STI/PUB/72)
1963, IAEA   Composite: E/F/S $10.00 cl.
Papers and discussions on environmental considerations with particular reference to atmosphere (8 papers), and to the ground (4); containment as it affects site selection (4); criteria for site selection (5); experience relating to site selection for nuclear research centers (4) and power reactors (4); and panel discussion on future trends in site selection criteria.

975 CONTAINMENT AND SITING OF NUCLEAR POWER PLANTS. Proceedings of a Symposium on the Containment and Siting of Nuclear Power Plants, held by IAEA in Vienna, 3-7 April 1967. 818 pp., fig., tables. (Proceedings Series; STI/PUB/154)
1967, IAEA   Composite: E/F/S $16.50
Papers and discussions on national practices in reactor siting (14 papers); reactor siting considerations (12); containment for particular reactor types (6); containment for reactor types—general considerations (6); and release and transport of pollutants (10); and panel discussion.

976 ENVIRONMENTAL ASPECTS OF NUCLEAR POWER STATIONS. Proceedings of a Symposium on Environmental Aspects of Nuclear Power Stations, held by IAEA in co-operation with the United States Atomic Energy Commission in New York, 10-14 August 1970. 970 pp., fig., tables. (Proceedings Series; STI/PUB/261)
1971, IAEA   Composite: E/F/R/S $25.00
Papers and discussions on activities in the production of nuclear power which might have an impact upon the environment, with attention to future needs for electric power and the role of public understanding in power plant siting studies or regulatory practices. Covers nuclear power as an energy source (5 papers); standards for the control of effluents (13 papers and a panel discussion); effluent control and monitoring, including thermal effects (20 papers); considerations affecting selection of sites for steam power stations (18 papers); benefit-risk assessments (2 papers); and a panel discussion on prospects for the future.

MONITORING AND MEASUREMENT

977 Methods of RADIOCHEMICAL ANALYSIS. Report of a Joint WHO/FAO Expert Committee. 116 pp. (WHO Technical Report Series, 173)
1959, WHO $1.00
————. FAO. (FAO Atomic Energy Series, 1) $1.00
Report on methods for determining radioactive materials in the environment, in certain biological materials and foods and in the human body.

978 METEOROLOGICAL FACTORS INFLUENCING THE TRANSPORT AND REMOVAL OF RADIOACTIVE DEBRIS. Papers presented by WMO Experts during the 7th Session of the United Nations Scientific Committee on the Effects of Atomic Radiation. Ed. by W. Bleeker. xiii, 171 pp., illus. (Technical Notes, 43)
1961, WMO   WMO-No.111.TP.49 $3.00
Deals with the meteorological aspects of worldwide radioactive

fallout; transport and removal of debris originating from nuclear detonations, and the possibility of forecasting the future distribution of such fallout.

979   RADIOACTIVE SUBSTANCES IN THE BIOSPHERE. Procedures Recommended by the IAEA's Panel on the Methods for Collection and Analysis of Samples for the Determination of Trace Amounts of Radioactive Substances in the Biosphere, held at Vienna, 7-12 September 1959. 43 pp. (STI/PUB/28)
1961, IAEA   Free on request to IAEA
Provides recommended procedures for measurement of radioactivity over large areas: water (rivers, lakes, oceans, well and tap water); rain, snow, and dry deposition; air; food (milk, fish, and meat; rice and vegetables); human beings.

980   AGRICULTURAL AND PUBLIC HEALTH ASPECTS OF RADIOACTIVE CONTAMINATION IN NORMAL AND EMERGENCY SITUATIONS. Papers, presented at the Seminar, Scheveningen, the Netherlands, 11-15 December 1961, sponsored jointly by FAO, WHO and IAEA. 421 pp., illus. (FAO Atomic Energy Series, 5)
1964, FAO $7.00
Papers (27) on sources of radionuclides in the environment; mechanisms of transfer of radionuclides from the environment through food, air, and water, to man; radiation protection standards and criteria; types and objectives of monitoring programs; emergency situations, responsibilities and interrelations of public health, agricultural, and atomic energy administrations.

981   ENVIRONMENTAL MONITORING IN EMERGENCY SITUATIONS. A manual based on the Work of a Panel of Experts. 122 pp., illus. (Safety Series, 18; STI/PUB/118)
1965, IAEA $3.00
Nature of, and hazards resulting from, accidental releases of radioactive materials to the environs; an emergency monitoring programme: its objectives and formulation, organization, staffing, procedures, and equipment. References and bibliography.

982   METEOROLOGICAL ASPECTS OF ATMOSPHERIC RADIOACTIVITY. Papers presented by WMO Experts during the 13th Session of the United Nations Scientific Committee on the Effects of Atomic Radiation. Ed. by W. Bleeker. 194 pp., illus. (Technical Notes, 68)
1965, WMO   WMO.No.169.TP.83 $6.50
Reviews state of knowledge relating to formation and transport in the atmosphere of artificially formed radioactive particles and removal mechanisms. Includes one paper on contamination of the oceans by natural and man-made radioactivity.

983   Methods of SURVEYING AND MONITORING MARINE RADIOACTIVITY. Report of an Ad Hoc Panel of Experts. 95 pp. (Safety Series, 11; STI/PUB/86)
1965, IAEA $2.00
Offers adequate technical methods for surveying and monitoring the sea and marine products with regard to the presence of radioactive substances, including the selection of required measurements and sampling, analytical procedures for radionuclides of interest, and examples of practices and facilities.

984   Manual on ENVIRONMENTAL MONITORING IN NORMAL OPERATION.
        70 pp., illus. (Safety Series, 16; STI/PUB/98)
        1966, IAEA $1.50
        Technical guidance for setting up programs of routine environmental
        monitoring in the vicinity of nuclear establishments.

985   Methods of RADIOCHEMICAL ANALYSIS. 163 pp., illus.
        1966, WHO $4.75 cl.
        Describes detailed analytical procedures for the determination of
        tritium, phosphorus, strontium, ruthenium, iodine, cesium, polo-
        nium, radon, radium, thorium, uranium, and plutonium. The meth-
        ods were selected by a Joint FAO/IAEA/WHO Scientific Meeting as
        those of the greatest utility in the evaluation of radioactive con-
        tamination. The first chapters review the various sources of radio-
        active contamination and discuss the general principles involved in
        the measurement of environmental and human contamination, with
        special attention to the importance of obtaining representative
        samples.

986   TECHNIQUES FOR CONTROLLING AIR POLLUTION FROM THE
        OPERATION OF NUCLEAR FACILITIES. Report of a Panel on Tech-
        niques for Preventing Atmosphere Pollution from the Operation of Nu-
        clear Facilities held in Vienna, 4-8 November 1963. 118 pp. and erra-
        tum slip; illus. (Safety Series, 17; STI/PUB/121)
        1966, IAEA $2.50
        Manual prepared by IAEA aided by a panel of experts and provided
        for the guidance of those persons or authorities responsible for the
        organization, control, and operation of ventilation systems and air-
        cleaning installations in nuclear establishments.

987   RISK EVALUATION FOR THE PROTECTION OF THE PUBLIC IN RA-
        DIATION ACCIDENTS. A Report published on behalf of IAEA and WHO.
        78 pp. (Safety Series, 21; STI/PUB/124)
        1967, IAEA $2.00
        Report of a panel of experts convened to define a range of reference
        doses of radiation that should be helpful in assessing the hazards of
        nuclear installations and in making policy decisions. References.

988   ASSESSMENT OF AIRBORNE RADIOACTIVITY. Proceedings of a
        Symposium on Instruments and Techniques for the Assessment of Air-
        borne Radioactivity in Nuclear Operations held in Vienna, 3-7 July
        1967. 766 pp., illus. (Proceedings Series; STI/PUB/159)
        1967, IAEA   Composite: E/F/R $15.50
        Papers (56) on characteristics and control of airborne radioactive
        pollutants; techniques of sampling; instruments.

989   The MEASUREMENT OF ATMOSPHERIC RADIOACTIVITY. By O.
        Suschny. 109 pp., illus. (Technical Notes, 94)
        1968, WMO   WMO-No.231.TP.124 $10.50
        Dr. Suschny (IAEA) was designated by the Commission for Instru-
        ments and Methods of Observation as rapporteur on the subject and
        prepared this outline of the origin and movement of atmospheric ra-
        dioactivity and comprehensive guide to the techniques and apparatus
        used for the measurement of natural and artificial radionuclides in
        the atmosphere. Bibliography.

990   ROUTINE SURVEILLANCE FOR RADIONUCLIDES IN AIR AND WATER.
        64 pp., illus.
        1968, WHO $2.75
        Guide to surveillance of environmental radioactivity, particularly in
        air and water.

991   ENVIRONMENTAL CONTAMINATION BY RADIOACTIVE MATERIALS.
        Proceedings of a Seminar on Agricultural and Public Health Aspects of
        Environmental Contamination by Radioactive Materials jointly orga-
        nized by FAO, IAEA and WHO and held in Vienna, 24-28 March 1969.
        746 pp., illus. (Proceedings Series; STI/PUB/226)
        1969, IAEA   Composite: E/F/S $20.00
        Papers (62) on: Review of sources of radionuclides in the environ-
        ment; mechanisms of transfer of radionuclides from the environ-
        ment through food and water to man; radiation protection standards
        and criteria; monitoring programs; emergency situations; adminis-
        trative responsibilities for protection of the public.

992   Handbook on CALIBRATION OF RADIATION PROTECTION MONI-
        TORING INSTRUMENTS. 95 pp. (Technical Reports Series, 133; STI/
        DOC/10/133)
        1971, IAEA $3.00
        Guide for the use of those who are establishing or operating calibra-
        tion facilities for their radiation monitoring instruments, in five
        chapters: (1) introduction; (2) requirements of a calibration labora-
        tory and the techniques to be used; (3) calibration facilities; (4)
        errors and quality control; and (5) instrument maintenance and re-
        pair.

993   RAPID METHODS FOR MEASURING RADIOACTIVITY IN THE EN-
        VIRONMENT. Proceedings of the International Symposium on Rapid
        Methods for Measurement of Radioactivity in the Environment, orga-
        nized by the Government of the Federal Republic of Germany in co-
        operation with the Gesellschaft für Strahlen- and Umweltforschung
        mbH München, with the co-sponsorship of the International Atomic En-
        ergy Agency, and held in Neuherberg bei München, 5-9 July 1971. 967
        pp., fig. (Proceedings Series; STI/PUB/289)
        1971, IAEA   Composite: E/F/R/German $25.00
        Papers (76) and discussions on rapid techniques and methods of
        monitoring and analyzing environmental radioactive contamination:
        basic considerations (4 papers), chemical laboratory methods (20),
        physical laboratory methods (14), field methods (11), normal sur-
        veillance (8), emergency surveillance (9), data evaluation (3), rapid
        methods in the U.S.S.R. (6), and future developments (1 paper and
        panel discussion).

CONTROL LEGISLATION

994   PROTECTION AGAINST IONIZING RADIATION: A SURVEY OF EX-
        ISTING LEGISLATION. 170 pp.
        1964, WHO $2.00
        General survey of legislation and information on national legislation
        of 26 countries, with bibliographical and legislative references. Re-
        print from International Digest of Health Legislation, vol. 15, no. 2,
        1964, pp. 209-376.

        International Digest of Health Legislation. See entry 1196.

CHEMICAL AND BIOLOGICAL WEAPONS

995   CHEMICAL AND BACTERIOLOGICAL (BIOLOGICAL) WEAPONS AND
THE EFFECTS OF THEIR POSSIBLE USE. Report of the Secretary-
General. xiv, 100 pp. (A/7575/Rev.1-S/9292/Rev.1)
1969, UN   Sales No. E.69.I.24 $1.00
Scientific appraisal of the characteristics of the chemical and bacteriological (biological) weapons (CBW) which could be used in war-
fare; the effects they could have on individuals (military and civilian) and populations, plants, and animals; the atmospheric factors
affecting the use of CBW; the possible long-term effects of the use of
CBW on human health and ecology; and the economic and security
implications of the development, acquistion, and possible use of CBW
and associated weapons systems. Bibliography. The report was
prepared with the assistance of a Group of Consultant Experts and
others.

996   HEALTH ASPECTS OF CHEMICAL AND BIOLOGICAL WEAPONS. Report of a WHO Group of Consultants. 132 pp.
1970, WHO $4.00
Analyzes the health effects of the possible use of chemical and biological weapons on civilian population groups at different levels of
social and economic development, and the resulting implications for
WHO and its Member States. Also makes qualitative and quantitative
estimates of the health effects of selected chemical and biological
agents employed under specific hypothetical conditions. Apart from
conclusions on health effects of CBW, also notes that large-scale use
of CBW could also cause lasting changes of an unpredictable nature
in man's environment, and that the possible effects of CBW are
subject to a high degree of uncertainty and unpredictability, owing to
the involvement of complex and extremely variable meteorological,
physiological, epidemiological, ecological, and other factors. Iso-
lated and sabotage attacks not requiring highly sophisticated weapons
systems could be effective against large civilian targets in certain
circumstances with some CB agents (e.g., sabotage of water supplies
by contamination).

WATER POLLUTION

GENERAL

997   WATER POLLUTION CONTROL; Report of a WHO Expert Committee.
32 pp. (WHO Technical Report Series, 318)
1966, WHO $0.60
Reviews most important problems of water pollution control in areas
where the density of population and degree of industrialization are
greatest; evaluates progress and trends; identifies areas requiring
further research; and makes recommendations.

998 WATER POLLUTION CONTROL IN DEVELOPING COUNTRIES; Report of a WHO Expert Committee. 38 pp. (WHO Technical Report Series, 404)
1968, WHO $1.00
Reviews most important problems of water pollution in developing countries, including interrelationship of water pollution and water resources, evaluation of health aspects; makes recommendations on planning of water use and pollution control, training of requisite personnel, and suggestions of areas where further research is needed.

999 "Polynuclear Aromatic Hydrocarbons in the Water Environment." By Julian B. Andelman and Michael J. Suess. Bulletin of the World Health Organization, vol. 43, no. 2, 1970, pp. 479-508.
1970, WHO $3.25
Reviews the literature on polynuclear aromatic hydrocarbons (PAH) in the water environment, both carcinogenic and noncarcinogenic— physico-chemical properties and analysis of PAH; origin, source, and vehicles of transmission of PAH; PAH in environmental waters; effect of water and waste-water treatment on PAH; human exposure to PAH; health considerations; conclusions.

Europe

1000 "Water Sanitation." Bulletin of the World Health Organization, vol. 14, no. 5/6, 1956, pp. 839-1107.
1956, WHO Out of print
Articles (6), mostly selected from papers prepared for the fourth and fifth European Seminars for Sanitary Engineers, on pollution of surface water and groundwater in Europe; safe disposal of radioactive wastes; treatment and disposal of toxic and infectious industrial effluents; wastes from pulp and paper mills; estimation of bacterial density of water samples.

1001 Conference on WATER POLLUTION PROBLEMS IN EUROPE held in Geneva from 22 February to 3 March 1961. Documents submitted to the Conference. 3 vol. (600 pp.) ([WATER POLL./CONF.])
1961, UN   Sales No. 61.II.E/Mim.24, vol. 1-3   Out of print
Papers on economic, technical, and administrative aspects of water pollution problems in Europe. Vol. 1 contains papers dealing with the most urgent problems; vol. 2, papers on administrative and legal aspects of control; vol. 3, papers on economic aspects and the possibilities for international action through the establishment of international control bodies and the exchange of information.

1002 ASPECTS OF WATER POLLUTION CONTROL; A Selection of Papers prepared for the Conference on Water Pollution Problems in Europe, Geneva, 1961. 115 pp. (Public Health Papers, 13)
1962, WHO $1.25
Contains papers (7) on organizational, legal, financial, and economic aspects of water pollution control and damages through water pollution.

1003 ECONOMIC COMMISSION FOR EUROPE. ANNUAL REPORT (9 May 1965-29 April 1966). 104 pp. (Economic and Social Council. Official Records, 41st session, 1966, Supplement No. 3 (E/4177-E/ECE/622)).
1966, UN   No sales number $2.00
Contains (pp. 61-62) ECE resolution 10 (XXI), "The ECE declaration of policy on water pollution control," and appended ten Principles recommended to ECE governments.

1004   WATER POLLUTION CONTROL. Report on a Conference...Budapest,
       11-15 October 1966. 39 pp. (EURO 159.4)
       1967, WHO Regional Office for Europe
           Discusses various aspects of a comprehensive approach to water
           pollution control applicable to the present situation in Europe.

1005   Proceedings of the SEMINAR ON THE PROTECTION OF GROUND
       AND SURFACE WATERS AGAINST POLLUTION BY CRUDE OIL
       AND OIL PRODUCTS, organized by the Committee on Water Problems
       of the United Nations Economic Commission for Europe and held in
       Geneva, 1-5 December 1969. 2 vols. (ST/ECE/WATER/2)
       1971, UN   Sales No. E.70.II.E/Mim.30 $4.00 the set
           Volume 1 contains a general account of the seminar, introductory
           statements by its Chairman and Vice-Chairman and members of the
           ECE secretariat, summaries of the discussions, and three main
           papers; volume 2 contains seven main papers, statistical data, the
           road sign to be used in areas threatened by dangerous substances,
           and a list of participants.

CONTROL LEGISLATION

1006   CONTROL OF WATER POLLUTION; A SURVEY OF EXISTING LEGIS-
       LATION. 210 pp.
       1966, WHO $4.00
           Surveys legislation on water pollution control of 13 countries, with
           bibliographical and legal references. Reprint from International
           Digest of Health Legislation, vol. 17, no. 4, 1966, pp. 629-834.

1007   WATER POLLUTION CONTROL: NATIONAL LEGISLATION AND POL-
       ICY; a Comparative Study. By E.R. Malakoff. 67 pp.
       1968, FAO
           Compares and analyzes national and subnational legislation of vari-
           ous countries which have dealt with problems of water pollution con-
           trol or of water quality management. A main document (reference
           no. 02132-68-MR) issued jointly by the Land and Water Development
           Division and the Legislation Branch of the FAO Secretariat.

       Food and Agricultural Legislation. See entry 1195.

       International Digest of Health Legislation. See entry 1196.

CONTAMINATION OF FOOD

PESTICIDE RESIDUES: GENERAL

1008   TOXIC HAZARDS OF CERTAIN PESTICIDES TO MAN. By J.M.
       Barnes. 129 pp. (WHO Monograph Series, 16)
       1953, WHO   Out of print
           Describes the risks of poisoning from pesticides during their
           manufacture and application, and from traces of them which may
           occur in food products. Indicates methods of avoiding these risks.
           Contains nearly 700 bibliographical references.

1009   TOXIC HAZARDS OF PESTICIDES TO MAN. Report of a Study Group.
        51 pp. (WHO Technical Report Series, 114)
        1956, WHO   Out of print
            Covers toxic properties of pesticides; incidence and nature of
            poisoning; measures for the protection of pesticide operators; con-
            tamination of food and water; effect on domestic animals and fish;
            control of pesticide hazards by means of regulations.

1010   PRINCIPLES GOVERNING CONSUMER SAFETY IN RELATION TO
        PESTICIDE RESIDUES. Report of a Meeting of a WHO Expert Commit-
        tee on Pesticide Residues held jointly with the FAO Panel of Experts on
        the Use of Pesticides in Agriculture. 18 pp. (WHO Technical Report
        Series, 240)
        1962, WHO $0.30
            Discusses the four basic requirements for adequate control of
            potentially dangerous residues of pesticides—supervision by quali-
            fied agricultural advisers; analytical facilities for determining
            residues; the establishment, with toxicological advice, of the amount
            of pesticide which may be ingested daily without ill effect; and joint
            consultation between all the authorities concerned. Considers vari-
            ous methods of controlling pesticides; makes recommendations.

1011   RADIOISOTOPES IN THE DETECTION OF PESTICIDE RESIDUES.
        Proceedings of a Panel on the Uses of Radioisotopes in the Detection of
        Pesticide Residues held in Vienna, 12-15 April, 1965. 116 pp., illus.
        (Panel Proceedings Series; STI/PUB/123)
        1966, IAEA $2.50
            Papers (17) and panel recommendations.

1012   "Endrin Food-poisoning. A Report on Four Outbreaks Caused by Two
        Separate Shipments of Endrin-contaminated Flour." By D.E. Weeks.
        Bulletin of the World Health Organization, vol. 37, no. 4, 1967, pp. 499-
        512.
        1967, WHO $3.25
            Describes the course of the outbreaks in June and July 1967 in Hoha
            (Qatar) and Hofuf (Saudi Arabia), during which 874 persons were
            hospitalized and 26 died. Describes also the measures taken to
            ascertain the cause of the outbreaks and to prevent their extension
            and recurrence.

1013   "Protein-deficient Diet and DDT Toxicity." By Eldon M. Boyd and
        Elvira S. De Castro. Bulletin of the World Health Organization, vol. 38,
        no. 1, 1968, pp. 141-150.
        1968, WHO $3.25
            Report on a study to investigate the hypothesis that a diet low in pro-
            tein would affect ability to detoxify pesticides. Within the parame-
            ters of toxicity measured, the results suggest that DDT toxicity is
            not augmented by a low-protein diet.

1014   Recommended International Tolerances for PESTICIDE RESIDUES. 10
        pp. (Joint FAO/WHO Food Standards Programme: CAC/RS 2-1969)
        1969, FAO/WHO $1.00
            Recommendations of tolerances, advisory in nature, adopted by the
            FAO/WHO Codex Alimentarius Commission.

1015   NUCLEAR TECHNIQUES FOR STUDYING PESTICIDE RESIDUE PROB-
        LEMS. Proceedings of a Panel on Isotopic Tracer and Radioactive
        Techniques for Studying Residue Problems with particular reference to
        Food Entering International Trade, organized by the Joint FAO/IAEA

Division of Atomic Energy in Food and Agriculture and held in Vienna, 16-20 December 1968. 84 pp., fig. (Panel Proceedings Series; STI/PUB/252)
1970, IAEA $3.00
Papers (8), report and recommendations of the panel and examples of coordinated experiments with labeled insecticides recommended for initial study.

PESTICIDE RESIDUES: FAO/WHO EVALUATIONS OF TOXICITY

1016   Evaluation of the Toxicity of Pesticide Residues in Food. REPORT OF A JOINT MEETING of the FAO Committee on Pesticides in Agriculture and the WHO Expert Committee on Pesticide Residues, Geneva, 30 September-7 October 1963. 172 pp. (FAO Meeting Report No. PL/1963/13; WHO/Food Add./23 (1964)).
1964, FAO   Available on request*

1017A Evaluation of the Toxicity of Pesticide Residues in Food. REPORT OF THE SECOND JOINT MEETING of the FAO Joint Meeting of the FAO Committee on Pesticides in Agriculture and the WHO Expert Committee on Pesticide Residues, Rome 15-22 March 1965. 15 pp. (FAO Meeting Report No. PL/1965/10; WHO/Food Add./26.65)
1965, FAO   Available on request*

1017B EVALUATION OF THE TOXICITY OF PESTICIDE RESIDUES IN FOOD. 194 pp. (FAO Meeting Report No. PL/1965/10/1; WHO/Food Add./27.65)
1965, FAO   Available on request*

1017C EVALUATION OF THE HAZARDS TO CONSUMERS RESULTING FROM THE USE OF FUMIGANTS IN THE PROTECTION OF FOOD. 71 pp. (FAO Meeting Report No. PL/1965/10/2; WHO/Food Add./28.65)
1965, FAO   Available on request*

1018A Pesticide Residues in Food. JOINT REPORT of the FAO Working Party on Pesticide Residues and the WHO Expert Committee on Pesticide Residues, Geneva, 14-24 November 1966. 19 pp. (FAO Agricultural Studies, 73)
1968, 2nd printing, FAO $0.60
———. 1967, WHO. (WHO Technical Report Series, 370) $0.60

1018B EVALUATION OF SOME PESTICIDE RESIDUES IN FOOD. 237 pp. (FAO PL:CP/15; WHO/Food Add./67.32)
1967, FAO   Available on request*

1019A Pesticide Residues. REPORT OF THE 1967 JOINT MEETING of the FAO Working Party and the WHO Expert Committee, Rome, Italy, 4-11 December 1967. 43 pp. (FAO Meeting Report No. PL: 1967/M/11)
1968, FAO
———. 1968, WHO. (WHO Technical Report Series, 391) $1.00

1019B 1967 EVALUATIONS OF SOME PESTICIDE RESIDUES IN FOOD. 242 pp. (FAO/PL:1967/M/11/1; WHO/Food Add./68.30)
1968, FAO   Available on request*

* From Food Additives, WHO (Geneva), or from Food Science and Technology Branch, FAO (Rome).

1020A Pesticide Residues in Food. REPORT OF THE 1968 JOINT MEETING
of the FAO Working Party of Experts on Pesticide Residues and the
WHO Expert Committee on Pesticide Residues, Geneva, 9-16 December
1968. 40 pp. (FAO Agricultural Studies, 78)
1969, FAO $1.00
————. 1969, WHO. (WHO Technical Report Series, 417) $1.00

1020B 1968 EVALUATIONS OF SOME PESTICIDE RESIDUES IN FOOD. 293
pp. (FAO/PL:1968/M/9/1; WHO/Food Add./69.35)
1969, FAO   Available on request*

1021A Pesticide Residues in Food. REPORT OF THE 1969 JOINT MEETING
of the FAO Working Party of Experts on Pesticide Residues and the
WHO Expert Committee on Pesticide Residues, Rome, 8-15 December
1969. 43 pp. (FAO Agricultural Studies, 84)
1970, FAO $1.00
————. 1970, WHO. (WHO Technical Report Series, 458) $1.00

1021B 1969 EVALUATION OF SOME PESTICIDE RESIDUES IN FOOD: the
Monographs. 243 pp. (FAO/PL:1969/M/17/1; WHO/Food Add./70.38)
1970, FAO/WHO   Available on request*

1022A Pesticide Residues in Food. REPORT OF THE 1970 JOINT MEETING
of the FAO Working Party of Experts on Pesticide Residues and the
WHO Expert Committee on Pesticide Residues, Rome, 9-16 November
1970. 44 pp. and corrigendum slip. (FAO Agricultural Studies, 87)
1971, FAO $1.00
————. 1971, WHO. (WHO Technical Report Series, 474) $1.00
The summary reports of the joint meetings listed above contain general
considerations, including the principles adopted for the evaluations of
consumer hazards arising from the use of pesticides, proposal of ac-
ceptable daily intakes for man of residues, estimation of tolerances to
pesticide residues, and a summary of a number of pesticide evaluations. The other FAO main documents are summaries of data considered at the joint meetings in arriving at the recommendations for ac-
ceptable daily intakes, tolerances and methods of analysis.

CONTROL LEGISLATION

1023 CONTROL OF PESTICIDES: a Survey of Existing Legislation. 150 pp.
1970, WHO $4.00
General survey of legislation on pesticide control; legislation of 11
countries; bibliography. Reprinted from International Digest of
Health Legislation, vol. 20, no. 4, 1969, pp. 579-726.

Current Food Additives Legislation. See entry 1194.

Food and Agricultural Legislation. See entry 1195.

International Digest of Health Legislation. See entry 1196.

* From Food Additives, WHO (Geneva), or from Food Science and
Technology Branch, FAO (Rome).

RADIONUCLIDES

1024   RADIOACTIVE MATERIALS IN FOOD AND AGRICULTURE. Report of an FAO Expert Committee, Rome, 30 November-11 December 1959. 123 pp. (FAO Atomic Energy Series, 2)
        1960, FAO $1.50
        Deals with problems of radioactive contamination of the food chain in terms of food, agricultural and fisheries experience and knowledge; contains a brief summary of the conclusions and recommendations of the committee, and five technical appendices.

    a. ————. Supplement to the Report...Working Papers submitted to the Committee. 315 pp., illus.
        1960, FAO   Mimeographed   Out of print

1025   DIETARY LEVELS OF STRONTIUM-90 AND CESIUM-137. A Summary of World Information. 78 pp., illus. (FAO Atomic Energy Series, 3)
        1962, FAO $1.00
        Summary of information from a review of all published sources to the end of 1960 on food surveys and sampling techniques for radionuclides. References.

1026   Organization of Surveys for RADIONUCLIDES IN FOOD AND AGRICULTURE. Report of an FAO Expert Committee, Rome, 12-18 July 1961. 103 pp. (FAO Atomic Energy Series, 4)
        1962, FAO $1.00
        Indicates methods by which various types of surveys of radioactivity in the human diet may be carried out.

DISPOSAL OF COMMUNITY AND INDUSTRIAL WASTES

1027   COMPOSTING. Sanitary Disposal and Reclamation of Organic Wastes. By H.B. Gotaas. 205 pp., illus. (WHO Monograph Series, 31)
        1956, WHO $5.00 cl.
        Covers decomposition of organic matter; sanitary and agricultural importance; historical development; raw material—quantity and decomposition; methods and planning for cities; methods for villages and small farms; manure and night-soil digesters for methane recovery on farms and in villages.

1028   EXCRETA DISPOSAL FOR RURAL AREAS AND SMALL COMMUNITIES. By Edmund G. Wagner and J.N. Lanoix. 187 pp., illus. (WHO Monograph Series, 39)
        1958, WHO $5.00
        Covers the privy method of excreta disposal, water-carried methods of disposal for rural areas, and excreta disposal programs.

1029   TREATMENT AND DISPOSAL OF WASTES. Report of a WHO Scientific Group. 30 pp. (WHO Technical Report Series, 367)
        1967, WHO $0.60
        Reviews present knowledge of, and the most important problems in, the treatment of waste waters and solid wastes, with recommendations on relevant research needs. Bibliography.

1030   "La Pollution microbienne, virale et parasitaire des eaux littorales et
        ses consequences pour la santé publique." By J. Brisou. Bulletin of
        the World Health Organization, vol. 38, no. 1, 1968, pp. 79-118.
        1968, WHO $3.25
            Summarizes information on pollution of coastal waters by pathogenic
            bacteria, parasites, and viruses; the survival of these organisms in
            sea water; and the health hazard presented by the problem of such
            pollution beyond the means of natural self-purification as the rapid
            rise in the world's population increases the pollution of coastal
            waters by human wastes. Bibliography.

1031   PROBLEMS IN COMMUNITY WASTES MANAGEMENT. by H.M. Ellis
        and others. 89 pp. (Public Health Papers, 38)
        1969, WHO $2.00
            The six chapters of this volume range from contributions dealing
            with the management of solid and liquid wastes to research needs
            and the controlled reuse of waste water.

1032   WASTE STABILIZATION PONDS. By Earnest F. Gloyna. 175 pp., fig.,
        tables. (WHO Monograph Series, 60)
        1971, WHO $6.00 cl.
            Summarizes available information on waste stabilization ponds, de-
            scribes past and present use of such ponds through the world, de-
            fines acceptable design criteria based on public health considera-
            tions, suggests alternative approaches to design, and deals with
            operational problems. Also provides useful information on the the-
            ory of biological waste treatment and on the control of disease
            transmission in ponds.

1032   SOLID WASTES DISPOSAL AND CONTROL. Report of a WHO Commit-
bis   tee. 34 pp. (WHO Technical Report Series, 484)
        1971, WHO $1.00
            Reviews the existing impact on health and on socioeconomic factors
            of improper handling of solid wastes, appraises current practices in
            solid wastes management, and identifies areas for further action.

                                     Europe

1033   "Aspects économiques de l'épuration des eaux useś." By H. Rohde.
        Bulletin of the World Health Organization, vol. 20, no. 4, 1959, pp. 509-
        533, illus.
        1959, WHO $3.25
            Describes construction and operating costs, and their calculation, in
            the Federal Republic of Germany for a variety of sewage treatment
            plants.

1034   "The Economics of the Disposal of Sewage and Trade Effluents." By
        C.B. Townsend. Bulletin of the World Health Organization, vol. 20, no.
        4, 1959, pp. 535-562.
        1959, WHO $3.25
            Touches on all aspects of the subject in England and Wales.

1035   ECONOMIC ASPECTS OF TREATMENT AND DISPOSAL OF CERTAIN
        INDUSTRIAL EFFLUENTS. 3 vol. (134; 132; 194 pp.) Mimeographed.
        (WATER/POLL/Econ/6, vol. 1-3)
        1967, UN   Sales No. 67.II.E/Min.56, vol. 1-3 $5.00 the set
            Papers and case studies submitted to a meeting of experts convened
            by ECE in November 1966 to consider measures for improving the

techniques of treatment and disposal of industrial effluents, and reducing costs. Vol. 1 covers papers on legislative and administrative standards and regulations in ECE countries. Vol. 2 contains case studies relating to the pulp, paper, and board industries; and vol. 3, case studies relating to the textile industry and to the mixed treatment of industrial and domestic waste products.

DISPOSAL OF RADIOACTIVE WASTES

1036 DISPOSAL OF RADIOACTIVE WASTES. Proceedings of the Scientific Conference on the Disposal of Radioactive Wastes sponsored by IAEA and UNESCO, with the co-operation of FAO, and held at the Oceanographic Museum of the Principality of Monaco, 16-21 November 1959. 2 vol. (612, 584 pp.); illus. (Proceedings Series; STI/PUB/18)
1960, IAEA   Composite: E/F/R/S
Papers (70) and discussions. Contents: Vol. I: Nature of radioactive wastes; treatment and processing; methods of disposal; administrative and general considerations—Vol. II: Biological, physical, and chemical aspects of sea disposal; advantages and disadvantages of sea disposal; oceanographic and fisheries research required for safe disposal; general considerations for ground disposal; advantages and disadvantages of disposal into geological structures.

1037 RADIOACTIVE-WASTE DISPOSAL INTO THE SEA. Report of the Ad Hoc Panel convened by the Director-General of IAEA.... 174 pp., illus. (Safety Series, 5; STI/PUB/14)
1961, IAEA $2.50
Report, completed in 1960, of a panel meeting in 1958 and 1959. Subjects include extent of and approach to problems, control measures, maximum permissible exposure to radiation, nature of the marine environment, data on wastes, mixing and exchange processes.

1038 PROCESSING OF RADIOACTIVE WASTES. By C.A. Mawson. 44 pp. (Review Series, 18; STI/PUB/15/18)
1961, IAEA $1.00
Review of techniques for processing gaseous, liquid, and solid radioactive wastes between 1955 and the end of 1960. References and classified, supplemental bibliography.

1039 TREATMENT AND STORAGE OF HIGH-LEVEL RADIOACTIVE WASTES. Proceedings of the Symposium on Treatment and Storage of High-level Radioactive Wastes held by the IAEA in Vienna, 8-12 October 1962. 666 pp., illus. (Proceedings Series; STI/PUB/63)
1963, IAEA   Composite: E/F/R/S $13.00
Papers (33) on: Problems and practices; calcination and storage; solidification and fixation of liquids; treatment of solid wastes; shipment of large quantities of radioactive materials.

1040 TECHNOLOGY OF RADIOACTIVE WASTE MANAGEMENT AVOIDING ENVIRONMENTAL DISPOSAL. Report of a Panel held in Vienna from 12-16 February 1962. 148 pp., illus. (Technical Reports Series, 27; STI/DOC/10/27)
1964, IAEA $3.00
Examines the relation of present radioactive waste management to

total containment and how closely total containment might be approached using present technology; assesses the effects of research and development work under way and points out what further technological developments are needed in the collection, segregation, treatment, and storage of radioactive waste. Bibliography.

1041   Proceedings of the Third International Conference on the PEACEFUL USES OF ATOMIC ENERGY, held in Geneva, 31 August-9 September 1964. Multilingual ed. 16 vol. (A/CONF.28/1, vol. 1-16)
1965, UN   Composite: E/F/R/S
   Vol. 14: Environmental Aspects of Atomic Energy and Waste Management. 380 pp., illus. (A/CONF.28/1, vol. 14)
   Sales No. 65.IX.14 $12.50 cl.
      Papers on safety aspects of large-scale use of atomic energy; measurement techniques (3 papers), fission products in the atmosphere (5) and in fresh and sea water (5), health and safety problems (7), radioactive waste management (16).

1042   The MANAGEMENT OF RADIOACTIVE WASTES PRODUCED BY RADIOISOTOPE USERS. A Code of Practice based on the Report of a Panel of Experts. 58 pp. (Safety Series, 12; STI/PUB/87)
1965, IAEA $1.50
      Code of practice intended for the guidance of radioisotope users and of other persons or authorities responsible for the individual or collective management of the relatively small quantities of radioactive waste arising from the use of radioisotopes in laboratories, hospitals and industry when no special facilities for disposal are available on the site. The code is based on the work of two panels meeting in 1961 and 1962.
   a. ———TECHNICAL ADDENDUM. 81 pp., illus. (Safety Series, 19; STI/PUB/119)
      1966, IAEA $2.00
      Contains detailed technical information on processes and procedures which are outlined in more general terms in the code of practice.

1043   RADIOACTIVE WASTE DISPOSAL INTO THE GROUND. 111 pp., illus. (Safety Series, 15; STI/PUB/103)
1965, IAEA $2.50
      Report based on discussions by an ad hoc panel of experts. Subjects: Site characteristics affecting ground disposal and its investigation; chemical reactions of wastes in the ground and their physical behavior; modes of release; evaluation of sites and methods of ground disposal; standards and control techniques. References and bibliography.

1044   PRACTICES IN THE TREATMENT OF LOW- AND INTERMEDIATE-LEVEL RADIOACTIVE WASTES. Proceedings of a Symposium on Practices in the Treatment of Low- and Intermediate-level Radioactive Wastes jointly organized by IAEA and ENEA and held in Vienna, 6-10 December 1965. 952 pp., illus. (Proceedings Series; STI/PUB/116)
1966, IAEA   Composite: E/F/R/S $19.00
      Papers (53) on general management; operating experience with existing facilities; treatment of solid wastes; special techniques.

1045   DISPOSAL OF RADIOACTIVE WASTES INTO SEAS, OCEANS AND SURFACE WATERS. Proceedings of the Symposium on the Disposal of Radioactive Wastes into Seas, Oceans and Surface Waters, held by

IAEA at Vienna, 16-20 May 1966. 898 pp., illus. (Proceedings Series; STI/PUB/126)
1966, IAEA   Composite: E/F/R $18.50
Papers (56) on physical and biological transport of radionuclides resulting from the disposal of radioactive wastes; evaluation of the resulting main exposure routes to man; possible effects of the disposal on marine and surface water resources.

1046   OPERATION AND CONTROL OF ION-EXCHANGE PROCESSES FOR TREATMENT OF RADIOACTIVE WASTES. 147 pp., illus. (Technical Reports Series, 78; STI/DOC/10/78)
1967, IAEA $3.00
Manual, compiled by L.A. Emelity, on disposal of large-volume, low-level radioactive wastes by concentration and containment, using technique of ion exchange. Reviews facilities currently using this method. Bibliography.

1047   DISPOSAL OF RADIOACTIVE WASTES INTO THE GROUND. Proceedings of a Symposium on the Disposal of Radioactive Wastes into the Ground jointly organized by IAEA and ENEA of OECD and held in Vienna, 29 May-2 June 1967. 666 pp., illus. (Proceedings Series; STI/PUB/156)
1967, IAEA   Composite: E/F/R/S $14.00
Papers (43) on operational experience; uptake and migration; site selection; buried solidified wastes; salt disposal and disposal into deep or porous formations. Symposium was held jointly by IAEA and ENEA.

1048   BASIC FACTORS FOR THE TREATMENT AND DISPOSAL OF RADIO-ACTIVE WASTES. 41 pp., illus. (Safety Series, 24; STI/PUB/170)
1967, IAEA $1.00
Manual prepared by the Panel on Selection Factors for Waste Management Systems, Vienna, 24-28 October 1968. Defines a waste management system and lists factors for selecting waste management systems.

1049   TREATMENT OF AIRBORNE RADIOACTIVE WASTES. Proceedings of a Symposium on Operating and Developmental Experience in the Treatment of Airborne Radioactive Wastes held by IAEA in New York, 26-30 August 1968. 816 pp., illus. (Proceedings Series; STI/PUB/195)
1968, IAEA   Composite: E/F/S $19.00
Papers (52) on monitoring air contaminants; characteristics of air contaminants from nuclear reactors; filters: design, development cost; testing high-efficiency filters; removal of noble gases; special problems related to heat and moisture; removal of iodine and its compounds; spray technology; airborne wastes from incineration; and operational experience in the treatment of airborne wastes.

1050   MANAGEMENT OF RADIOACTIVE WASTES AT NUCLEAR POWER PLANTS. 225 pp. and 2 fold. fig. in pocket, illus. (Safety Series, 28; STI/PUB/208)
1968, IAEA $6.00
Manual on factors to be considered in design and operations (sources and character of wastes, standards and criteria, environmental factors, process capabilities, performance verification and monitoring). Practice in Canada, France, the United Kingdom, and the U.S.A. summarized.

1051   TREATMENT OF LOW- AND INTERMEDIATE-LEVEL RADIOACTIVE
       WASTE CONCENTRATES. Report of a Panel. 110 pp., illus. (Techni-
       cal Reports Series, 82; STI/DOC/10/82)
       1968, IAEA $2.50
           Report of a panel convened in Vienna, 2-6 May 1966. Reviewed cur-
           rent operating practices and experiences. Developed detailed cost
           data on typical operations in various countries for treatment of con-
           centrates. Deals with concentration by evaporation; filtration;
           insolubilization in bitumen and in cement or cement-vermiculite.

1052   ECONOMICS IN MANAGING RADIOACTIVE WASTES. Report resulting
       from two panels of experts on the economics of radioactive waste man-
       agement. 46 pp. (Technical Reports Series, 83; STI/DOC/10/83)
       1968, IAEA $1.00
           Panel agreed on a definition of waste management for accounting
           purposes and devised a relatively simple system of cost reporting,
           which the participants tested by applying it to waste management
           operations at their 15 nuclear research establishments in 12 coun-
           tries.

1053   DESIGN AND OPERATION OF EVAPORATORS FOR RADIOACTIVE
       WASTES. Comp. by Y. Yamomoto, assisted by N. Mitsuishi and S.
       Kadoya. 115 pp., illus. (Technical Reports Series, 87; STI/DOC/10/87)
       1968, IAEA $2.50
           Manual, compiled under the direction of Frank N. Browder, on the
           application of evaporators to the treatment of radioactive wastes,
           including the design and description of evaporator types and as-
           sociated equipment, operational procedures, disposal of evaporator
           condensates and concentrates, costs. Bibliography.

1054   CHEMICAL TREATMENT OF RADIOACTIVE WASTES. Comp. by P.E.
       Pottier. 80 pp., illus. (Technical Reports Series, 89; STI/DOC/10/89)
       1968, IAEA $2.50
           Manual on the concentration of radioactive liquid wastes by chemical
           precipitation. Deals with the principles of coagulation-flocculation
           and sedimentation and associated processes, equipment, conditioning
           and disposal of flocculation sludge, sampling, laboratory work, and
           the factors governing the selection of processes. Organized by L.H.
           Keher, ed. by Frank N. Browder.

1055   "Disposal of Radioactive Wastes into Deep Geological Formations."
       By H. Krause. Atomic Energy Review, vol. 7, no. 1, April 1969, pp.
       47-70, illus.
       1969, IAEA $4.00
           References.

1056   STANDARDIZATION OF RADIOACTIVE WASTE CATEGORIES. Report
       of a Panel held in Vienna, 6-10 November 1967. 18 pp. (Technical Re-
       ports Series, 101; STI/DOC/10/101)
       1970, IAEA $1.00
           Simple international standard classification of categories of radioac-
           tive wastes, to be used as a common language between people work-
           ing in the field of waste management. While the waste categories
           are not suitable for regulatory purposes of for use in health and
           safety evaluations, a future panel may assess the experience gained
           from the use of the present standard and consider expanding the con-
           cept of an international standard to cover these purposes also.

1057   The VOLUME REDUCTION OF LOW-ACTIVITY SOLID WASTES. Report of a Panel. Prepared by J. Pradel in collaboration with P.J. Parsons and E. Malášek. 44 pp., illus. (Technical Reports Series, 106; STI/DOC/10/106)

1970, IAEA $2.00

Describes methods for reducing the volume of solid radioactive wastes in order to transport them in regulation packages to storage places away from populated areas where local storage might hinder local development. Considers reduction by mechanical methods (compacting, fragmentation), by incineration, and by other methods.

1058   BITUMINIZATION OF RADIOACTIVE WASTES. Review of the Present State of the Development and Industrial Application. 135 pp., illus. (Technical Reports Series, 116; STE/DOC/10/116)

1970, IAEA $4.00

Report, in 15 chapters, arising from the Research Co-ordination Meeting on the Incorporation of Radioactive Wastes in Bitumen, Dubna, U.S.S.R., 1968. Reviews research and development in many countries of the incorporation of solid radioactive wastes in bitumen, the advantages and disadvantages of the process, and its economic aspects.

1059   MANAGEMENT OF LOW- AND INTERMEDIATE-LEVEL RADIOACTIVE WASTES. Proceedings of a Symposium on Developments in the Management of Low- and Intermediate-level Radioactive Wastes, jointly organized by IAEA and the European Nuclear Energy Agency and held in Aix-en-Provence, 7-11 September 1970. 814 pp., fig., tables. (Proceedings Series; STI/PUB/264)

1970, IAEA   Composite: E/F/R/S $22.00

Review paper and papers on national waste management policies and experience (16), operational experience and site policies (25), and developments (11).

1060   DISPOSAL OF RADIOACTIVE WASTES INTO RIVERS, LAKES AND ESTUARIES. Report of a Panel of Experts sponsored by IAEA and WHO. 77 pp. (Safety Series, 36; STI/PUB/283)

1971, IAEA $3.00

Presents principles and practices of radioactive waste management which will ensure that use of fresh-water systems will not be jeopardized by pollution. Revises and brings up to date the 1963 publication, Disposal of Radioactive Wastes into Fresh Water (Safety Series, 10; STI/PUB/44).

1061   "Leach Testing of Immobilized Radioactive Solid Wastes. A Proposal for a Standard Method." Ed. by E.D. Hespe. Atomic Energy Review, vol. 9, no. 1, 1971, pp. 195-207.

1971, IAEA $4.00

An important treatment preliminary to the disposal of radioactive waste is its solidication and incorporation into a stable medium from which the radioactive components cannot readily be leached by waters which may come into contact with the waste after disposal. In order to test the quality of any solidified form of waste and to intercompare various products from different processes or different laboratories it is desirable to have a standard method of measuring the leach rate by water of the radioactive components in the solid waste. Two leach rate tests are presented which were the result of recommendations by an IAEA panel of experts (1969). One method is for intercomparison purposes and the other to approximate the behavior of an immobilized solid under the conditions of the storage environment.

1062　"Solidification of Low- and Intermediate-level Wastes." By B.H.
　　　　Burns. Atomic Energy Review, vol. 9, no. 3, 1971, pp. 547-599.
　　　　1971, IAEA　　　　　　　　　　　　　　　　　　　　$4.00
　　　　Surveys present processes for the incorporation of concentrated
　　　　radioactive wastes in cement or concrete or into bitumen, and other
　　　　processes; assesses leaching of these incorporates; and compares
　　　　costs of cementation and bituminization processes.

1063　"Principles for Limiting the Introduction of Radioactive Wastes into the
　　　　Sea." Ed. by C.M. Slansky. Atomic Energy Review, vol. 9, no. 4, 1971,
　　　　pp. 853-868.
　　　　1971, IAEA　　　　　　　　　　　　　　　　　　　　$4.00
　　　　Discusses the basic principles governing the marine disposal of
　　　　radioactive wastes; shows how these principles can be applied in
　　　　practice; and indicates the basic methodology for establishing pro-
　　　　cedures for limiting the releases of radioactive wastes into the seas
　　　　and oceans. Draws attention to monitoring both waste and the en-
　　　　vironment and suggests a need for both national and international
　　　　records of waste discharge. Mentions areas where further research
　　　　is required.

MARINE POLLUTION

1064　POLLUTION OF THE SEA BY OIL. Results of an Inquiry Made in 1963.
　　　　104 pp., 2 fold. charts.
　　　　1964, IMCO　　　　Sales No. IMCO.1964.2　　　　　$1.10
　　　　Contains information from 37 countries on prevention of pollution of
　　　　the sea by oil. Brings up to date the information in the out-of-print
　　　　United Nations inquiry of 1956. Information received on port facili-
　　　　ties was separately published (entry 1065).

1065　FACILITIES IN PORTS FOR THE RECEPTION OF OIL RESIDUES. Re-
　　　　sults of an Inquiry Made in 1963. 59 pp.
　　　　1964, IMCO　　　　Sales No. IMCO.1964.3　　　　　$0.70
　　　　Surveys situation with regard to oil pollution and progress made in
　　　　the provision of reception facilities in the ports of 27 countries.
　　　　Brings up to date the information in the 1956 United Nations survey
　　　　(out of print).

1066　INTERNATIONAL CONVENTION FOR THE PREVENTION OF POLLU-
　　　　TION OF THE SEA BY OIL, 1954, including the Amendments Adopted in
　　　　1962.... 47 pp.
　　　　1967, IMCO　　　　Sales No. IMCO.1967.4　　　　　$1.00
　　　　　　　　　　　　Bilingual: E/F
　　　　Text of 1954 Convention, as amended by the 1962 Conference and the
　　　　resolutions adopted by the Conference.

1067　CHARTS OF PROHIBITED ZONES. 6 pp., 5 fold. maps.
　　　　1967, IMCO　　　　Sales No. IMCO.1967.5(E)　　　$0.70
　　　　Contains texts of Annex A to the 1954 Convention and Annex A to the
　　　　Convention as modified by the 1962 Conference, and related maps of
　　　　zones in which the discharge of oil or oily mixtures is prohibited.
　　　　Also contains the text of Annex B (Form of Oil Record Book).

1068 INTERNATIONAL LEGAL CONFERENCE ON MARINE POLLUTION
DAMAGE, 1969. FINAL ACT of the Conference with Attachments in-
cluding the Texts of the Adopted CONVENTIONS.... 88 pp.
1970, IMCO   IMCO.1970.3 $2.20
Bilingual: E/F
In addition to the Final Act of the Conference, contains texts of the
International Convention relating to Intervention on the High Seas in
Cases of Oil Pollution Casualties, and the International Convention
on Civil Liability for Oil Pollution Damage.

1069 Reference Methods for MARINE RADIOACTIVITY STUDIES. Sampling
Techniques and Analytical Procedures for the Determination of Selected
Radionuclides and Their Stable Counterparts. I: Strontium, Caesium,
Cerium, Cobalt, Zinc and Others. 284 pp., illus. (Technical Reports
Series, 118; ST/DOC/10/118)
1970, IAEA $7.00
Sets of analytical methods for the detection of selected radionuclides
and corresponding stable elements, including the collection, storage,
and preparation of samples for chemical and radiochemical analyses. The report results from a study by a 1968 panel on strontium,
cesium, cerium, cobalt, and zinc, and aims at improving techniques
for measuring radioactivity in marine organisms and their environment and ensuring comparability of results. References and ten
supporting technical papers.

NOISE

1070 NOISE; an Occupational Hazard and Public Nuisance. By Alan Bell.
131 pp. (Public Health Papers, 30)
1966, WHO $2.00
Short account of the effects of noise upon the human body and mind
and of possible methods of controlling noise. Contains an extensive
list of references.

FILMSTRIPS AND OTHER VISUAL AIDS

1071 AFRICA'S WILDLIFE IS IN OUR HANDS. Filmstrip. Color, 87 frames,
Single frame. With commentary (bilingual: E/F; 24 pp.)
1967, FAO $5.00
Teaches the importance of saving Africa's wildlife. Text by Florita
Botts.

1072 BETTER HOMES AND BETTER COMMUNITIES IN THE SOUTH PA-
CIFIC. Filmstrip. Color, 52 frames. Single frame. With commentary
(8 pp.)
1965, FAO $5.00
Illustrates the work of the South Pacific Community Education Training Centre, in Fiji. Script by Joyce Meyer and Florita Botts. Adults
and youth.

1073   BILL, PEPITO AND COMPANY. Cartoon strip with commentary, 47
        drawings in full color. 8 pp.
        1967, FAO   50 copies $1.75*
        Text by Brian Taylor, illustrations by Jack Maxwell. Contrasts the
        life of four children and their families in rural North America, Af-
        rica, Asia, and Central America. Shows how FAO works to improve
        rural conditions in the developing countries, and how Bill, the Amer-
        ican boy, can help in this work. *Available in bulk only.

1074   CLOUDS AND METEORS. Filmstrip, 35 mm. Color, 35 frames. With
        commentary (44 pp.)
        1960, UNESCO
        Filmstrip is intended to help teachers instruct their pupils in cloud
        classification and weather observation, and was produced in cooper-
        ation with WMO. Text by W. Bleeker. For senior high schools and
        adults.

1075   DISCOVERING THE OCEAN. Filmstrip, 35 mm. Black and white, 41
        frames. With commentary (trilingual: E/F/S; 77 pp.)
        1965, UNESCO $3.50
        Illustrates modern techniques in oceanographic research aboard
        ships and in laboratories on land. Explains the need for international
        cooperation in this field and the importance of the scientific findings
        in everyday life. Text by Jean Lévigne and Jean Vidal. For junior
        and senior high schools and adults.

1076   THE EXTENSION AGENT. Filmstrip. Color, 61 frames. Double
        frame. With commentary (8 pp.)
        1970, FAO $5.00
        Shows how a typical agricultural extension agent goes about his daily
        rounds in his district. Filmed in the Andean highlands of Ecuador.
        For training extension agents.

1077   EXTENSION WORK AND TEACHING AIDS. Filmstrip. Black and white,
        46 frames. Single frame. With commentary (trilingual: E/F/S; 19 pp.)
        1965, FAO $4.00
        Cartoon drawings tell the story of two extension workers—one a
        specialist in agriculture, the other a home economist—at work in a
        rural village. Text by Florita Botts, drawings by Jack Maxwell.
        For training rural community leaders and extension agents.

1078   FAO IN THE FIELD. Filmstrip. Color, 90 frames. With commentary
        (8 pp.)
        1970, FAO   Double frame version $5.00
                                               Single frame version $5.00
        Illustrates FAO's programs of technical assistance in developing
        countries. Useful for lecturers.

1079   FERTILIZERS AND THEIR USE. Filmstrip. Color, 62 frames. Double
        frame. With commentary (9 pp.)
        1965, FAO $5.00
        Based on the FAO pocket guide, Fertilizers and Their Use (entry
        430). A teaching aid for students of agronomy, agricultural exten-
        sion trainees, and for farmers' training courses.

1080   FERTILIZERS IN THE FIGHT AGAINST HUNGER. Filmstrip. Color,
        50 frames. Double frame. With commentary (8 pp.)
        196 , FAO $5.00
        Shows how the FAO-FFHC Fertilizer Program is working and its
        aims and accomplishments.

1081　FISH—FOOD FOR BILLIONS. Filmstrip. Color, 41 frames. Single
　　　frame. With commentary (trilingual: E/F/S; 16 pp.)
　　　1965, FAO　　　　　　　　　　　　　　　　　　　　　　$5.00
　　　　Illustrates the significance of fish as an inexpensive protein-rich
　　　　food; surveys primitive methods of fishing and fish curing around the
　　　　world; shows various ways of improving boats and nets, introducing
　　　　mechanization, and new methods of preservation; shows the farming
　　　　of freshwater fish; and suggests new ways of making fish palatable to
　　　　the public. Text by Armand Defever and Florita Botts. Adults and
　　　　youth.

1082　FOCUS ON THE FAMILY. Filmstrip. Color, 45 frames. Single frame.
　　　With commentary (8 pp.)
　　　1965, FAO　　　　　　　　　　　　　　　　　　　　　　$5.00
　　　　Shows how home economics programs help women to achieve better
　　　　living standards for the family. Text by Mary Elizabeth Keister,
　　　　Florita Botts, and Armand Defever. Adults and youth.

1083　A GIFT IS NEVER SMALL. Filmstrip. Color, 43 frames. Single
　　　frame. With commentary (4 pp.)
　　　1965, FAO/UNESCO　　　　　　　　　　　　　　　　　　$5.00
　　　　Illustrates the Freedom from Hunger Campaign (FFHC) sponsored
　　　　by FAO, indicating the types of projects to which gifts can be made
　　　　through the UNESCO Gift Coupon Scheme for technical education,
　　　　agricultural tools and equipment. Adults and youth.

1084　GOOD FOOD WINS THE GAME. Filmstrip. Color, 62 frames. Double
　　　frame. With commentary (8 pp.)
　　　1970, FAO　　　　　　　　　　　　　　　　　　　　　　$5.00
　　　　Aims to teach children of elementary school age the basic facts of
　　　　nutrition by showing two boys in Chile, both eight years of age, but
　　　　different in size and strength because one boy eats the right kinds of
　　　　food but the other does not.

1085　HOW TO FEED YOUR BABY. Filmstrip. Color, 45 frames. Double
　　　frame. With commentary (7 pp.)
　　　1969, FAO　　　　　　　　　　　　　　　　　　　　　　$5.00
　　　　Prepared for teaching mothers in Africa the right foods for the baby
　　　　during breast feeding and weaning, how to cook them, and the proper
　　　　way to feed the baby. Text and photographs by Florita Botts.

1086　A LITTLE BIT MORE. Filmstrip. Color, 56 frames. Double frame.
　　　With commentary (7 pp.)
　　　196-, FAO　　　　　　　　　　　　　　　　　　　　　　$5.00
　　　　Shows women (filmed in Chile) practical ways of feeding their fami-
　　　　lies more nourishing food.

1087　MILK FROM POWDER TO LIQUID. Filmstrip. Color, 46 frames.
　　　Double frame. With commentary (4 pp.)
　　　　Designed to convince women that growing children need milk, and
　　　　showing how to prepare powdered milk. Made in Latin America.

1087　MORE THAN FOOD. Filmstrip. Color, 71 frames. Single frame. With
bis　　commentary (trilingual: E/F/S, 21 pp.)
　　　1971, UN/FAO　　　　　　　　　　　　　　　　　　　　$5.00
　　　　Illustrates use of food aid under the World Food Program as a capi-
　　　　tal investment in improvement projects in the developing countries.
　　　　Text by Bruno de Marco.

1088 THE NEW RICE GIVES YOU A CHOICE. Filmstrip. Color, 66 frames. Double frame. With commentary (7 pp.)
1970, FAO $5.00
 A teaching aid for Asian farmers, to show them why the improved rice varieties are advantageous and how they should be grown.

1089 OUR AFRICAN HERITAGE. Filmstrip. Color, 39 frames. With commentary (bilingual: E/F; 53 pp.)
 A presentation of the threats to wildlife in Africa and what can be done about them. Text by Jean Vidal and Alain Gille. Suitable for junior and senior high schools.

1090 PEASANTS INTO FARMERS. Filmstrip. Color, 43 frames. Single frame. With commentary (8 pp.)
1965, FAO $5.00
 Tells the story of a group of United Nations experts who established the first model training center for farmers and future teachers of farmers in Upper Volta. Text by Armand Defever and Florita Botts. Adults and youth.

1091 THE PROTEIN GAP. Filmstrip. Color, 98 frames. With commentary (10 pp.)
1971, FAO   Double frame version $5.00
      Single frame version $5.00
 Focuses attention on the widespread problem of protein deficiency in diet, points to its main causes, and suggests some ways of solving the problem.

1092 THE SEA. Filmstrip. Color, 43 frames. Commentary (trilingual: E/F/S; 72 pp.)
1964, UNESCO $4.50
 Illustrates the challenges and promises inherent in the study of oceanography. Text by Jean Lévigne and Jean Vidal. For junior and senior high schools and adults.

1093 THAT ALL THE WORLD MAY EAT. Filmstrip. Color, 47 frames. Single frame. With commentary (bilingual: E/F; 23 pp.)
1966, FAO $5.00
 A series of cartoon drawings in color with a text and story line made for children of 8-12 years. Shows how FAO works to increase food production and the income of farmers, foresters, and fishermen of Africa, Asia, and Latin America. Drawings by Jack Maxwell, text by Brian Taylor.

1094 TO SURVIVE. Filmstrip. Color, 41 frames. Single frame. With commentary (4 pp.)
1963, FAO $5.00
 Presents the drama of hunger and some of its causes: soil erosion, drought, floods, poverty, disease, pests, and ignorance. Illustrates some means of relieving many of these urgent problems: reforestation, irrigation, improved farming tools and methods, breeding better plant varieties and animals, education of the farmers, technical assistance, and community development.

1095 WATER IN ARID ZONES. Filmstrip. Black and white, 47 frames. With commentary (22 pp.)
1959, UNESCO $3.50
 Presents the problem of procuring and conserving water in arid

lands. Describes the water cycle, hydrology and utilization of surface water, underground water, rainmaking, dew, and the problems
of saline water. Text by J. Guilloteau. For junior and senior high
schools and adults.

1096 WHY DO WE EAT? Filmstrip. Black and white, 23 frames. Single
frame. With commentary (4 pp.)
1970, FAO $4.00
Cartoon drawings explain the basic elements of nutrition and why the
body needs certain foods for growth, energy, and protection from
diseases.

Catalogue

1097 FAO CATALOGUE OF FILMS AND FILMSTRIPS 1968. 108 pp.
1968, FAO   Free on request to FAO
Lists films available for borrowing from FAO Film and TV Section;
includes instructions for the purchase of FAO filmstrips. Lists
names and addresses of distributors from whom films may be pur-
chased (or borrowed) directly. Among the subjects covered are food
supply, processing, and storage; land and water development; and
nutrition.

DICTIONARIES, GLOSSARIES, ETC.

1098 Multilingual DEMOGRAPHIC DICTIONARY. Prepared by the Demographic Dictionary Committee of the International Union for the Scientific Study of Population. (Population Studies, 29; ST/SOA/SER.A/29)
1958, UN
a. English Section. 1958. 77 pp. Sales No. E.58.XIII.4 $1.00
b. Volume français. 1958. 105 pp.   Sales No. F.58.XIII.4 $1.00
c. Russian volume. 1958. 113 pp.   Sales No. R.58.XIII.4 $1.00
d. Volumen español. 1958, reprinted 1965. 108 pp. Sales No.
S.58.XIII.4 $1.00
Terms are fully explained and are numbered in order to facilitate reference to other language editions.

1099 Glossary of HYDROLOGIC TERMS USED IN ASIA AND THE FAR EAST.
38 pp. (ECAFE. Flood Control Series, 10; ST/ECAFE/SER.F/10)
1956, UN   Sales No. 56.II.F.7   Out of print
Text adopted by a working group of experts in hydrologic data and
terminology in the region convened by the UN Economic Commission
for Asia and the Far East and WMO in Bangkok, 1955.

1100 International METEOROLOGICAL VOCABULARY.... 291 pp.
1966, WMO   WMO-No.182.TP.91 $14.00
Quadrilingual: E/F/R/S
Consists of three main parts: (1) Multilingual meteorological no-
menclature in four languages; (2) international meteorological definitions in English and French; (3) alphabetical indexes in four languages. Terms and definitions are grouped in accordance with the
divisions and subdivisions of the meteorology section of the Universal Decimal Classification (UDC)—included as an appendix. Includes also as an appendix the Abridged International Ice Nomen-
clature, with terms in the four languages, and definitions in English
and French.

1101   The International Standard Classification of OCCUPATIONS. Rev. ed.
1968. 355 pp.
1969, ILO $5.00
    Provides a basis for the international comparison of occupational
data, affords basis for developing the systems of occupational classification, and serves as a means of identifying specific national occupations of international interest. Designed as a comprehensive,
multipurpose instrument for the organization of occupational information, it is composed of definitions of occupations and of groups of
occupations embodied in a corresponding classification structure
with code numbers. Over 1,800 occupational titles are defined. In
addition to the text of the classification with definitions there is a list
of titles by group and also an alphabetical index of titles used. A
completely revised edition of the work originally issued in 1958.

    Saharan nomads: local terms. See entry 13.

    Scientific and technical dictionaries: bibliographies. See entries 1141,
1142.

1102   Multilingual Vocabulary of SOIL SCIENCE. By G.V. Jacks, R.
Tavernier and D.H. Boalch. 430 pp.
1968, 3rd printing, FAO   Multilingual $4.50
    Covers nine languages: English, French, Spanish, German, Portu-
guese, Italian, Dutch, Swedish, Russian. First issued in 1960.

BIBLIOGRAPHIES

1103   Bibliography on the ANALYSIS AND PROJECTION OF DEMAND AND
PRODUCTION. 1963. 279 pp. (Commodity Reference Series, 2)
1963, FAO $5.00
    Annotated entries covering postwar papers, published or unpub-
lished, on analysis and projection of the demand for, and production
of, foodstuffs and agricultural raw materials, classified in five
major groups: methodology, commodities, projections, statistical
sources, bibliographies. Contains indexes by author, country, and
international organization.

    ANIMALS. See entry 1111.

1104   CURRENT BIBLIOGRAPHY FOR AQUATIC SCIENCES AND FISHER-
IES. Compiled by Biological Data Section, Fish Stock Evaluation
Branch, Fishery Resources and Exploitation Division, FAO.
1959- , Taylor & Francis, Ltd.(London) Annual 3 vols. $17.20 per vol.
    References cover: general (oceanography, limnology, and fisheries),
physical oceanography and limnology, plankton, benthos, fishing,
aquatic stocks, miscellaneous, and auxiliaries. List of meetings;
indexes: author, geographic, taxonomic, subject, and citation.

1105   INIS ATOMINDEX. Vol. 1, No. 1, May 1970- . (STI/PUB/245)
1970- , IAEA   Annual subscr. $25.00 (12 issues and including semi-
annual cumulative indexes)
    Contains complete bibliographic descriptions and keywords for all
items reported to IAEA's International Nuclear Information System
(INIS). Items are grouped according to a subject classification sys-

tem. The first cumulative index covers Vol. 1, No. 1-8, May-December 1970 (report nos. RN 000001-004053) and contains indexes of personal authors, report numbers, and corporate entries. Semiannual and annual cumulations begin with Vol. 2 (1971); multivolume cumulations are planned to appear at appropriate intervals in the future. INIS formally began to operate in April 1970. It was established cooperatively by IAEA and its member states to construct a data basis for identifying publications relating to nuclear science and its peaceful applications. The member states (and cooperating international and regional organizations) scan the scientific and technical literature and report the input data for INIS to IAEA. In addition to a magnetic tape service available only to participating governments and organizations, IAEA issues the following services: (1) INIS Atomindex. (2) Abstracts-on-microfiche: an abstract for every item reported in INIS Atomindex is available at an annual subscription in 1971 of $45.00. (3) Full texts of "nonconventional" literature, i.e., all items other than journal articles and commercially published books, reported in INIS Atomindex are available on standing order at $0.20 per microfiche; individual orders, at $0.65 per document irrespective of the number of microfiches contained therein. Prices cited above include airmail delivery to destinations outside Europe. All output services of INIS should be ordered from IAEA (INIS Section) direct; they are not available through sales agents.———INIS Atomindex supersedes the twice-a-month Atomindex: International List of the Documents That Are Available from the IAEA Microfiche Service, vol. 1, no. 1-vol. 11, no. 30, October-4 May 1970 (STI/DOC/12), and the quarterly Nuclear Medicine: A Guide to Recent Literature, no. 1-32, 1962-1966.

1106 Bibliography on BILHARZIASIS, 1949-1958.... 158 pp.
1960, WHO   Bilingual: E/F $2.00
Contains over 2,700 entries, a combined subject and geographical index referring to item numbers, and a list of principal sources consulted.

COMMODITIES. See entry 1112.

1107 CURRENT BIBLIOGRAPHICAL INFORMATION.... vol. 1, no. 1, 1 January 1971- . (ST/LIB/SER.K/1-)
1971- , UN   Bilingual: E/F   Annual subscr. $12.00 (12 issues)
                                                  Single issue $1.00
Subject list of recently received books, publications of governments and of national and international organizations, and selected periodical articles related to topics considerd by organs of the United Nations system, new reference works, books of general interest in the fields of geography, history, biography etc., and journals acquired for the first time by the Dag Hammarskjold Library, United Nations. Replaces Current Issues (no. 1-11, December 1965-Autumn 1970; ST/LIB/SER.G/1-11) and New Publications in the Dag Hammarskjold Library (vol. 1-21, September 1949-December 1970; ST/LIB/SER.A/1-255).

1108 Bibliography of Recent Official DEMOGRAPHIC STATISTICS.... 80 pp. (Statistical Papers, Series M, No. 18; ST/STAT/SER.M/18)
1953, UN   Sales No. E/F.53.XIII.14   Out of print
Reprint of comprehensive, cumulated bibliography published in the Demographic Yearbook 1953 (pp. 345-421), covering the most recent census returns, sources of periodic and other demographic statistics

since 1920, and life tables since 1900. Bibliography was prepared by the Census Library Project of the U.S. Library of Congress, and was continued in the <u>Demographic Yearbook 1954</u>, <u>1955</u> (entry 1170).

ECONOMIC ANALYSIS. See entry 1113.

ENVIRONMENT. See entry 1114.

1109   FAMILY PLANNING, INTERNAL MIGRATION AND URBANIZATION IN ECAFE COUNTRIES; a Bibliography of Available Materials. Part One: Family Planning; Part Two: Internal Migration and Urbanization. 66 pp. (ECAFE. Asian Population Studies Series, 2; E/CN.11/807)
1968, UN   Sales No. E.68.II.F.13 $1.50
Bibliography (869 entries), primarily from 1963 onward but including some earlier material in the case of countries in which material is scarce, although no material earlier than 1956; country and author indexes. Supplements "A Select Annotated Bibliography on Population and Related Questions in Asia and the Far East," an unpublished background paper prepared for the Asian Population Conference, 1963 (entry 875).

1110   FAO DOCUMENTATION: Current Index. January 1967- (DC 67/1-)
1967- , FAO   Free on request to FAO (12 issues a year, with annual cumulations)
————Cumulative, January-June 1967. 277 pp. (DC 67/6)
    Accession nos. 00001-00650.   Out of print
————Cumulative, July-December 1967. 425 pp. (DC 67/12)
    Accession nos. 00651-01890.   Out of print
————Cumulative, January-June 1968. 451 pp. (DC 68/6)
    Accession nos. 01891-03440.   Out of print
————Cumulative, July-December 1968. 450 pp. (DC 68/12)
    Accession nos. 03441-04925.   Out of print
————Cumulative, January-December 1969. 2 vols. (DC 69/12)
    Accession nos. 04926-08430.
————Cumulative, January-December 1970. 2 vols. (DC 70/12)
    Accession nos. 08431-12276.
————January 1971- . (DC 71/1-)
    Accession nos. 12277- .
In three parts: an analytical index, an index of authors, and a bibliographical list of publications and documents (including separate entries for each section or report contained in a publication or document and for each article in a periodical issue).

FAO has published also a number of special indexes to publications and documents issued from 1945 to date:

1111   Animals, 1945-1966. 1967. 408 pp. (DC/Sp.6) $4.00
    Covers 1,617 items.

1112   Commodities, 1945-1966. 1969. 297 pp. (DC/Sp.11) $3.00
    Covers 945 items.

1113   Economic Analysis, 1945-1966. 1969. 260 pp. (DC/Sp.12) $3.00
    Covers 873 items.

1114   Environment, 1945-1970. 1971. 782 pp. (DC/Sp.21) $8.00
    Covers 1,982 items.

1115   FAO Technical Assistance Reports, 1951-1965. 1967. 244 pp. (DC/
       Sp.1)   Out of print
       Covers 2,120 items.

1116   FAO/UNDP (SF) Projects, 1963-1966. 1967. 82 pp. (DC/Sp.3) $1.00
       Covers 139 items.

1117   FAO/UNDP (SF) Project Reports and Documents:
       a. 1: [Guyana, Kenya, Korea, Sudan, Syria, Tunisia, Zambia, Interre-
          gional Desert Locust Project.] 1969. 189 pp. (DC/Sp.13) $2.00
          Covers 517 items.
       b. 2: [Afghanistan, Brazil, British Guiana, Chile, China (Taiwan),
          Colombia, Cyprus, Ecuador, Honduras, Peru, Philippines, Saudi
          Arabia, Turkey, U.A.R.] 1970. 172 pp. (DC/Sp.15) $2.00
          Covers 284 items.
       c. 3: [Chile, Costa Rica, Ethiopia, Greece, Iran, Lebanon, Mexico,
          Nicaragua, Somalia, Sudan, Togo, Turkey.] 1970. 184 pp. (DC/Sp.
          16) $2.00
          Covers 359 items.
       d. 4: [Bolivia, Ecuador, El Salvador, Ghana, Greece, Haiti, Israel,
          Peru, Syria, Thailand, Tunisia, U.A.R., Regional Africa, Regional
          Latin America.] 1970. 209 pp. (DC/Sp.18) $2.00
       e. 5: Sebou Project. 1971. 350 pp. (DC/Sp.19) $4.00
          Covers 789 items. The Sebou River is in Morocco.

1118   Fisheries, 1945-1969. 1969. 2 vols. (DC/Sp.14) $12.00
       Covers 4,504 items. Supersedes Fisheries 1945-1966 (DC/Sp.2).

1119   Food and Agricultural Industries, 1945-1970. 1970. 504 pp. (DC/Sp.
       17) $5.00
       Covers 1,512 items.

1120   Forestry, 1945-1966. 1967. 656 pp. (DC/Sp.4) $7.00
       Covers 2,326 items.

1121   Land and Water, 1945-1966. 1968. 424 pp. (DC/Sp.8) $5.00
       Covers 1,308 items.

1122   Land Reform, 1945-1970. 1971. 257 pp. (DC/Sp.20) $3.00
       Covers 909 items.

1123   Nutrition, 1945-1966. 1968. 372 pp. (DC/Sp.7)   Out of print
       Covers 1,354 items.

1124   Plants, 1945-1966. 1967. 606 pp. (DC/Sp.5) $6.00
       Covers 2,423 items.

1125   Rural Institutions, 1945-1966. 1968. 320 pp. (DC/Sp.9) $4.00
       Covers 1,185 items.

1126   Statistics, 1945-1966. 1968. 371 pp. (DC/Sp.10) $4.00
       Covers 1,320 items.

FAO Catalogue of FILMS AND FILMSTRIPS. See entry 1097.

FISHERIES. See entry 1118.

1127   World FISHERIES ABSTRACTS: a Quarterly Review of Technical
        Literature on Fisheries and Related Industries. vol. 1, no. 1, January/
        February 1950-
        1950-   , FAO   Annual subscr. $4.00 (4 issues)
                                                        Single copy $1.25
        Covers field of fisheries technology, processing methods, boat de-
        sign, fishing methods and gear, chemical examination of fishery
        products, production of fish meal, etc. The abstracts are published
        in such a form that they may be cut out and filed.

1128   FOOD AID: a Selective Annotated Bibliography on Food Utilization for
        Economic Development. By Elizabeth Henderson. 203 pp. (P/WFP:B1)
        1964, UN/FAO $2.50
        Contains 422 entries, subject index and author index.

        FOOD AND AGRICULTURAL INDUSTRIES. See entry 1119.

        FORESTRY. See entry 1120.

1129   Bibliography of African HYDROLOGY. By J. Rodier. 166 pp. (Natural
        Resources Research, 2)
        1963, UNESCO   Bilingual: E/F $6.50 cl.
        Excludes references dealing with applied hydraulic engineering.

1130   Isotope Techniques in HYDROLOGY. Vol. 1 (1957-1965). 228 pp. (Bib-
        liographical Series, 32; STI/PUB/21/32)
        1968, IAEA $6.00
        Most references (985 entries) are provided with an abstract in
        English. Main bibliographical source is Nuclear Science Abstracts,
        but other scientific journals were also consulted and abstracted.
        Abstracts of relevant papers given at IAEA symposia are also in-
        cluded. Author and keyword indexes.

1131   Analytical Bibliography of INTERNATIONAL MIGRATION STATISTICS,
        SELECTED COUNTRIES, 1925-1950. 195 pp. (Population Studies, 24;
        ST/SOA/SER.A/24)
        1955, UN   Sales No. 56.XIII.1   Out of print
        Sources of international migration statistics for 24 countries (7 in
        Europe), selected with a view to facilitating studies of emigration
        from Europe.

1132   INTERNATIONAL MIGRATION AND ECONOMIC DEVELOPMENT; A
        Trend Report and Bibliography. By Brinley Thomas. 85 pp. (Popu-
        lation and Culture)
        1961, UNESCO $1.50
        General survey prepared in consultation with the International Com-
        mittee for Social Sciences Documentation.

        LAND AND WATER. See entry 1121.

1133   Bibliography on LAND AND WATER UTILIZATION AND CONSERVA-
        TION IN EUROPE. By C.H. Edelman and B.E.P. Eeuwens. 347 pp.
        1955, FAO   Out of print
        Covers general and agricultural geography, maps; climate; soils;
        vegetation; hydrology and hydrography; land reclamation, drainage,
        irrigation, consolidation of fragmented holdings; soil erosion and
        soil conservation. Includes Cyprus and Turkey; excludes Albania,
        Bulgaria, Czechoslovakia, Hungary, Iceland, Poland, Romania, and
        the U.S.S.R. Subjects extensively dealt with in the Bibliography on
        Land Tenure (1955) are omitted.

LAND REFORM. See entry 1122.

1134   MARINE SCIENCE CONTENTS TABLES.... vol. 1, no. 1, May 1966-
        1966- , FAO/UNESCO   Trilingual: E/F/S Free on request (12 issues
                                                                     a year)
        Reproduces the tables of contents of core journals (59) in marine
        science. Prepared jointly (since January 1970) by FAO and
        UNESCO. Distributed free to governmental and academic institu-
        tions receiving International Marine Science (entry 1200). FAO
        (Biological Data Section, Fishery Resources and Exploitation Divi-
        sion) and UNESCO (Office of Oceanography) will be glad to place the
        names of institutions concerned with marine science on the mailing
        list for either or both of these publications.

1135   List of Bibliographies on NUCLEAR ENERGY. vol. 1, no. 1, June
        1960- .
        1960- , IAEA   Free on request to IAEA (3 issues a year)
        Disseminates information on bibliographies published or in prepara-
        tion. It is compiled mainly from information made available by
        member states and from sources indicated by them. Titles shown
        as "in preparation" are reported in current issues until published;
        bibliographies shown as "published" in an issue are then dropped
        from subsequent issues. Includes also a section on the IAEA
        documentation program, with detailed information on bibliographies
        recently published by IAEA and the program of forthcoming publica-
        tions in the IAEA Bibliographical Series (STI/PUB/21/1-).

1136   Bibliografía sobre NUTRICIÓN, ALIMENTACIÓN, SALUD PÚBLICA
        Y CIENCIAS Afines para Uso en las Escuelas de Nutrición y Dietética
        de Latinoamérica. 457 pp. (Publicaciones Científicas. 174)
        1969, PAHO $2.00
        Selected list of reference materials, primarily in Spanish, Portu-
        guese, and English, classified by subject fields: basic sciences, eco-
        nomic and social sciences, public health, food sciences and nutrition,
        and education.

NUTRITION. See entry 1123.

PLANTS. See entry 1124.

1137   Disposal of RADIOACTIVE WASTES INTO MARINE AND FRESH
        WATERS. Comp. by Sophie Stephens and Mrs. J.M. O'Leary; ed. by
        I.C. Roberts. 365 pp. (Bibliographical Series, 5; STI/PUB/21/5)
        1962, IAEA $3.00
        Covers literature on the subject for the period 1955-1960 (1,918
        entries). References (with abstracts) are arranged in eight major
        subdivisions. Author index.

1138   RADIOISOTOPE APPLICATIONS IN INDUSTRY. A Survey of Radio-
        isotope Applications Classified by Industry or Economic Activity with
        Selected References to the International Literature. (STI/PUB/70)
        1963, IAEA $2.50
        A listing by industry of uses by radioisotopes gathered from the
        world literature up to 1960. In the listings by use extensive refer-
        ences are made to a bibliography at the end. The industrial cate-
        gories include canal, dock, and harbor construction (investigation of
        movement of water bed materials); electric light and power (water
        and steam flow studies; investigation of steam-water systems);
        water supply; sanitary services (sewage).

1139   RADIOISOTOPE INSTRUMENTS IN INDUSTRY AND GEOPHYSICS. 411
        pp. (Bibliographical Series, 20; STI/PUB/21/20)
        1966, IAEA $8.50
        Approximately 1800 entries (1960-1964), classified by technique and
        field of application. Author index.

        RURAL INSTITUTIONS. See entry 1125.

1140   List of Annual Reviews of Progress in SCIENCE AND TECHNOLOGY...
        2nd ed. 22 pp.
        1969, UNESCO   Bilingual: E/F
        Consists mainly of annual series containing papers which review
        fairly narrow topics within the broader subject designated by the
        serial title. Arranged in seven groups: information sciences; ag-
        ricultural and veterinary sciences (including food and nutrition);
        biological sciences; engineering and technology; medical sciences;
        physical and mathematical sciences (including astronomy and space
        sciences, geology and geophysics); yearbooks of science and tech-
        nology. Free on request to UNESCO Department of Science Policy
        and Promotion of Basic Sciences.

1141   Bibliography of Monolingual SCIENTIFIC AND TECHNICAL DICTIO-
        NARIES. 2 vol. (Documentation and Terminology of Science)
        Vol. 1: National Standards. 219 pp.
        1955, UNESCO $2.50
        Based upon standardized technical glossaries published by na-
        tional standards associations. Entries cover 27 languages and
        are arranged in accordance with the Universal Decimal Classifi-
        cation.
        Vol. 2: Miscellaneous Sources. 134 pp.
        1959, UNESCO $2.50
        Covers 1,043 publications published privately, among which 26
        languages are represented.

1142   Bibliography of Interlingual SCIENTIFIC AND TECHNICAL DICTIO-
        NARIES. 5th ed. 250 pp. (Documentation and Terminology of Science)
        1969, UNESCO $9.00 cl.
        Lists 2,491 dictionaries under 263 subject headings and in 75 lan-
        guages. Includes author, language, and subject indexes.

1143   SCIENTIFIC MAPS AND ATLASES and Other Related Publications;
        Catalogue 1971-1972. 57 pp., maps.
        1971, UNESCO   Free on request
        Catalogue of maps (geology, minerals, hydrogeology, climate, soils,
        vegetation, oceanography) published by UNESCO, and of related pub-
        lications in ecology and natural resources, geology and geophysics,
        geography, soil science, hydrology, and oceanography.

1144   SOCIAL FACTORS IN ECONOMIC GROWTH; Trend Report and Bibliog-
        raphy. [65] pp. (Current Sociology, vol. 6, no. 3, pp. 173-237)
        1958, UNESCO   Out of print
        Contains material on demographic characteristics as impediments to
        economic growth, and on the labor force.

1145   World List of SOCIAL SCIENCE PERIODICALS. 3rd ed., rev. and en-
        larged; 448 pp. (Documentation in the Social Sciences)
        1966, UNESCO $12.00
        Lists scientific periodicals containing original material written by
        specialists (2,193 journals from 90 countries and 24 journals is-

sued by international agencies). Data include title, name and address of publisher, periodicity, year publication commenced, special index issues, pages per issue, regular features, description of a typical issue. Alphabetical index of journals includes standard abbreviation for each; subject indexes in English and French; also index of scientific institutions publishing periodicals.

1146 International Bibliography of SOCIOLOGY. Vol. 1 (1951)- . Prepared by the International Committee for Social Sciences Documentation. (International Bibliography of the Social Sciences)
1952-61, vol. 1-9, UNESCO
1962- , vol. 10- , Tavistock Publications, Ltd. (London) Annual; bilingual: E/F $4.75
This annual bibliography includes entries on population, migration, and urbanization. Each volume has an author index and subject indexes in English and French. Volumes 1-9, covering the years 1951-1959 were published by UNESCO, during which time volumes 1-4 formed numbers of Current Sociology.

1147 National STATISTICAL PUBLICATIONS ISSUED IN 1966. 120 pp. (Conference of European Statisticians. Statistical Standards and Studies, 19; ST/CES/19)
1969, UN   Sales No. E.69.II.E/Mim.36 $1.35
Lists statistical publications of 30 European countries (including both the Federal Republic of Germany and the German Democratic Republic) and Cyprus and the U.S.A.; includes statistical publications of the European Economic Community (EEC). Publications include census and other demographic titles.

STATISTICS. See entry 1126.

1148 A Guide to Basic STATISTICS IN COUNTRIES OF THE ECAFE REGION. 2nd ed. 217 pp. (E/CN.11/881)
1970, UN   Sales No. E.70.II.F.6 $3.00
Guide to official statistical publications considered basic to the formulation and implementation of plans for economic and social development in countries of the region. Includes, among other publications, those concerning population and labor statistics (number of persons and households, birth and death rates, infant mortality, life expectancy, migration) and statistics of services dealing with water supply, sewerage, sanitation, garbage disposal, and irrigation.

1149 UNITED NATIONS DOCUMENTS INDEX. vol. 1, no. 1, January 1950- . (ST/LIB/SER.E/1-)
1950- , UN   Annual subscr. $25.00 (11 issues and cumulative subject index and cumulative checklist; single issue $1.50; single cumulation, prices vary)
Lists and indexes all documents and publications of the United Nations (except restricted material and internal papers) and all printed publications of the International Court of Justice. Each monthly issue includes in general the documents and publications processed by the Documentation Division of the United Nations Headquarters Library, New York, during the month. Issues for July and August are combined in one issue. Periodicals are listed annually in the Cumulative Checklist. Beginning with volume 14 (1963), the monthly issues are superseded at the end of the year by two separate annual cumulations, the Cumulative Checklist and the Cumulative Index.

The Cumulative Index consolidates the monthly indexes and also contains a consolidated list by symbol of all documents issued during
the year, a consolidated list of documents republished in the Official
Records or elsewhere, a consolidated list of sales publications, a
list of new document series symbols, and a list of libraries and
United Nations Information Centres receiving United Nations material.

1150 Geology of URANIUM AND THORIUM. Comp. by E.M. El Shazly.
1962- , IAEA
a. [Vol. 1(1940-1960)]. 1962. 134 pp. (Bibliographical Series, 4; STI/
PUB/21/4) $1.50
b. Vol. 2 (1961-1966). 1968. 102 pp. (Bibliographical Series, 31; STI/
PUB/21/31) $2.50
Covers abstracts in Nuclear Science Abstracts with a few important ad-
ditions; includes a section giving the main estimates of the reserves
and resources of uranium and thorium. Author index. Volume 1 contains 892 entries; volume 2 contains 902 entries.

1151 Selected Bibliography on URBAN CLIMATE. Prepared by Dr. T.J.
Chandler. 383 pp.
1970, WMO   WMO-No.276.TP.155 $12.50
Covers a wide range of books and articles on urban climatology, including air pollution. Incorporates many new references proposed
for inclusion after the first draft had been circulated to participants
in the WHO Symposium on Urban Climate and Building Climatology,
Brussels, 1968 (see entries 4, 5) and to member states of WMO.

1152 "Summary of Information Published by National Authorities on VITAL
AND HEALTH STATISTICS." Epidemiological and Vital Statistics Report, vol. 20, no. 3, 1967, pp. 169-310.
1967, WHO   Bilingual: E/F $3.25
Summary of demographic and health statistics published; available
mortality statistics published annually; published statistics of notified cases of infectious and parasitic diseases; published statistics
of notified cases of diseases other than infectious and parasitic ones;
list of publications.

1153 WASTE MANAGEMENT RESEARCH ABSTRACTS. No. 1- .
1965- , IAEA   Annual   Free on request to IAEA
Abstracts of recently completed research in radioactive waste management; compiled and issued annually by IAEA from abstracts provided by the governments of member states. Abstracts are published in the language in which they are submitted. Russian titles,
authors' names, and names of institutes are translated into English.

1154 Publications of the WORLD HEALTH ORGANIZATION...a Bibliography.
1958- , WHO
————. 1947-1957. 1958. 128 pp. $3.25 cl.
————. 1958-1962. 1964. 125 pp. $4.00 cl.
————. 1963-1967. 1969. 152 pp. $6.00 cl.
Covers technical articles and publications, administrative and
general articles and publications. Contains author index, country
index, and list of WHO publications by series.

DIRECTORIES

1155 Institutions Engaged in ECONOMIC AND SOCIAL PLANNING IN AF-
RICA. Prepared under the responsibility of Michèle Cser. 155 pp.
(Reports and Papers in the Social Sciences, 22)
1967, UNESCO   Composite: E/F $1.75
Lists 300 public and private organizations, both alphabetically and
grouped by country. Four types of activity are covered: project execution, consultation, applied research, and training of personnel.

1156 SOCIAL SCIENTISTS SPECIALIZING IN AFRICAN STUDIES.... 375
pp.
1963, UNESCO/École Pratique des Hautes Études
                  Bilingual: E/F $11.50
Groups information on 2,072 social scientists, either African or
specializing in African studies. Gives data on experts whose work is
concerned partially or entirely with the social sciences and the hu-
manities, including certain cultural or ecological activities affecting
the socioeconomic development of Africa.

1157 International Directory of AGRICULTURAL ENGINEERING INSTITU-
TIONS. By H.J. Hopfen. 462 pp. (AgS:MISC/68/2)
1968, FAO   Composite: E/F/S $3.50
Replaces 1962 ed. In two principal parts: (1) List of names and addresses of central governmental services, as well as international
and national institutes dealing with land and water development, farm
power and machinery, rural electrification, farm buildings, and farm
work organization. For each institution information is provided on
scientific staff, training and research, recent publications, and languages of correspondence. Information on each country is presented
in the language chosen by that country for official correspondence
with FAO. (2) Lists of national associations of farm machinery
manufacturers; exhibitions; periodicals; reference books; and films
dealing with agricultural engineering. Contains an index of special
institutions, and an index of the names of specialists.

1158 Directory of Institutions Engaged in ARID ZONE RESEARCH. 110 pp.
(Arid Zone Programme, 3)
1953, UNESCO   Out of print
Data on staff organization, facilities, and programs of some 90
specialized institutions. Obsolete.

1159 Organization of HYDROMETEOROLOGICAL AND HYDROLOGICAL
SERVICES. [59 pp., including tables] (Reports on WMO/IHD Projects,
10)
1969, WMO $2.00
Reviews replies received from 91 countries to WMO questionnaire.
Gives information on whether or not a central body responsible for
the coordination of hydrometeorological and hydrological services
exists in each country. Where no central body is responsible for
such coordination, lists the principal hydrometeorological and hydrological services. Indicates also whether there are repositories
of data and national catalogs and where the latter are obtainable.

1160 METEOROLOGICAL SERVICES OF THE WORLD.... Ed. 1971. [277]
pp., loose-leaf, with binder.
1971- , WMO   Composite: E/F   WMO-No.2 $13.00
Gives information for each country on the state meteorological ser-

vice; other state meteorological organizations dependent on that service; other state meteorological organizations independent of the state service; meteorological institutes forming part of the state service; meteorological institutes not forming part of the state service; meteorological publications (regular; bibliographical; addresses from which occasional or nonserial publications may be obtained); responsible authority for aeronautical meteorological services. Kept up to date by irregularly issued Supplements.

1161 Yearbook of INTERNATIONAL ORGANIZATIONS. The Encyclopaedic Dictionary of International Organizations, Their Officers, Their Abbreviations. 12th (1968-69) ed. Ed. by Eyvind S. Tew. 1,220 pp. (UIA. Publication No.210)
Brussels, Union of International Associations $24.00 cl.
   Provides information on a total of 4,252 international organizations, intergovernmental and nongovernmental: aims, titles, structure, officers, activities, finance, publications, meetings, acronyms.

1162 Directory of METEORITE COLLECTIONS AND METEORITE RESEARCH. 50 pp.
1968, UNESCO   Bilingual: E/F $3.50
   Information on conservation and cataloguing of meteorites; existing arrangements for international exchange of meteoritic material between scientific institutions and museums; analytical methods and techniques. Data from 55 contributing countries.

1163 International Repertory of Institutions Conducting POPULATION STUDIES.... 240 pp. (Reports and Papers in the Social Sciences, 11)
1959, UNESCO   Composite: E/F $2.50
   Gives for each institution: address: structure, administrative directorate, professional staff, finance, research activities, teaching and training, use of basic demographic data, and publications.

1164 World Guide to SCIENCE INFORMATION AND DOCUMENTATION SERVICES. 211 pp. (Documentation and Terminology of Science)
1965, UNESCO   Composite: E/F   Out of print
   Guide to the location of services providing information and documentation on the natural sciences, covering 144 services in 65 countries. (See also entry 1166.)

1165 World Directory of National SCIENCE POLICY-MAKING BODIES. 4 vols. 1966- , UNESCO   Composite: E/F
   a. Vol. 1: Europe and North America. 1966. 356 pp. $13.00 cl.
   b. Vol. 2: Asia and Oceania. 1968. 157 pp. $6.50 cl.
   c. Vol. 3: Latin America. 1968. 187 pp. $6.50 cl.
   d. Vol. 4: Africa and the Arab States. In preparation.
   Directory of those bodies in each country which play a leading part at the national level in the planning, coordination, financing, or organization of scientific and technological research. The selection of those bodies to be included—difficult in the case of countries where there is yet no centralized organization of research—includes provisionally bodies which clearly exert some influence over the development of national activities in the realm of science. Information given was supplied by the countries themselves and submitted, before publication, to the organization concerned for their approval. Covers 70 countries (vol. 1-3 inclusive). Information English or French depending on language chosen by each country for correspondence with UNESCO. Vol. 1 was published by Francis Hodgson Ltd., London.

1166   World Guide to TECHNICAL INFORMATION AND DOCUMENTATION SERVICES. 287 pp. (Documentation and Terminology of Science)
1969, UNESCO   Composite: E/F $6.00 cl.
A companion volume to the World Guide to Science Information and Documentation Services (entry 1164). Lists the principal centers offering technical information and documentation services. Includes 273 centers in 73 countries and territories, 60 of them—marked with an asterisk—dealing at the same time with scientific and technical documentation, were included also in the first of these two guides.

1167   International Repertory of SOCIOLOGICAL RESEARCH CENTRES OUTSIDE OF THE U.S.A.... Prepared by the International Committee for Social Sciences Documentation. 125 pp. (Reports and Papers in the Social Sciences, 20)
1965, UNESCO   Bilingual: E/F $1.25
Lists public and private institutions exclusively or primarily devoted to sociological research and employing a salaried scientific staff.

1168   Directory of National Bodies Concerned with URBAN AND REGIONAL RESEARCH. 134 pp. and corrigenda sheet. (ST/ECE/HOU/34 and Corr.1)
1968, UN   Sales No. E.69.II.E.8 $2.00
Covers 27 European countries and the U.S.A. Lists national bodies concerned with physical planning. The entry for each country is in two parts: (A) General information on the organization of and financial resources allotted to regional research; (B) a list of bodies, each accompanied by a description of its organization (identification, status, date of establishment, directorate, staff, financing) and its research activities, including publications.

1169   Directory of ZOOLOGICAL (AND ENTOMOLOGICAL) SPECIMEN COLLECTIONS OF TROPICAL INSTITUTIONS. 31 pp. (Humid Tropics Research)
1962, UNESCO $1.00
Lists collections in Africa, Latin America, and Asia; material collected or exchanged and its identification.

PERIODICALS

ANNUALS

1170   DEMOGRAPHIC YEARBOOK 1970.... ix, 830 pp., tables.
1971, UN   Sales No. E/F.71.XIII.1 $22.00 cl.; $15.00 pa.
Twenty-second issue. Contains statistics of area, population, natality, mortality, expectation of life, nuptiality, divorce, and international migration. The special text in this issue (chap. I) is entitled "How Well Do We Know the Present Size and Trend of the World's Population?" and reexamines, on the basis of the experience of the last ten years, this same question discussed in the Demographic Yearbook 1960. Technical notes for each table are included. A cumulative subject index covers this issue and previous issues. The first twenty-one issues, for the years 1948-1969, were published during 1949-1970. The Demographic Yearbook is supplemented by the quarterly Population and Vital Statistics Report (entry 1207). (See also entry 1108.)

1171   Segundo COMPENDIO ESTADÍSTICO CENTROAMERICANO. 62 pp.,
        tables. (E/CN.12/597)
        1962, UN   Sales No. S.63.II.G.11   Out of print
        In Spanish only. Statistics cover the countries of Central America
        and Panama for the period 1950-1961: population estimates (1 table);
        agriculture; industry; transportation; external trade; balance of pay-
        ments; prices and cost of living; national accounts; money and bank-
        ing. The first issue (1957; 125 pp.; E/CN.12/487; in Spanish only;
        out of print) covered the period 1950-1955 for the same topics, but
        contained more demographic data (18 tables) and, in addition, data
        on public finance and education.

1172   Annual FERTILIZER REVIEW 1970.... xviii, 176 pp., tables.
        1971, FAO   Trilingual: E/F/S $5.00
        Based on data received from governments up to 31 May 1971. Re-
        view covers a period of six years—1964/65-1969/70, the fertilizer
        year being 1 July-30 June. Issued since 1946 under various titles.

1173   Yearbook of FISHERY STATISTICS....
        a. Vol. 28: Catches and Landings...1969. [322] pp., tables.
           1970, FAO   Trilingual: E/F/S $7.00
           Annual statistics by country, by species, and by major fishing
           areas of catches or landings of all fish, crustaceans, molluscs or
           other aquatic animals, residue and plants, marine and freshwater.
           Published under this title from 1964 as volumes 16, 18, 20, 22,
           24, 26, and 28 for the years 1963-1969.
        b. Vol. 29: Fishery Commodities...1969. [334] pp., tables.
           1970, FAO   Trilingual: E/F/S $7.00
           Annual statistics on the disposition of the catch, production, and
           international trade by countries of fish, fish products and prepa-
           rations, crustacean and mollusc products and preparations, oils
           and fats, meals, solubles, and similar animal feedingstuffs.
           Published under this title from 1964 as vols. 17, 19, 21, 23, 25,
           27, and 29 for the years 1963-1969.

1174   The State of FOOD AND AGRICULTURE 1971. World Review; Review
        by Regions; Water Pollution and Its Effects on Living Aquatic Re-
        sources and Fisheries. xii, 234 pp. and corrigendum slip, tables.
        1971, FAO $10.00
        Twenty-fifth issue. An authoritative annual review of developments
        in the world food and agricultural situation, with a year-by-year re-
        port on the outlook for the future. Beginning with the issue of 1957,
        detailed studies on special subjects are also a feature. The first
        twenty-four issues were published during 1947-1970; issues for
        1957, 1958, 1963, 1968, 1969, and 1970 are still in print. (For re-
        prints of special chapters separately listed, see entry 429 (from
        1963), 707 (from 1966), 501 (from 1967), 510 (from 1968), 552 (from
        1969) and 538 bis (from 1970).)

1175   Yearbook of FOREST PRODUCTS 1969-70.... L, 230 pp. and corrigen-
        dum sheet, tables.
        1971, FAO   Trilingual: E/F/S $5.00
        Twenty-fourth issue. Contains statistics from 170 countries and
        territories concerning production, consumption, and international
        trade covering roundwood, sawnwoods and wood-based panels, pulp,
        paper, and paperboard; also forest products other than wood (bam-
        boo, raw cork, bark and other tanning materials, materials for
        plaiting, natural gums exclusive of natural rubber, oilseeds and oil-

nuts, vegetable oils, waxes), land and forest areas, and population. Published under this title since 1968, beginning with the issue for 1967. The issues for 1947-1966 were published during 1948-1966 under the title Yearbook of Forest Products Statistics. Supplemented for Europe by the quarterly Timber Bulletin for Europe (entry 1214).

1176   World HEALTH STATISTICS ANNUAL, 1968....
        1971, WHO   Bilingual: E/F
        Vol. 1: Vital Statistics and Causes of Death.... xxviii, 675 pp.,
        tables.
        Vol. 2: Infectious Diseases: Cases, Deaths and Vaccinations....
        xviii, 207 pp., tables $6.75
        Vol. 3: Health Personnel and Hospital Establishments....
        xvi, 217 pp., tables $6.75
        Contains the information collected by WHO on vital statistics of
        notifiable diseases in a large number of countries and territories, to-
        gether with information on medical and paramedical personnel, hospi-
        tal establishments, and numbers of vaccinations performed. Present
        title begins with fifteenth issue, for the year 1962, also publication in
        three separate volumes (published in 1965-1966); previous title of first
        to fourteenth issues, covering the years 1939-1946 through 1961 (pub-
        lished during 1951-1964): Annual Epidemiological and Vital Statistics.

1177   Annual Bulletin of HOUSING AND BUILDING STATISTICS FOR
        EUROPE.... Vol. 14: 1970. 87 pp., tables
        1971, UN   Sales No. E/F/R.71.II.E.7 $2.00
        Includes general data (estimates of midyear population and its rate
        of change, fixed capital formation in Western European countries and
        the U.S.A., and investments in Eastern European countries); dwelling
        construction; labor; price indices; building materials. Volumes 1-13
        for the years 1957-1969 were issued during 1958-1970.

1178   Year Book of LABOUR STATISTICS 1970.... 30th issue. xxiii, 919
        pp., tables.
        1971, ILO   Trilingual: E/F/S $11.50
        Presents a summary of the principal labor statistics in more than
        170 countries and territories. Wherever possible, data cover the
        last ten years (1960-1969) and, for some tables, one month or one
        period near the middle of 1970. Subjects are grouped into ten
        chapters: (1) total and economically active population, (2) employ-
        ment, (3) unemployment, (4) hours of work, (5) labor productivity,
        (6) wages, (7) consumer prices, (8) household budgets, (9) industrial
        accidents, and (10) industrial disputes. Supplemented by the monthly
        Bulletin of Labour Statistics (entry 1199). The first twenty-nine
        issues for the years 1935-1936 to 1969 were published during the
        years 1936-1970, and were preceded by the I.L.O. Year-book, 1933
        (Appendix II: Labour Statistics) and 1934-35 (vol. 2: Labour Statis-
        tics), issued in 1934 and 1935.

1179   Annual Summary of Information on NATURAL DISASTERS. No. 4: 1969.
        73 pp., map.
        1971, UNESCO $2.50
        Fourth number. Summary data on scientific and technical aspects of
        earthquakes, tsunamis, storm surges, and volcanic eruptions occur-
        ring during 1969. Nos. 1-3, covering the years 1966-1968, were
        issued in 1970.

1180   PRODUCTION YEARBOOK. Volume 24: 1970. xvi, 822 pp., tables.
        1971, FAO   Trilingual: E/F/S $10.00 cl.
        Twenty-fourth issue. Contains information available as of 1 Febru-
        ary 1971. Statistics cover: land (land use, irrigation in specified
        countries, number and area of agricultural holdings); population
        (total and agricultural); index numbers of agricultural production;
        crops; livestock numbers and products; food supply per capita
        (quantity, calories, proteins, fats); means of production (commercial
        fertilizers, farm machinery, pesticides); prices (agricultural com-
        modities, export prices and freight rates, fertilizers, farm ma-
        chinery, index numbers of prices); wages. The first to twenty-third
        issues for 1947-1969 (appearing for the years 1947-1957 under the
        title Yearbook of Food and Agricultural Statistics) were published
        during 1947-1970. (See also entry 1185.)

1181   STATISTICAL YEARBOOK, 1970.... 814 pp., tables.
        1971, UN   Sales No. E/F.71.XVII.1 $22.00 cl.
        Twenty-second issue. Contains tables (212) grouped into two sec-
        tions, a world summary by region (17 tables), and the remaining
        tables, containing data country by country arranged in chapters:
        population; manpower; land; agriculture; forestry; fishing; industrial
        production; mining and quarrying; manufacturing; construction;
        energy; internal trade; external trade; transport; communications;
        consumption; balance of payments; wages and prices; national ac-
        counts; finance; public finance; international capital flow; health;
        housing; education; culture. Country index. The first to twenty-first
        issues, covering the years 1948-1969, were published in 1949-1970.
        (See also entry 1213.)
        a. 1967 Supplement to the Statistical Yearbook and the Monthly Bulletin
           of Statistics. Methodology and Definitions. (1st issue) 408 pp.
           1968, UN   Sales No. E.68.XVII.9 $5.00
           The Supplement provides more detailed definitions and explana-
           tory notes for the statistical series published in the Statistical
           Yearbook (entry 1181) and the Monthly Bulletin of Statistics
           (entry 1213). Includes a bibliography of the most important gov-
           ernment and central bank publications which are sources for sta-
           tistical data.

1182   STATISTICAL YEARBOOK [FOR AFRICA]...1970. 1971- . 5 vols.
        1971- , UN   Bilingual: E/F
        Part 1: National Accounts.... xxi, 219 pp., tables.
        1971
        Part 2:
        In preparation
        Part 3: Agriculture. xi, 332 pp., tables.
        1971
        Part 4:
        In preparation
        Part 5: Transports, Communications. xi, 105 pp., tables.
        1971
        First issue. Replaces the former Statistical Bulletin for Africa (entry
        1211). This unpriced Statistical Yearbook has been prepared by the
        secretariat of the Economic Commission for Africa and has been
        planned for a number of years. Three parts had been received at
        United Nations Headquarters by January 1972, but none states the total
        number of parts envisaged or the titles of the parts still to appear.

1182 DEMOGRAPHIC HANDBOOK FOR AFRICA.... 127 pp., tables.
bis   1971, UN   Bilingual: E/F
     A collection of demographic data from various sources in seven
     chapters: (1) world summary, (2) population, size, density, dis-
     tribution, structure, and growth, (3) fertility and mortality, (4) in-
     ternational migration, (5) economically active population, (6)
     population projections, (7) rates of increase of gross national prod-
     uct and of population.

1183 STATISTICAL YEARBOOK FOR ASIA AND THE FAR EAST, 1970....
     xxiv, 400 pp., tables. (E/CN.11/942)
     1971, UN   Sales No. E/F.71.II.F.3 $7.00
     Third issue. Statistics are arranged by country (29 countries and
     territories) and subdivided into twelve groupings: population; man-
     power; national accounts; agriculture; industry; consumption; trans-
     port and communications; internal trade; external trade; wages,
     prices and household expenditures; finance; social statistics. In
     many cases these are more detailed than the statistics in the Statis-
     tical Yearbook of the United Nations (entry 1181). The first and sec-
     ond issues, for 1968 and 1969, were published in 1969 and 1970.

    a. STATISTICS ON CHILDREN AND YOUTH. Supplement to the Statis-
      tical Yearbook for Asia and the Far East, 1968.... 198 pp. (E/
      CN.11/879)
      1969, UN
      Presents available data (for the period 1960-1967) in whatever
      form they are available, irrespective of their limitations and of
      factors affecting their international comparability. Tables for
      each of 18 countries and territories in the ECAFE region are
      grouped under five headings: (1) demographic and vital statistics;
      (2) housing, nutrition, and health; (3) education; (4) employment;
      and (5) welfare. There are also 20 summary tables for the
      region.

1184 Unesco STATISTICAL YEARBOOK 1970.... 786 pp., tables.
     1971, UNESCO   Bilingual: E/F $35.00 cl.; $29.00 pa.
     Eighth issue. Contains statistics from over 200 countries and ter-
     ritories, covering: Population (5 tables); education (22); science and
     technology (11); libraries and museums (6); book production (13);
     newspapers and other periodicals (2); paper consumption (2); films
     and cinema (2); radio broadcasting (2); television (2); and cultural
     expenditure (1). The first to seventh issues, covering the years
     1963-1969, were published during 1964-1970.

1185 TRADE YEARBOOK. Volume 24: 1970. xxxii, 375 pp., tables.
     1971, FAO   Trilingual: E/F/S $7.00 cl.
     Twenty-fourth issue. Contains annual external trade information
     available as of 31 December 1970. Covers trade in agricultural
     commodities, 1964-1969, trade in agricultural requisites, 1964-
     1969 (tractors, crude and manufactured fertilizers, seeds and
     spores for planting, insecticides and fungicides); value of agricul-
     tural trade, by countries, 1964-1969. The first twenty-three issues,
     for the years 1947-1969 (appearing for the years 1947-1957 under
     the title Yearbook of Food and Agricultural Statistics), were pub-
     lished during 1947-1969. (See also entry 1180.)

OTHER PERIODICALS

1186   AGRICULTURAL ECONOMICS BULLETIN FOR AFRICA. No. 1,
September 1962-   (E/CN.14/AGREB/1-)
1962- , UN   Single issue $1.50 (2 issues a year)
Articles on agricultural development, food processing, and related
subjects. Issued by the ECA/FAO Joint Agriculture Division, Economic Commission for Africa, Addis Ababa.

1187   Monthly Bulletin of AGRICULTURAL ECONOMICS AND STATISTICS.
Vol. 1, no. 1, May 1952- .
1952- , FAO   Annual subscr. $6.00 (12 issues)
                                 Single issue $0.60
Contains statistics of production, prices, and external trade concerning agricultural products (other than forest products), and special features such as statistics of production and consumption of
commercial fertilizers, trend reports and projections, and articles on agricultural economics.

Current Bibliography for AQUATIC SCIENCES AND FISHERIES. See
entry 1104.

1188   ATOMIC ENERGY REVIEW. Vol. 1, no. 1, January 1963- . (STI/
PUB/66)
1963- , IAEA   Annual subscr. $18.00 (4 issues)
                                 Single issue $5.00
Contains, reports, and reviews articles in the general field of nuclear energy and reports of IAEA conferences and symposia. Articles are published in English, except for those submitted in French,
and each is provided with an abstract in E/F/R/S. Special issues
are published occasionally. Cumulative Index Vols. 1-7 (1963-1969),
1970, $2.00.

ATOMINDEX, INIS. See entry 1105.

1189   CERES; the FAO Review. Vol. 1, no. 1, January/February 1968- .
1968- , FAO   Annual subscr. $5.00 (6 issues)
                                 Single issue $1.00
Covers the multiple aspects of agricultural, economic, and social
development in the developing countries as well as the work of FAO;
illustrated; articles, book reviews.

CURRENT BIBLIOGRAPHICAL INFORMATION. See entry 1107.

1190   ECONOMIC BULLETIN FOR AFRICA. Vol. 1, no. 1, January 1961- .
1961- , UN   Price per issue varies (2 issues a year)
Covers recent economic developments in Africa and its international
trade. Articles are included on economic developments in individual
African countries, on specific industries, demographic trends,
balance of payments, public finance, etc. (See also entry 1211.)

1191   ECONOMIC BULLETIN FOR ASIA AND THE FAR EAST. Vol. 1, no. 1,
August 1950- .
1950- , UN   Price per issue varies (4 issues a year)
Issued quarterly in June, September, December, and March. The
March issue contains the annual Economic Survey of Asia and the
Far East. The June and September issues contain articles and
notes on subjects related to the Asian economy. The December

issue features special studies and reports related to economic development and planning. All four issues include a compendium of Asian economic statistics. Population and social development are also included in the subjects dealt with.

1192 ECONOMIC BULLETIN FOR EUROPE. Vol. 1, no. 1, July 1949- .
1949- , UN   Price per issue varies (2 issues a year)
In recent years one issue deals with recent changes in Europe's trade, with a statistical annex and notes to the statistics, while the other issue contains special studies on particular aspects of Europe's foreign trade and economic development.

1193 ECONOMIC BULLETIN FOR LATIN AMERICA. Vol. 1, no. 1, January 1956- .
1956- , UN   Price per issue varies (2 issues a year)
Presents a resumé of the economic situation in Latin America designed to supplement and bring up to date the information published in the annual Economic Survey of Latin America, and contains articles on different subjects related to the economy of the region, including informative and methodological notes. (See also entry 1212.)

1194 Current FOOD ADDITIVES LEGISLATION. No. 1-150, June 1956-December 1971.
1956-71, FAO   Ten issues a year
Issued by the FAO Legislation Research Branch under the joint FAO/WHO program on intentional nonnutritive food additives, this bulletin lists current acts and regulations in the field and covers legislation on pesticide residues in foods. Listings are country by country in the following groups: Colors; antimicrobial preservatives (excluding antibiotics); antioxidants; artificial sweeteners; emulsifiers and stabilizers; flavorings; enrichments; antibiotics; packaging materials and other food-contact substances; plant treatment substances—postharvest; animal treatment substances (including feed additives); others. Discontinued at end of 1971.
————. Index to CFAL 1956-1961, covering abstracts 1-484 in issues 1-50. 37 pp.
————. Index to CFAL 1962-1965, covering abstracts 485-1385-8, issues 51-90. 19 pp.

1195 FOOD AND AGRICULTURAL LEGISLATION. Vol. 1, no. 1, January 1952- .
1952- , FAO   Annual subscr. $2.50 (2 issues)
                                          Single issue $1.50
Contains a selection of food and agricultural laws of international importance. Legislative texts are, according to their special interest, reproduced in full or in extract, or summarized. They are issued in leaflets and indexed according to a classification scheme published in the first number in each volume. Annual subject and country indexes are published with the last number in each volume. Subjects include food additives regulation; food laws in general, including handling, hygiene, processing, storage, and transport of foodstuffs; standards for plant, animal, and fish products; food rationing; special nutritional programs; manufacture and use of pesticides; pollution control (air and water); rural hygiene; agricultural and commercial statistics including censuses; water laws in general; wildlife, national parks, and natural resources. Issued four times a year during 1952-1969.

1196   International Digest of HEALTH LEGISLATION. Vol. 1, no. 1, 1948- .
        1948- , WHO   Annual subscr. $12.00 (4 issues)
                                                    Single issue $4.00
        Contains a selection of health laws and regulations and, from time to
        time, studies in comparative health legislation as well as a biblio-
        graphical section. Subjects include environmental hygiene, air and
        water pollution control, control of pesticides, food hygiene, protec-
        tion against ionizing radiations.
        ——————Cumulative Index 1948-1954 (Vol. 1-5). 54 pp.
        1958, WHO $0.70
        ——————Cumulative Index 1955-1959 (Vol. 6-10). 27 pp.
        1961, WHO $0.60
        ——————Cumulative Index 1960-1964 (Vol. 11-15). 36 pp
        1965, WHO $0.60
        ——————Cumulative Index 1965-1969 (Vol. 16-20).
        In preparation

1197   World HEALTH STATISTICS REPORT.... Vol. 1, no. 1, 1947- .
        1947- , WHO   Bilingual: E/F   Annual subscr. $16.00 (12 issues)
                                            Single issue: Price varies
        Contains statistics on notifiable diseases, general mortality, infant
        mortality, and birth rates. Also contains statistical tables on
        epidemiological and demographic topics as well as on causes of
        mortality. Entitled Epidemiological and Vital Statistics Report
        from 1947 through 1967 (vol. 1-20).

1197   Freedom from Hunger Campaign/Action for Development. IDEAS AND
bis   ACTION BULLETIN. No. 1, May 1965- .
        1965- , FAO   Free on request to FAO (12 issues a year)
        Newsletter concerning activities of nongovernmental organizations
        in behalf of the developing countries through projects aiding agricul-
        tural development, rural communities, and nutrition education.
        Available from the Office of the Coordinator, Action for Develop-
        ment/FFH, FAO (Rome).

1198   International LABOUR REVIEW. Vol. 1, no. 1, January 1921- .
        1921- , ILO   Annual subscr. $7.25 (12 issues)
                                                    Single issue $0.75
        Concerned with social aspects of economic life, in particular with
        development of human resources, improvement of labor standards
        and conditions, and promotion of sound industrial relations in the
        interest of economic and social progress. Contains international
        studies, articles analyzing experience of international interest in
        different countries, factual information on measures of policy taken
        in various parts of the world, and reviews and notes on new publi-
        cations in fields of concern to ILO.

1199   Bulletin of LABOUR STATISTICS. Vol. 1, no. 1, First Quarter 1965- .
        1965- , ILO   Annual subscr. $5.50 (4 main issues and 8 supplements)
                                            Single main issue $1.75
        Supplements the annual data given in the Year Book of Labour Sta-
        tistics (entry 1178). Contains indices of general level of employ-
        ment and of employment in nonagricultural sectors; indices of num-
        bers employed and total hours worked in manufacturing; numbers
        and percentages unemployed; average number of hours worked in
        nonagricultural sectors and in manufacturing; average earnings or
        wage rates in nonagricultural sectors and in manufacturing; general

indices and food indices of consumer prices. Notes on methods of computation and other technical material are included. A more detailed inquiry into wage rates, hours of work, and consumer prices is carried out annually, and the results of this are published in the Bulletin. Information in the latest issue available is brought up to date by means of a trilingual Supplement, published in January, February, April, May, July, August, October, and November. The interest of the Supplement is therefore limited to the month of its publication since the information given therein is incorporated in the following quarterly issue of the Bulletin.

1200   International MARINE SCIENCE. Vol. 1, no. 1, April 1963- .
       1963- , UNESCO/FAO   Annual subscr. (4 issues)
       Contains information on international, regional, and national activities of international significance to marine science, including meeting and training facilities, notes on new periodicals and publications. Publication suspended with vol. 7, no. 4, February 1970.

       MARINE SCIENCE CONTENTS TABLES. See entry 1134.

1201   NATURAL RESOURCES, SCIENCE AND TECHNOLOGY NEWSLETTER.
       No. 1, October 1963- .
       1963- , UN   Irregular   Free on request
       News notes on activities of international organizations in the field of science and technology and development of natural resources in Africa; lists relevant documents of the United Nations Economic Commission for Africa. Name through no. 16 (December 1969): Natural Resources Newsletter. Free on request to United Nations Economic Commission for Africa.

1202   NATURAL RESOURCES FORUM.... Vol. 1, no. 1, 1971- .
       1971, UN   Trilingual: E/F/S Irregular; price per issue varies
       Deals with technical, economic, legal, and institutional aspects of natural resources exploration and development, featuring articles by United Nations and other experts. The first issue (ST/ECA/142; sales no. E/F/S.71.II.A.13) is priced at $2.00.

1203   NATURE AND RESOURCES; Newsletter about Scientific Research on Environment, Resources and Conservation of Nature; Bulletin of the International Hydrological Decade. Vol. 1, no. 1/2, June 1965- .
       1965- , UNESCO   Free on request to UNESCO (4 issues a year)
       Covers all aspects of the work of the program of the Natural Resources Research Division, Department of Environmental Sciences and Natural Resources Research in the fields of hydrology, geology, soil sciences, ecology, and conservation of nature, with particular emphasis on news of the International Hydrological Decade (1965-1974), including articles and notes on new publications. Continues information on arid lands, and replaces the former Arid Zone (No. 1-26, 1958-1964) and Humid Tropics Newsletter (No. 1-4, 1962-1964).

1204   NUTRITION NEWSLETTER. Vol. 1, no. 1, January/March 1963- .
       1963- , FAO   Free on request to FAO (4 issues a year)
       Contains notes and articles; notices of FAO priced publications and of reports by experts and consultants; notes on meetings; book reviews. Covers food science and technology, food consumption and planning, applied nutrition, and home economics.

1205   Boletín de la OFICINA SANITARIA PANAMERICANA. Vol. 1,
        no. 1, mayo de 1922-   .
        1922-   , PAHO   Annual subscr. $5.00 (12 issues in 2 vols.)
                                                        Single issue $0.50
                                            Annual English ed. (selections) Free
        Articles are primarily in Spanish but there are occasionally articles
        in English, French, or Portuguese. In addition to articles on pre-
        ventive medicine, public health, and other subjects within the inter-
        ests of the Panamerican Sanitary Bureau, there are sections devoted
        to medical and health news; information items of public health inter-
        est; a calendar of meetings; book reviews, and a library list of cur-
        rent publications. Title of vol. 1-2, no. 6 (mayo de 1922-junio de
        1923): Boletín Panamericano de Salud.

1206   BOLETÍN DE LA OFICINA SANITARIA PANAMERICANA. English
        Edition: Selections from 1966-   .
        1967-   , PAHO   Free on request to PAHO (1 issue a year)
        Selections of articles, reports, and news items from the other edi-
        tion of the Boletín, translated into (or reprinted in) English.

1207   POPULATION AND VITAL STATISTICS REPORT. . . . Vol. 1, no. 1, 1
        January 1949-   . (Statistical Series A, vol. 1, no. 1-   ; ST/STAT/
        SER.A/1-   )
        1949-   , UN   Annual subscr. $4.00 (4 issues)
                                                        Single issue $1.25
        Contains the latest census returns and official estimates of popula-
        tion, the population estimate for a fixed (recent) year, and the
        latest birth, death, and infant death statistics. World and continental
        aggregates are also included. The reference date of the fixed-year
        population for each country, as well as the aggregate estimate for
        the world and each continent, is updated in the 1 July issue. Supple-
        ments data published in the Demographic Yearbook (entry 1170).
        Each issue supersedes all previous issues.

1208   POPULATION BULLETIN OF THE UNITED NATIONS. No. 1-7.
        1952-63, UN   Irregular
        a. No. 1: 1951. 57 pp., illus. (ST/SOA/SER.N/1)
           1952, UN   Sales No. 52.XIII.2   Out of print
           Articles (4) on: past and future growth of world population; inter-
           national migration in the Far East during recent times (the coun-
           tries of emigration); dynamics of age structure in a population
           with initially high fertility and mortality; some quantitative as-
           pects of the aging of Western populations.
        b. No. 2: 1952. 79 pp., illus. (ST/SOA/SER.N/2)
           1952, UN   Sales No. 52.XIII.4   Out of print
           Articles (4) on: an analysis of infant mortality; the assimilation of
           expellees in the Federal Republic of Germany; international mi-
           grations in the Far East (the countries of immigration); accuracy
           tests for census age distributions tabulated in five-year and ten-
           year groups.
        c. No. 3: 1953. 70 pp., illus. (ST/SOA/SER.N/3)
           1953, UN   Sales no. 53.XIII.8   Out of print
           Articles (3) on: population structure as a factor in manpower and
           dependency problems of developing countries; census statistics
           of population dependent on various types of economic activities;
           the measurement of infant mortality.

d. No. 4: 1954. 40 pp., illus. (ST/SOA/SER.N/4)
1955, UN   Sales No. 55.XIII.1   Out of print
Articles (4) on: gaps in existing knowledge of the relationship between population trends and economic and social conditions; fertility according to size of family: application to Australia; differential fertility in Ceylon; the cause of aging in populations: declining mortality or declining fertility?

e. No. 5: 1956.   pp., illus. (ST/SOA/SER.N/5)
1956, UN   Sales No. 56.XIII.4   Out of print
Articles (6) on: cultural values, social structure, and population growth; growth of population and public health programs in Asia and the Far East; demographic aspects of manpower in the Far East; the sample survey: its uses and problems; recent demographic trends in Cuba, Haiti, and the British Caribbean; analysis and calculation of the fertility of populations of developing countries.

No. 6: 1962. See entry 822.
No. 7: 1963. See entry 832.

1209   POPULATION NEWSLETTER. No. 1, April 1968- .
1968- , UN   Free on request to UN (4 issues a year)
Contains notes of United Nations population activities, including demographic statistics. Prepared by the Population Division, United Nations, New York.

1210   Impact of SCIENCE on Society. Vol. 1, no. 1, April/June 1950- .
1950- , UNESCO   Annual subscr. $4.00 (4 issues)
Single issue $1.25
Reports on science as a major force for social change. Written for the educated layman and the scientist by outstanding natural scientists, social scientists, and other specialists. Each issue is entirely devoted to a subject of significant interest and importance for the citizen today.

1211   STATISTICAL BULLETIN FOR AFRICA.... Vol. 1, parts 1, 2—vol. 2, parts 1, 2, 3, November 1965-March 1967.
1965-67, UN   Bilingual: E/F
Part 1: Statistical tables of population and vital statistics; employment; national accounts: summary tables; agriculture; industry; transport and communications; external trade; balance of payments: summary tables; prices. Part 2: Household budgets and consumption expenditure; finance; public finance; social statistics (physicians, dentists, midwives, pharmacists); education and culture; input-output: country tables; national accounts: country tables. Part 3: Balance of payments: country tables. Replaced the Statistical Annexes to the Economic Bulletin for Africa (entry 1190), formerly included in various issues or issued as separate annexes. The SBA itself was replaced by the Quarterly Statistical Bulletin for Africa (vol. 1, no. 1, January 1968- , bilingual: E/F, unpriced) and the Statistical Yearbook [for Africa] (entry 1182). The Quarterly Statistical Bulletin for Africa provides monthly data for two years in respect of external trade, transport, money, banking and finance, price indexes, and production, in two parts: (1) general statistics (prices, transport, money and finance, values of imports and exports), allowing comparisons between countries; and (2) country tables (production, foreign trade, airport traffic).

1212   STATISTICAL BULLETIN FOR LATIN AMERICA.... Vol. 1, no. 1,
       March 1964- .
       1964- , UN   Twice a year   Price per issue varies
                         Bilingual: E/F
       Contains regional statistics: selected economic indicators, total
       reserves, foreign exchange reserves, balance of payments with the
       U.S.A., price indexes of main export products in the world market,
       quantum indexes of production (agriculture, mining, manufacturing);
       and national statistics: population (including births, deaths, econom-
       ically active population), agriculture, mining, manufacturing, con-
       struction, electricity, price indexes, transport, and international
       trade. Supersedes the Statistical Supplement formerly appearing as
       part of the Economic Bulletin for Latin America (see entry 1193),
       either as part of issues of the Bulletin (1958/60) or as a bilingual
       (E/S) separate (1960/62).

1213   Monthly Bulletin of STATISTICS.... Vol. 1, no. 1, January 1947- .
       1947- , UN   Monthly; bilingual: E/F   Annual subscr. $25.00
                                                       Single issue $3.00
       Provides monthly statistics on 70 subjects from over 200 countries
       and territories together with special tables illustrating important
       economic developments. Quarterly data for significant world and
       regional aggregates are also published in the Bulletin. Supplements
       the statistics published in the Statistical Yearbook (entry 1181).
       ————. 1967 Supplement. (See entry 1181a.)

1214   TIMBER BULLETIN FOR EUROPE. Vol. 1, no. 1. 1948- .
       1948- , UN/FAO   Bilingual: E/F   Annual subscr. $20.00 (4 issues
       and supplements) Single issue $1.50   Supplements: prices vary
       Contains data on production and trade of the principal forest prod-
       ucts as well as selected series of price statistics. Supplements
       comprise an annual market review, a triennial survey of the pro-
       duction capacity of the raw material consumption by the wood-
       based panel products industries, special studies, and reports of
       seminars.

1215   UNASYLVA; an International Review of Forestry and Forest Industries.
       Vol. 1, no. 1, July/August 1947-vol. 25, no. 4, 1971.
       1947-71, FAO   Quarterly
       Contains articles on world conditions and developments; reports, re-
       gional and technical conferences; commodity reports on forest prod-
       ucts entering international trade; news items and reviews of current
       technical literature. Bimonthly 1947-1949. Ceased publication at
       the end of 1971.

       UNITED NATIONS DOCUMENTS INDEX. See entry 1149.

1216   WATER RESOURCES JOURNAL. September 1949- . (Series ST/
       ECAFE/SER.C/1-)
       1949- , UN   Mimeographed   Free (4 issues a year)
       Issued in Bangkok, Thailand, by the UN Economic Commission for
       Asia and the Far East. Contains articles, extracts from reports,
       etc., on floods, water resources development, hydrological services,
       etc. Former title: Flood Control Journal, no. 1-53, September 1949-
       September 1962; present title from no. 54, December 1962.

1217   WHO CHRONICLE. No. 1, no. 1/2, January/February 1947-   .
       1947-   , WHO   Annual subscr. $4.00 (12 issues)
                                                      Single issue $0.60
       Provides a monthly record of the principal health activities under-
       taken in various countries with WHO assistance. Also contains sum-
       maries and detailed accounts of the other publications issued by
       WHO. Former title (vol. 1-12): Chronicle of the World Health
       Organization.

1218   Bulletin of the WORLD HEALTH ORGANIZATION. Vol. 1, no. 1, 1947/
       48-   .
       1948-   , WHO   Bilingual: E/F Annual subscr. $30.00 (12 issues in
       2 vols.)   Single issue $3.25
       Contains technical articles reporting the results of laboratory, clin-
       ical, or field investigations of international significance on subjects
       within the wide scope of WHO fields of interest; review articles.
       Subjects include environmental hazards to health, physical, chemi-
       cal, and biological; sanitary engineering, including air and water
       pollution control, community sanitation and housing, water and
       wastes; protection against ionizing radiation; nutrition; epidemiology
       and health statistics.

1219   WMO BULLETIN. Vol. 1, no. 1, April 1952-   .
       1952-   , WMO   Annual subscr. $6.00
                                                      Single issue $1.75
       Summarizes the work of WMO and developments in international
       meteorology concerned with the application of meteorology to human
       activity. Includes occasional articles, notes on WMO publications
       and reviews of other publications in the field.

Author Index

References are to entry numbers, not to page numbers. Included in this index are the corporate names of issuing bodies. Corporate entries which include the elements <u>Committee</u>, <u>Conference</u>, <u>Congress</u>, <u>Group</u>, <u>Meeting</u>, <u>Mission</u>, <u>Panel</u>, <u>Seminar</u>, <u>Symposium</u>, <u>Working Party</u>, and <u>Workshop</u> are indexed under the appropriate element, e.g.,

Committee...
    Air Pollution (WHO), Expert C. on
    Application of Science and Technology to Development (UN), Advisory C. on the, ACAST
    Arctic Research (ICSU), Scientific C. on, SCAR

The organization of which the body is a part or by which it was convened is indicated parenthetically by acronym.

Aalsmeer, W.C., 706
Abbott, J.C., 543
Abdussalam, M., 584
Abensour, E.S., 9
Abercrombie, K.C., 535, 539
Acha Jamet, Pedro, 577, 579
Adams, F., 386
Alaka, M.A., 165
Albertsen, V.E., 578
Allen, G.R., 563
Allison, Stephen V., 396
Almeida, F.F.M., 47
Amiran, D.H.K., 300
Andelman, Julian B., 999
Andersen, K.L., 11
Angot, Alfred, 192
Applied Research and Engineering, Ltd., AREL, Washington, (Durham, United Kingdom) 350, 351
Araten, Y., 441
Association of African Geological Surveys, AAGS, 44-46, 48
Aten, A., 609, 615, 625, 626
Auger, Pierre, 22
Aung Din, U, 480
Austin Bourke, P.M., 258
Autret, Marcel, 742, 743
Aykroyd, W.R., 708, 711, 733

Bairoch, P., 881
Barker, J.E., 953
Barnea, Joseph, 120
Barnes, J.M., 1008
Barnoya de Asturias, Concha, 765
Bataillon, Claude, 13

Baum, Samuel, 879
Bear, J., 302
Beekman, Evert, 627
Behar, Moise, 743
Behrman, Daniel, 128
Bell, Alan, 1070
Ben-David, Joseph, 299
Benson, M.A., 369
Beresford-Peirse, Henry, 467
Berk, Zeki, 611
Biraud, Y., 811
Bleeker, W., 978, 982, 1074
Blombergsson, Helge, 621
Boalch, D.H., 1102
Bohdal, M., 734
Bolin, Bert, 204
Booher, L.J., 391
Borrie, W.D., 930
Botts, Florita, 1071, 1072, 1077, 1081, 1082, 1085, 1090
Bouquiaux, J., 959
Bourgeois-Pichat, Jean, 765 bis
Boyd, Eldon M., 1013
Bradbury, J.C.C., 119
Brasnett, N.V., 462
Brisou, J., 1030
Brock, J.F., 742
Brosh, R., 441
Browder, Frank N., 1053, 1054
Bruce, J.P., 228, 284
Buck, W.Keith, 74
Budowski, Gerardo, 7
Bundesanstalt für Bodenforschung, Hannover, 55, 56
Bureau Central International de Séismologie, BCIS, 66

Stephen, Joan M.L., 696
Stephens, Sophie, 1137
Stevens, R.A., 20
Subrahmanyam, V.P., 301
Suess, Michael J., 999
Suri, P.N., 473
Suschny, O. 989
Swaroop, S. 811
Symonds, Richard, 841
Symposium ... (See also Simposio ...)
  Agroclimatological
    (UNESCO), S. on, Reading,
    1966, 214
  Application of Radioisotopes in
    Hydrology (IAEA), S. on the,
    Tokyo, 1963, 263
  Arid Zone Hydrology (UNESCO),
    S. on, Ankara, 1952, 292, 293
  Arid Zone Plant Ecology
    (UNESCO), S. on, Montpellier,
    1953, 452, 453
  Brain Research and Human Be-
    haviour (UNESCO/IBRO), S.
    on, Paris, 1968, 728
  Changes of Climate (UNESCO/
    WMO), S. on, Rome, 1961, 218
  Climatology and Microclimatology
    (UNESCO), S. on, Canberra,
    1956, 294, 295
  Containment and Siting of Nuclear
    Power Plants (IAEA), S. on
    the, Vienna, 1967, 975
  Criteria for Guidance in the
    Selection of Sites for the
    Construction of Reactor and
    Nuclear Research Centres
    (IAEA), S. on, Bombay, 1963,
    974
  Dams and Reservoirs (UN), Re-
    gional S. on, Tokyo, 1961, 381
  Development of Petroleum Re-
    sources in Asia and the Far
    East (UN), S. on the 1st, New
    Delhi, 1958, 107, 107a; 2nd,
    Tehran, 1962, 108, 109; 3rd,
    Tokyo, 1965, 51, 111, 112; 4th,
    Canberra, 1969, 113
  Developments in the Management
    of Low- and Intermediate-
    level Radioactive Wastes
    (IAEA/ENEA), S. on, Aix-en-
    Provence, 1970, 1059
  Disposal of Radioactive Wastes
    into Seas, Oceans and Surface
    Waters (IAEA), S. on the, Vi-
    enna, 1966, 1045
  Disposal of Radioactive Wastes
    into the Ground (IAEA/ENEA),
    S. on the, Vienna, 1967, 1047

Ecology of the Subarctic Regions
  (UNESCO), S. on the, Otaniemi,
  1966, 17
Education and Training in Nutri-
  tion in Europe (FAO/WHO), S.
  on, Bad Homburg, 1959,
  755
Environmental Aspects of Nuclear
  Power Stations (IAEA/USAEC),
  S. on, New York, 1970,
  976
Epidemiology of Air Pollution
  (WHO), S. on the, Copenhagen,
  1960, 965
Family Living Studies (ILO), S.
  on, 867 [Not a meeting but a
  collection of papers]
Flood Control, Reclamation,
  Utilization and Development of
  Deltaic Areas (UN), Regional
  S. on, Bangkok, 1963,
  378
Floods and Their Computation
  (UNESCO/IASH/WMO), S. on,
  Leningrad, 1967, 370
Food Irradiation (FAO/IAEA), In-
  ternational S. on, Karlsruhe,
  1966, 598
Functioning of Terrestrial Eco-
  systems at the Primary Pro-
  duction Level (UNESCO), S. on
  the, Copenhagen, 1965,
  449
Gondwana Stratigraphy (IUGS), S.
  on, Buenos Aires, 1967, 41
Granites of West Africa
  (UNESCO), S. on the, 1965, 47
Health Effects of Air Pollution
  (WHO), S. on the, Prague,
  1967, 950
Higher Education and Training,
  WMO/IAMAP S. on, Rome,
  1970, 174
Hydrological Forecasting (WMO/
  UNESCO), S. on, Surfers'
  Paradise, 1967, 371
Hydrology of Deltas (UNESCO/
  IASH), S. on the, Bucharest,
  1969, 289
Hydrology of Fractured Rocks
  (UNESCO/IASH/FAO), S. on
  the, Dubrovnik, 1965, 285
Industrial Development (UNIDO),
  International S. on, Athens,
  1967, 589
Industrial Feeding and Canteen
  Management in Europe (FAO/
  ILO/WHO), Joint S. on, Rome,
  1963, 747

Symposium ... (continued)

Instruments and Techniques for the Assessment of Airborne Radioactivity in Nuclear Operations (IAEA), S. on, Vienna, 1967, 988

Investigations and Resources of the Caribbean Sea and Adjacent Regions (UNESCO/FAO), S. on, Willemstad, 1968, 130

Iron Metabolism and Anemia (PAHO), S. on, Washington, 1969, 736

Isotopes in Hydrology (IAEA/IUGS), S. on, Vienna, 1966, 267

Kuroshio, S. on the, Tokyo, 1963, 133

Land Subsidence (UNESCO/IAHS), S. on, Tokyo, 1969, 233

Land Use in Semi-arid Mediterranean Climates (UNESCO/IGU), S. on, Iraklion, 1962, 300

Meteorite Research (IAEA), S. on, Vienna, 1968, 40

Methodology of Plant Eco-physiology (UNESCO), S. on, Montpellier, 1962, 457

Methods of Study of Soil Ecology (UNESCO), S. on, Paris, 1967, 426

Nuclear Desalination (IAEA), S. on, Madrid, 1968, 354

Nuclear Energy Costs and Economic Development (IAEA), S. on, Istanbul, 1969, 357

Numerical Weather Prediction (WMO/IUGG), S. on, Tokyo, 1968, 162

Oceanography and Fisheries of the Tropical Atlantic (UNESCO/FAO/OAU), S. on the, Abidjan, 1966, 129

Operating and Developmental Experience in the Treatment of Airborne Radioactive Wastes (IAEA), S. on, Vienna, 1968, 1049

Peaceful Uses of Atomic Energy in Africa (OAU/IAEA), S. on the, Kinshasa, 1969, 103

Physical Oceanography, Unesco S. on, Tokyo, 1955, 121

Physiology and Psychology in Arid and Semi-arid Conditions (UNESCO), S. on, Lucknow, 1962, 14, 15

Planning and Development of New Towns (UN), S. on the, Moscow, 1964, 901

Plant Protein Resources: Their Improvement through the Application of Nuclear Techniques (FAO/IAEA), S. on, Vienna, 1970, 710

Plant-Water Relationships in Arid and Semi-arid Conditions (UNESCO), S. on, Madrid, 1959, 454, 455

Polar Meteorology (WMO/SCAR/ICPM), S. on, Geneva, 1966, 277

Positive Contribution by Immigrants (UNESCO/ISA/IEA), S. on the, 931 [Not a meeting but a collection of papers]

Practices in the Treatment of Low- and Intermediate-level Radioactive Wastes (IAEA/ENEA), S. on, Vienna, 1965, 1044

Problems of the Arid Zone (UNESCO), S. on, Paris, 1960, 298

Problems relating to the Environment (UN), ECE S. on, Prague, 1971, 8 bis

Radiation including Satellite Techniques (WMO/ICSU), S. on, Bergen, 1968, 193, 201

Radioactive Dating (IAEA), S. on, Athens, 1962, 38

Radioactive Dating and Methods of Low-level Counting (IAEA), S. on, Monaco, 1967, 39

Radioisotope Instruments in Industry and Geophysics (IAEA), S. on, Warsaw, 1965, 59

Radioisotope Tracers in Industry and Geophysics (IAEA), Prague, 1966, 60

Rapid Methods for Measurement of Radioactivity in the Environment (F.R. Germany/Gesellscaft für Strahlen- und Umweltforschung/IAEA), International S. on, Neuherberg bei München, 1971, 993

Reactor Safety and Hazards Evaluation Techniques (IAEA), S. on, Vienna, 1962, 973

Recovery of Uranium from Its Ores and Other Sources (IAEA), S. on the, São Paulo, 1970, 102

Series and Serials Index

References after numbers in series and after titles are to entry numbers, not to page numbers.

UN, General Assembly, Official
Records (continued)

147a; 25th sess., 1970, Supp.
No.21, 148; 26th sess., 1971,
Supp. No.21, 149
International Social Development
Review, 1: 902
Manuals on Methods of Estimating
Population, 1: 784; 2: 785;
3: 802; 4: 816; 5: 882 bis; 6: 790
Mineral Resources Development
Series. See subheading Eco-
nomic Commission for Asia
and the Far East (ECAFE)
Monthly Bulletin of Statistics,
1213; 1967 Supplement, 1181a
Natural Resources Forum, 1202
selected articles, 43, 73, 74,
106, 119, 120, 304, 315, 374
Natural Resources, Science and
Technology Newsletter. See
subheading Economic Commission for Africa (ECA)
Official Records. See subheading
Economic and Social Council
and General Assembly
Population and Vital Statistics
Report, 1207
Population Bulletin of the United
Nations, 1-5: 1208; 6: 822;
7: 832
Population Newsletter, 1209
Population Studies, 1: 783; 2: 778;
3: 770; 4: superseded, 793;
5: 924; 6: superseded, 793;
7: 782; 8: superseded, 793;
9: superseded, 793; 10: 784;
11: 925; 12: 928; 13: 820;
14: 779; 15: 777; 16: superseded by no. 41; 17: 863;
19: superseded by no. 29;
20: 864; 21: superseded by
no. 41; 22: 821; 23: 785;
24: 1131; 25: 802; 26: 866;
27: 830; 28: superseded by
no. 41; 29: 1098; 30: superseded by no. 41; 31: superseded by no. 41; 32: 887;
33: 878; 34: 780; 35: 786;
36: 795; 37: 884; 38: 803;
39: 788; 40: 894; 41: 804;
42: 816; 43: 880; 44: 903;
45: 842; 46: 882 bis; 47: 791;
48: 805
Quarterly Statistical Bulletin for
Africa. See subheading Economic Commission for Africa
(ECA)

Recursos Hildráulicos de Amér-
ica Latina. See subheading
Economic Commission for
Latin America (ECLA)
Regional Plan Harmonization and
Integration Studies. See sub-
heading Economic Commission
for Asia and the Far East
(ECAFE)
Statistical Bulletin for Africa.
See subheading Economic Com-
mission for Africa (ECA)
Statistical Bulletin for Latin
America. See subheading Eco-
nomic Commission for Latin
America (ECLA)
Statistical Papers, Series C,
12: 787
Statistical Papers, Series K,
2: 870; 3: 870
Statistical Papers, Series M,
4/Rev.2: 872; 18: 1108;
19: 814; 20: 926; 27: 792;
28: 920; 44: 799; 45: 921;
47: 915; 50: 818; 51: 791
Statistical Standards and Studies.
See subheading Conference of
European Statisticians (CES)
Statistical Yearbook, 1181
1967 Supplement, 1181a
Statistical Yearbook [for Africa].
See subheading Economic
Commission for Africa (ECA)
Statistical Yearbook for Asia and
the Far East. See subheading
Economic Commission for
Asia and the Far East (ECAFE)
Studies in Methods, 5/Rev.1: 793;
7: 815; 10: 869; 12: 914;
13: 913; 15: 789; 16: 798C
Timber Bulletin for Europe. See
subheading Economic Commis-
sion for Europe (ECE)
United Nations Documents Index,
1144
Water Resources Journal. See
subheading Economic Com-
mission for Asia and the Far
East (ECAFE)
Water Resources Series. See
subheading Economic Com-
mission for Asia and the Far
East (ECAFE)
United Nations Educational, Scien-
tific and Cultural Organization
(UNESCO). See also Intergovernmental Oceanographic Commission (IOC); International
Bureau of Education (IBE),
Geneva

WHO, Technical Reports Series
(continued)

386: 854; 387: 838; 391: 1019A;
397: 855; 404: 998; 406: 938;
410: 956; 417: 1020A; 420: 327;
424: 856; 430: 646; 435: 839;
439: 940; 440: 829; 442: 843;
445: 648; 451: 607; 452: 699;
453: 583; 458: 1021A; 462: 651;
471: 840; 473: 857; 474: 1022A;
476: 844; 484: 1032 bis

WHO Chronicle. See subheading
Chronicle, WHO

World Health
selected articles, 943

World Health Statistics Annual,
1176 (continues Annual Epidemiological and Vital Statistics)

World Health Statistics Report,
1197 (continues Epidemiological and Vital Statistics Report)
selected article, 737

World Meteorological Organization
(WMO)
Basic Documents, 2: 154
Bulletin, WMO, 1219
selected articles, 8, 957
GARP Publications Series [ICSU/
WMO series], 1: 195; 2: 196,
3: 197, 4: 198; 5: 201; 6: 202;
7: 203
GARP Special Reports [ICSU/
WMO series], 1: 199; 2: 200

Reports on Marine Science Affairs, 1: 127; 2: 140; 3: 141
bis; 4: 141 ter
Reports on WMO/IHD Projects,
1: 252; 2: 301; 3: 253; 4: 284;
5: 254; 6: 255; 7: 272; 8: 244;
9: 231; 10: 1159; 11: 136;
12: 245; 13: 261; 14: 416;
15: 249; 16: 257 bis

Special Environmental Reports,
1: 6; 2: 164 bis

Technical Notes, 7: superseded by
no. 91; 10: 258; 11: 258;
13: superseded by no. 105;
25: 242; 26: 280; 43: 978;
46: 142; 47: 251; 50: 168;
53: 521; 54: 223; 56: 216;
58: 186; 59: 212; 60: 187;
62: 165; 63: 118; 65: 1; 68: 982;
69: 224; 70: 188; 76: 229;
77: 189; 82: 158; 83: 260;
86: 220; 90: 369; 91: 192;
92: 371; 94: 989; 96: 954;
97: 413; 98: 372; 101: 523;
102: 145; 103: 125; 104: 193;
105: 257; 106: 959; 107: 482;
108: 4; 109: 5; 111: 163;
113: 484; 114: 960; 117: 237

WMO Bulletin. See subheading
Bulletin, WMO
World Weather Watch
Planning Reports, 1; 190; 4: 178;
17: 159; 18: 182; 22: 215;
27: 160; 30: 183; 31: 184

Title Index

References are to entry numbers, not to page numbers.

Subject Index

References are to entry numbers, not to page numbers.

Anemias, nutritional, 669, 723, 733-
737, 758
Animal biology, nuclear research
techniques, 489, 490, 494-496,
509
Animal ecology
arid zones, 12, 294, 295, 297
humid tropics, 367
polar and subarctic regions, 17
Animal health, 10, 11, 103, 212, 407,
481, 482, 484, 494
application of existing knowledge,
26
pesticides and, 1009
Animal nutrition, 423, 435, 489-496,
717
Animal resources, 3, 23c, 27, 481-
502, 505
arid zones, 29, 488
biblio., 1111
CBW's possible long-term effects, 995, 996
development, Indicative World
Plan, IWP, 536a
marine mammals, 498A-B
statistics, 1180
wildlife conservation, 27, 279,
485-487, 497
filmstrips, 1071, 1089
legislation, 1195
wildlife forest influence, 459, 467
Antarctic. See Polar and subarctic
regions
Aquatic plants. See Plant resources
Arab States
FAO/UNDP (SF) projects, biblio.,
1117a-e
migration, international, 929
nomads, 16
salt industries, 151
science policy-making bodies,
1037d
urbanization, 911
water laws, 322
Arctic region. See Polar and subarctic regions
Argentina
agricultural projections (IWP),
534b
migration, international, 931
nutrition deficiency, 741
urbanization, 909
Arid zones
animal resources, 29, 488
drainage, 385
drought, 301, 454, 455, 457
ecology, 12-15, 294, 295
environmental health, 29
forest resources, 29, 303, 471

geography, 30, 31
geology, 29, 297
hydrometeorology and hydrology,
29, 292-303, 341, 452, 454,
455, 457
biblio., 1129
filmstrip, 1095
irrigation, 29, 298, 299, 340,
341, 376, 454
land tenure, 29
land use, 297-300, 454
locust, desert, 12, 223-225, 1117a
map symbols, 235
natural resources surveys, 29
nomads, 12, 13, 16, 297, 298
nutrition, 12, 14, 15, 29
ore concentration, 342
periodical, 1203
plant resources, 29, 222,
452-457, 471
medical plants, 456
research, 26
directory, 1158
sheep breeds, 488
soils, 29, 294
tree-planting practices, 471
urbanization, 300, 911
water resources, 29, 30, 286-303,
385
filmstrip, 1095
saline water use, 340-342
water-use efficiency studies, 392
weather and climate, 14, 15, 29,
214, 216-225, 292-300
Artificial satellites
in meteorology, 177, 182-184,
193, 194, 201
in snow hydrology, 272
Ascorbic acid requirements in diet,
699
Asia and the Far East
aerial surveys, 24, 77
agricultural censuses (1970),
860B
agricultural projections (IWP),
534a, d
agricultural statistics, 1183
bauxite ores, 79
children and youth, statistics,
1183a
civil registration, 818
coal and lignite resources, 52,
90-92
coconut industry, 613
community and industrial wastes,
1148
copper ore resources, 78
dams and reservoirs, 381
Economic Bulletin, 1191

Pesticide residues, 673
  detection by nuclear techniques,
    1011, 1015
  legislation, 1009, 1023
    periodical, 1194
  tolerances for human ingestion,
    1010, 1014
  toxic hazard evaluations (FAO/
    WHO), 1016-1022A
Pesticides
  biological effects in soil, 425
  legislation, 1009, 1023
    periodicals, 1194, 1196
  statistics, 1180, 1185
  toxic hazards, 1009
Petroleum and natural gas. See also
    Marine pollution, oil and oily
    mixtures
  exploration, 88, 104, 106-113
    geochemical methods, 80
    nuclear techniques, 70
  and fertilizer production from
    natural gas, 438
  legislation, 83, 107
  maps, 32, 49, 51, 52
  oil-shale utilization, 105
  production, 104, 107-113
Philippines
  agricultural projections (IWP),
    534d
  educational planning, 894
  filled milk, 715
  labor force, 877, 887
  petroleum legislation, 83
  rice enrichment, 706
  water laws, 318a
  water resources development,
    361B
Physical sciences, research trends,
    22
Pines
  Honduras survey, 478
  Pinus radiata, 462, 463
Plant breeding, 23c, 407, 423
  nuclear techniques, 103, 422,
    423, 710
Plant ecology, 1, 448-451
  arid zones, 294, 295, 297, 452-
    457, 471
  humid tropics, 458, 479
  subarctic regions, 17
Plant nutrition. See Fertilizers;
    Plant/Soil relations
Plant protection, 23e, 212, 407, 954
  nuclear techniques, 103
Plant resources, 27
  aerial surveys, 24
  air pollution injury, 954
  algae, 497

aquatic plants, 435, 497
  harvest statistics, 1173a
arid zones, 29, 222, 452-457
  biblio., 1124
  CBW's possible long-term ef-
    fects, 995, 996
  grasses, 407, 448
  medical plants, 456
  tropics, 367, 408, 458
  vegetation maps
    catalogue, 1143
    Mediterranean zones, 222
Plant/Soil relations, 417-427, 449,
    452, 509. See also Fertilizers
Poland
  labor force, 877
  pollution, 8 bis
Polar and subarctic regions
  ecology, 17
  environmental health, 10, 11
  glaciology, 278
  Gondwana stratigraphy, 41
Pollution, 8 bis, 934, 938, 942. See
    also Air pollution; Community
    and industrial wastes; Food
    contamination; Land pollution;
    Marine pollution; Noise; Pesti-
    cide residues; Pesticides;
    Radioactive pollution of the
    atmosphere; Radioactive
    wastes; Thermal pollution; Vi-
    bration; Water pollution
  biblio., 1114
  control legislation, 8 bis, 961
    periodicals, 1195, 1196
  industrial areas, 8 bis, 856, 951-
    953, 959, 961, 962 bis
    periodical, 1218
  recreation and tourist areas,
    8 bis
  urban areas, 4, 8 bis, 934, 956-
    959
  zones of historical value and in-
    terest, 8 bis
Poplars, 461
Population. See also Demography
  age structure, 770, 785, 816, 866,
    1208a-c
  arid zones, 29
  changes
    and distribution of genetic fac-
      tors, 772f
    social aspects, 772f
  distribution, 772c
  economic characteristics, 793b
  and food supply. See Food pro-
    duction and supply, population
    growth and
  social characteristics, 793c, 1208c